The international cotton trade

The international cotton trade

JULIAN ROCHE

WOODHEAD PUBLISHING LIMITED
Cambridge England

Published by Woodhead Publishing Ltd,
Abington Hall, Abington,
Cambridge CB1 6AH, England

First published 1994

British Library Cataloguing in Publication Data
A catalogue record for this book is available from the British Library

ISBN 1 85573 104 5

Designed by Andrew Jones (text) and Chris Feely (jacket)
Typeset by BookEns Ltd, Baldock, Herts.
Printed by St Edmundsbury Press, Suffolk, England.

To Angela Clark

Contents

Preface

Cotton is the world's most important textile fibre, forming almost half of all textile fibre used - and it is one of the world's most important agriculturally produced raw materials. About 83-96 million bales (480 lb/bale) are produced annually - different organisations produce slightly different statistics. Over 250 million people worldwide derive some or all of their cash income from it.

Everyone knows that it is used to make a wide variety of clothes and fabrics, from underwear to overcoats and tarpaulins. Rugs, towels, sheets, handkerchiefs and bandages are a few other uses of cotton, but in the bulk form in which it is usually traded, cotton has only limited uses. A single cotton fibre is a little less than the diameter of a human hair, and anything from 2 to 6 cm long. Thousands or millions of these fibres, wadded and matted together, are used for such products as mattresses and pillows, batting for insulation and quilted products as well as for industrial uses. The more sophisticated use of the processed fibres is for clothing and apparel, and an entire industry, the textile industry, has grown up around the creation of garments and other products from cotton yarn. The basic principles of prehistoric spinning and weaving are those used to produce yarn, threads and fabrics - albeit with increasingly sophisticated machinery - throughout the world in the late 20th century. It is a highly technical industry, one subject to many competitive pressures, and widely dispersed geographically.

Around this industry and the geographical dispersion of cotton production itself grew the cotton trade. It lays just claim to being the most sophisticated and complex of all the international commodity trades: the first to use telegraphs and telephones; the first to have reliable pricing; and the first to use futures contracts. Now it is a worldwide and deeply professional business involving dozens of companies - some of very considerable size and all with tremendous expertise in the numerous specialised markets for cotton of all the many different origins, approximately 23 million bales of which - about 25% of world production - is traded internationally each year from the 75 countries that produce cotton. World trade in cotton has risen steadily but undramatically, from 17 million bales in 1960/61 to 22 million bales in 1992/93. The increase in world trade has not kept pace with world production increases for important reasons of national textile industry development. As a result, the percentage of cotton traded internationally has fallen from 38% in 1960/61 to 23% in 1991/92 (an all-time record cotton production year) and 27% the following year.

But in this almost $7 billion worldwide industry, the key figures are all known to one another. The few dozen top cotton traders in the world all meet, for example at the annual Liverpool Cotton Dinner. Learning about cotton is a lifetime's work. Experts such as those at Tom Bell Associates, Cotton Outlook, the International Cotton Advisory Committee (ICAC) or US Department of Agriculture (USDA) and indeed cotton traders are astounding in the range and depth of their understanding of their subject which is for them, as for many others from farmers to futures analysts, a life's work.

For an author the cotton market presents special difficulties. It is not impenetrable for lack of information; on the contrary there is a superfluity of excellent analyses - principally from three sources: daily from Cotton Outlook in Liverpool, monthly from the USDA and the ICAC. The problem lies in interpreting and deciphering the immense amount of technical and detailed information that is presented. It would be possible to write an entire book on one national cotton industry since 1980, for example, or on one cotton trading firm's history.

This book, therefore is intended as a general guide for the uninitiated and a quick reference work for the expert. It is not, and within one volume could never be, a complete analysis of cotton or the international cotton trade. Experts in the cotton trade, for example traders and analysts, will find the comparison between markets, the evaluation of index numbers and prices, the statistical compendium, and the view of the future all of interest. Those approaching the daunting and information-rich world of cotton for the first time will find this book both comprehensive and compact, with directions for further in-depth study of particular areas evident throughout the text.

Acknowledgements

Without help from the USDA - in particular Russ Barlowe - and the New York Cotton Exchange - in particular Tom Bertolini - and all the staff at the ICAC and USDA this book would not have been written. They have helped both in providing their data and analysis and in scrutinising my rough treatment of it. I am especially grateful to a number of cotton analysts and traders for agreeing to review chapters of this book before I eventually submitted it. Especial thanks to Peter Scott of Amcot for pointing out directions of research. It goes without saying that any mistakes which remain are entirely my own responsibility.

Thanks are also due to Angela, who endured living with me during the production of this book and encouraged me throughout.

I should finally like to thank Chris Kennedy, formerly Director of Business Development at London Fox, for first encouraging my interest in cotton.

Conversion table

Bales vary in weight internationally. The standard US bale is 480 lb.

1 lb	= 0.436 kg
1 hectare	= 2.471 acres
1 inch	= 25.4 mm

Units have been retained as either metric or imperial according to the standard used by the source.

CHAPTER

1

History and background of cotton

Physical characteristics of cotton

Cotton fibres are single plant cells, lengthenings of the outer layer of cotton seed cells which come from the seed pod, or boll, of the cotton plant. The cotton plant thrives in the warm and moist tropical and subtropical regions of the world. The plant is perennial, but cotton cultivated for agriculture is an annual herbaceous plant which can grow from 3 to 160 cm (1–8 in.) tall, depending on the variety grown. The process from germination has several stages. First, the radicle appears from the hilum end of the seed. The radicle grows downwards, thereby securing the seed and then the hypocotyl grows upwards. The cotlyedons grow above the ground, and straight. The branches are more or less widespreading, especially on the lower portions of the stem, and tapering to a pointed top. When closely planted, the development of the branching is reduced. The plant has a tap root system, which begins as a continuation of the stem root, and then thins for the rest of its depth. The root system also differs greatly, depending on the nature of the soil and the water supply. In Egypt, for example, the main tap root has been shown to search over 3 m into the ground to find water. The root system can also stretch up to 3 m sideways. Whilst, therefore, cotton does grow well on shallow soils, it achieves better results in deeper soils. The plant is not well suited to soils that are either too hard or too loose.

The stem of the plant has vegetable branches (monopodia) which in turn produce fruiting branches (sympodia), which have a flower bud at each node. At what point the fruiting branches start depends on the environment - water supply and planting distances are both determinants. The plant responds to damage by growing new branches or, later into the growth cycle, by growing existing branches more vigorously. The leaves, occurring after about eight nodes up the plant, are large and more or less deeply divided into three or five lobes, the size, shape and colour differing greatly in the different species and varieties. Thus Sea Island cotton has very deeply cut lobes, whilst American upland represents the other extreme. The Indian leaf is smaller, with characteristically rounded lobes. The leaves have a typically hairy undersurface.

The flower also differs considerably in colour from one species to another. It is surrounded by three large bracts (bracteoles), or outside leaves. The flower of the *Gossypium barbadense*, or Egyptian type, is lemon or golden yellow, with crimson spots at the base of the five petals inside, and a golden brush of stamens; that of *G. hirsutum*, or American type, is creamy, usually without markings, and has buff-coloured stamens. The flowers of *G. arboreum*, or Asiatic type, are also yellow, but smaller than the Egyptian, and the spots are smaller and darker. When the plant first flowers is an important indicator of how early the crop will be. There is great variation internationally: the average is probably 50–55 days but it can range from 45 to 70 days. The expanding petals push open the ring of bracteoles when the flower opens. The bud is fully developed by this stage: it stays green, and develops the fruit. The plant fertilises on its opening day, as the anthers burst and the stigma becomes receptive. The flower is large and attractive to bees and other insects, but most seeds result from self-pollination. Both the vertical flowering interval between the sympodia and the horizontal flowering interval between flowers on a branch are quite stable during the plant's growth and across a given crop. The ratio between the intervals is approximately >2 : 1. The sheer size of the cotton plant when grown can pose problems in growing it: in the USA principal varieties can grow between 1 and 1.2 m high, but under tropical growing conditions the same varieties can reach 2 m. Efforts to reduce plant height through breeding have had only limited success. A tall crop can be knocked over by a storm, rendering it difficult to harvest and liable to damage.

About half the cotton crop is produced on the non-fruiting branches with wide spacing; at close spacing – over 100 000 plants/ha

– it can be less than 10%. The crop size in terms of the percentage of flowers that actually produce bolls varies over time and between crops, depending on growing conditions. Usually over half of the first sympodia produce bolls, a third of the second and a tenth of the third.

After anthesis, the petals wither and fall away to expose the green boll. Before maturity, the bolls, or fruits, are of different sizes, and their colours are various shades of green. The diameter can vary from 19 to 40 mm and the shape can vary too. Egyptian type bolls are narrow and pointed; upland cotton is much rounder and appears much shorter. Some US cotton has large, pointed bolls. The boll is divided into three to five loculi, or compartments, with 5–11 seeds in each, to which the lint – cotton itself – is attached. The fruit ripens some 50–70 days (the boll period) after fertilisation, depending on the variety, the weather, and even from plant to plant because of access to light and soil conditions. The boll becomes full size about half way through its development. By then the seeds are also full size, and the lint hairs are fully grown. As the boll dries out when it is mature, the outer wall of each of the loculi comes open, revealing both lint and seeds. The cotton mass expands as it dries, staying attached to the dry carpels but still liable, in the less storm-resistant varieties, to be detached. The cotton seed is about 10–16 mm in size, weighing about 5–10 mg/100 seeds, with two layers covering the embryo and seed hairs. Again, there is a marked difference between varieties. Seed fuzz develops from the epidermal cells during the 5–11 days after the cotton flower opens, as opposed to the lint development which begins on the day the plant flowers.

Commercial upland cotton has relatively uniform fuzz over the entire seed. Sea Island, Egyptian and Tanguic have only part of their seed fuzz-covered. Egyptian seeds are black or brown, and the lint comes easily and completely away from the seed in the process of ginning – separating the lint from the seed. For US and Indian cotton, by contrast, there are two types of hair: one, the actual cotton or lint, is about 25 mm long and comes away from the seed easily in the gin: the other is a short fuzz which adheres closely to the seed and is only partially removed, even by the delinting process. The lint results from a much reduced deposit of cellulose in the cuticle, leaving a central cavity (lumen) open at maturity. When the capsule opens and the lint hairs dry, the cavity left by the drying of the cell contents is so large that the hair collapses, forming a ribbon. Each fibre consists of an outer or primary cell wall and an inner layer of cellulose making up the secondary wall. Examination of a fibre's cross section can show daily growth rings forming a series of concentric circles. The inside of each fibre, during development, contains the living protoplasm of the cell. When mature and exposed to the atmosphere, the fibre dies, and in drying it flattens and twists. The cellulose is deposited in spirally arranged fibrillae, the spirals reversing direction at intervals so that they are present in the same hair and in both clockwise and anti-clockwise directions. This causes the dry ribbon to twist, and these irregularly twisted ribbons cling together when spun into thread. The convolutions of lint are important for spinning, but both species and environmental factors affect length, coarseness and fibre strength.

The length of the cotton fibres varies according to the variety of plant. The best cotton tends to have longer fibres. The finest cotton (Sea Island) is about 50 mm long, Indian may be less than 12 mm but the common varieties of cotton have fibres of under 40 mm. The diameter of the fibre varies from 0.018 to 25 mm, and both the diameter and the number of twists that the respective fibres have is dependent on the species: the more twists, the easier the cotton is to spin. The lint's appearance in the open boll is very different between varieties; also in some cases it hangs out of the boll and in others it clings closely to the boll.

The seed is usually about two-thirds of the weight of the seed cotton, and the lint one-third, but the proportion varies from 25 to 20% between varieties. Breeding has produced longer, finer and stronger fibre, but the cotton fibres from bits of cloth and yarn found in the Indus valley among relics dating back to 3000 BC, like the fibres in current cottons, are flattened, twisted tubes, the walls of which consist of fibrils of cellulose built up in a number of concentric layers: the textile character known as 'cottonlike' is the result of this persistent fibrous structure. The spirality of a cotton fibre distinguishes it from all other fibres, gives it elasticity and makes it especially suitable for spinning; a strong thread can be made from cotton, although the individual fibres are much shorter than either wool or flax. Efforts to produce a man-made fibre with a comparable set of characteristics have proved only partially successful.

From first cultivation to the end of World War II

It is not known exactly how old cotton cultivation is, nor where it originated exactly. The word cotton is derived from the Arabic *qutun*, or *kutun*, and the arts of spinning and weaving are amongst the oldest in the world. It is alleged that the remains of cotton fabrics discovered in a cave near Tehuacan in Mexico date back to about 5800 BC, although fibre decay makes these dates difficult to establish with certainty. Cotton has been grown since 3500 BC, at least: one of the oldest-known specimens of spun cotton yarn was found in the ruins of Mohenjo-Daro, a city in the Indus valley, which had become deserted by about 3000 BC (Hutchinson *et al.*, 1947). The '*Schuking*', the ancient history of China, indicates that cotton was being grown and processed in that region at about the same time. Other ancient cotton fabrics still in existence are the grave cloths of pre-Inca Peru found at Huaca Prieta, dated about 2500 BC and a specimen of a cotton plant has been found in Mexico dating from 3500 BC; cotton has also been found in the prehistoric pueblo ruins of Arizona. In the ancient city of Dacca, the finest and lightest cotton yarns and fabrics ever known were made. Cotton was mentioned in a Hindu Rig-Veda hymn 15 centuries before Christ.

It has, however, only been found in Middle Eastern civilisations such as Mesopotamia, Babylon and Egypt from 500 BC. Herodotus (*c.* 450 BC) mentions how Indian women plucked lint from seed, carded it and spun yarn for weaving on their crude hand looms (see the section on 'the cotton cycle' in Chapter 2 for these terms). Strabo referred to the flowered cottons or chintzes of the Indians, of their lustre and of the vivid quality of the dyes with which they figured their cloths. In *The Circumnavigation of the Erythraean Sea*, Arrian, an Egyptian Greek, referred to the Arab traders bringing Indian cottons to Aduli on the Red Sea – the first reference to such trade. Ports beyond the Red Sea had an established trade in calicoes, muslins and other cottons, both plain and with floral ornamentation. Yet cotton spinning lingered 1300 years on the Mediterranean shores before arriving in Greece and then Italy, despite the potential demand for it in Rome and Constantinople. Alexander the Great's Admiral Nearchus commented on Indian cotton garments about 327 BC. The term 'calico' comes originally from the Indian port of export, Calicut.

The Egyptians grew and spun cotton from about AD 6–700. Mohammedan Spain manufactured cloths from cotton in the tenth century: the Saracens and Arabs brought cotton fabrics with them on their conquests of Sicily and southern Spain. Cotton was the subject of legend and myth during the Middle Ages in Europe, and it was for an additional supply of cotton that many of the naval explorations of the fifteenth and sixteenth centuries were made. Christopher Columbus bartered with the inhabitants of the West Indies in 1492 for skeins of their cotton yarn. Yucatan Indians presented cotton garments to Hernan Cortes, and the Spanish conquistadors encountered widespread cotton-based textile production amongst the Incas in Peru – where spinning of cotton for nets and fishing lines apparently predates cloth weaving - and the Mayas and Aztecs in Mexico - Mexicans wore cotton when Cortes established Veracruz. Spinning and weaving were carried out by the earliest New World settlers. Vasco da Gama sailed around Africa in 1497 reaching Calicut (Kozhikode) and discovered that centuries before, Arab traders had taught Africans to cultivate and spin cotton. From Calcutta, Vasco da Gama brought back to Lisbon *pintados*, or calicoes. Venice exported a variety of cottons in 1560, and Milan sent out fustians and dimities of many types.

The first record of cotton in English trade is by Hakluyt in the late fifteenth century, who mentions Genoese cotton imports; by 1600 cotton was imported fron Antwerp, Sicily, the Levant and Lisbon. The primary port of import was London, but in 1709 the pattern of imports began to change in favour of Liverpool. In 1757 there was the first recorded cotton auction in Liverpool and by 1795 the majority of British cotton imports (some 17 000 bales) passed through Liverpool. The distribution of cotton was a gradual process occurring on a small scale up to the eighteenth century. Current major producers were still importers: for example Egypt imported cotton from Syria and Cyprus in the late sixteenth century. As the use of cotton increased, selection occurred, so that the tall, unproductive perennial cotton species were replaced by varieties which were more compact, higher yielding and annual. Seed dormancy and photoperiodicity were other characteristics which were reduced or eliminated.

Asia has seen a cotton industry for centuries. In Korea, for example, cotton seeds were first introduced in 1363. China concentrated on silk. But for over 100 years after the great merchant traveller Roberts wrote his book *Traffic* in 1641, India kept ahead of Europe in cotton manufactured goods, the Eastern trading companies gaining great wealth through the monopoly. By the eighteenth century woven goods imports from India had created such a demand in Europe that much of the spinning and weaving industry of the UK, which was based in Lancashire, switched to produce cotton. In 1775 a patriotic association was formed in Edinburgh to discourage the wearing of Indian cotton robes in favour of native calicoes and lawns.

This was a forlorn hope: a number of factors combined to increase the efficiency of the industry. Already by 1727 Defoe said of Manchester that 'the grand manufacture which has so much raised this town is that of cotton in all its varieties.' Firstly, before 1733, weavers had to pass the shuttle containing the weft thread by hand through the parted warp threads. For widths of cloth of more than a metre, two men were needed, one each side, to pass the shuttle back and forth. The flying shuttle used a simple, hand-operated device to drive the shuttle through the warp threads: this enabled a weaver to double his output and increase the width of cloth woven substantially. Textile workers did initially oppose the introduction of the machine, and the inventor, John Kay (1704–64) was driven into a penurious exile.

The invention of the flying shuttle created a demand for more yarn. Until the mid-1760s the yarn was spun either by hand on the distaff or on the single spindle spinning wheel. In 1764 James Hargreaves (1722–78) began experimenting with a multi-spindled spinning machine. By 1768 he had advanced to an eight-spindled machine. This included a moving carriage for drawing out the roving, and was called a spinning jenny. Hargreaves escaped Luddite persecution and established a mill with 30 spindles, patenting his machine in 1770. At the same time, another Lancastrian, Richard Arkwright (1732–92) was building his roller-spinning frame. The roving was drawn out into yarn by sets of rollers rotating at different speeds. Arkwright patented the spinning frame in 1769: his first model was horse-powered, but later larger models were water-powered. His machine became known as the water frame.

Yet another Lancastrian, Samuel Crompton – the area was and remains the centre for the British cotton industry – was dissatisfied with the yarn quality produced by the spinning machines of the day. In 1779 he produced the spinning mule, which combined the best features of the jenny and the water frame. With the mule, the yarn was first drawn out by rollers and then drawn out further by a travelling carriage. With this machine Crompton began to produce thread of exceptional fineness. As the newly invented spinning machines came into widespread use, yarn was produced much more quickly than it could be woven on the hand looms of the day. A Nottinghamshire clergyman, Dr Edmund Cartwright (1743–1823), patented the first mechanical loom in 1785, and despite Luddite opposition, the machines went into substantial use during the 1790s.

The result was that in the late eighteenth century England held a textile industry monopoly worldwide.

Cotton was spreading geographically, too. The New World (tetraploid) cottons were distributed to many Asian and Africa countries, some as ornamental plants such as the perennial tree cottons and kidney cotton, and some as crop plants like the punctatum cottons in the dry belt of West Africa south of the Sahara. Conversely, upland cottons reached India, Cambodia and the Philippines. These introductions frequently met with no commercial success but continued to grow as weeds and wild.

The earliest European settlers in the southern US states planted cotton – reputedly at Jamestown – and from that beginning grew the vast US cotton industry. The greatly increased pace of spinning and weaving led in turn to a greater demand for raw cotton. North American planters began large-scale cotton growing in the South in the late 1700s, and imported slaves to work on the cotton plantations. The largest bottleneck in large-scale production was the separation of the fibres from the seeds. Throughout the eighteenth century, the extraction and processing of cotton fibre demanded vast manual labour, so it remained a relatively minor crop until the US inventor Eli Whitney (1765–1825) built in 1793 the cotton gin, which separated the seeds mechanically. This could do the work of 50 men separating the seeds by hand. At the same time, advanced English yarn-spinning methods were introduced into the USA by Samuel Slater, breaking the UK laws on the export of machines, plans or parts for them, and the emigration of textile mechanics. The 1789 Almy Brown mill was the first; in 1813 Lowell's mill was the first where all operations from the opening of bales to the production of finished cloth were all conducted under the same roof. Production of raw cotton expanded rapidly, from 10 000 bales in 1793 to 126 000 bales in a decade. From then, the total size of the crop and the percentage exported increased dramatically, so that by 1820 it had spread westwards into the Midsouth, Texas and Oklahoma, and US production dominated the world industry.

Weavers who wanted to introduce patterns into their weaving had quite a complex task. Each individual warp thread had to be lifted in a certain sequence, while the different coloured weft threads were threaded through. The more elaborate the pattern, the greater were the weavers' problems. Joseph Marie Jacquard (1752–1834) succeeded in 1801 in producing the famous Jacquard loom; a pattern of holes in a punched card determined which warp threads were to be raised. The Jacquard loom was soon widely adopted.

In 1808 the cotton business of Liverpool transferred from the quayside, streets and nearby coffee-houses to the New Exchange Building. Merchants transacted business on the 'Exchange Flags' outside, which were to be the site of the Cotton Market for almost a century thereafter. From 1832 reliable information about the Market was provided in a weekly and monthly *General Circular*, which were publications produced by cotton traders. Seeing benefits from this type of co-operation, in 1841 the Cotton Brokers' Association was formed. By 1863, members had agreed to make and enforce common 'best practice' trading rules. In 1882 brokers entered into dispute with merchants about their respective spheres of influence and it was agreed to expand and rename the Brokers' Association to include merchants, hence the Liverpool Cotton Association.

In Egypt, cotton assumed importance when tetraploid varieties were introduced from 1820 onwards. This Jumel cotton and other varieties introduced were of a longer staple than the existing commercial crop, and by the late nineteenth century Egypt was producing quick-maturing annual hybrids.

By the late 1850s, world cotton production was dominated by the USA: over 75% of the world crop was grown there and 70% of this was exported to the UK. Some 90% of US lint production was exported, and these earnings offset the costs of almost two-thirds of all US imports. As early as the late 1840s, more people worked in the US textile industry than in any other. Cotton thread was produced for the first time.

The US Civil War (1861–65) prevented cotton supplies from reaching Europe. The significance of the US Civil War is much debated. The revisionist position is that

> The decade of 1850–1860 was one of unparalleled prosperity for the 'slavocrats' of the South. Cotton production, although increasing annually, was barely sufficient to meet the demand. Thus, the price tended to rise. But in 1860 a record crop was produced. This caused a veritable glut on the market and the price of cotton began to decline and went on declining almost up until Lincoln's proclamation of blockade in April of 1861. The British market was glutted to the extent that many mills closed down in 1861, not from a shortage of cotton but from overproduction ... Britain and France had several millions of bales on hand when the blockade was imposed. This was more than enough to keep the mills going through 1862, although some were forced to close. By 1862–63 the British mill owners were locating alternate sources of supply in India and Egypt, sources which had always been there but which has been of greater cost than the American cotton. So in the end, the blockade benefited many British mill owners. It permitted them to get rid of surplus stocks of cotton without a significant decrease in the price of finished goods. And it was these very same mill owners upon whom the Confederacy relied to pressure the British government into going to war with the USA. (Nofi, 1974, p. 19).

Nofi adds that the Lincoln Administration actually relieved poverty in Lancashire to express thanks for the anti-slavery position taken by millhands there.

General opinion remains, however, that whether because of shortage of supply or overproduction, the effect was devastation to the Lancashire industry and indeed the encouragement of production elsewhere, much of it based on tetraploid varieties, for example in Turkey, Greece and in southern Africa. Only in the Indian subcontinent were these varieties successful, and required hybridisation in the twentieth century to succeed. India's exports of cotton to Europe increased from 515 000 bales between 1856–60 to 1.3 million between 1860–65 and 1.5 million bales between 1866–70. High prices achieved during these years were a major incentive to Indian cotton growing and to the expansion of Bombay – the end of the US Civil War brought financial ruin to many in the city. Australian cotton production also increased during the US Civil War, and also decreased again until it revived in the mid-twentieth century. In fact, revived US dominance was only arrested by rising US labour costs in the early part of the twentieth century and the depredation of the boll weevil.

At the same time, the need to regulate international cotton trading was increasing. The pace of business was increasing: improved means of information meant crucially that information arrived in advance of the vessels bringing the cargo. 1840 saw the start of the Cunard trans-Atlantic mail service, which allowed traders to buy and sell cotton in the 'Arrivals Market'. Orders were placed at the best possible price. On invoice, shipping and other expenses were added to the purchase price as well as a fixed commission of 5% for unloading. Price increases owing to freight fluctuations or exchange rates were all borne by the purchaser. Such dealings were speculative and difficult in a falling market. However, in 1864 the Indo-European Telegraph Company opened lines via Russia, Persia and cables in the Persian Gulf to India; in 1865 a link via Asian Turkey was established; in 1866 the first underwater cable to North America was laid and in 1870 a direct line to India was laid. These enabled merchants to hedge their sales commitments by immediate purchases for shipment. Purchasers' demands now adjusted quickly to changes in trading conditions; risks were passed back to sellers. Shipping dates were cut from the earliest possible outgoing shipment to three months, then two and finally down to 15 days. It became usual to request deliveries to European ports at monthly or even weekly intervals.

Hedging principles developed by Rew were introduced to Liverpool in 1866, at the same time as the first Transatlantic cable. Cotton could now be imported at a time of falling prices without inevitable losses; and this ensured ample cotton supplies to the mills of Lancashire just as they were recovering from the US Civil War. The revival of the US cotton industry after the Civil War was dramatic – in Mississippi pre-war there were fewer than 600 farms of under 10 million acres – by 1870 there were over 11 000. By 1876 the crop was larger than 1860. In the north, the textile spinning industry grew rapidly. The success of the Chicago Board of Trade made a major impact on New York cotton traders and they responded by organising a cotton futures market by 1870; in 1871 an Exchange exclusively for trading cotton was established at New Orleans and cotton exchanges were established shortly thereafter in Liverpool, New York and Bremen - still vitally important cities for the cotton trade. Finally, at the end of the century the telephone speeded up the whole process of buying, hedging and selling.

Up until the 1870s, England occupied a pre-eminent position in the cotton industry, cotton spinning and textile manufacture being the principal source of national employment, but this entirely changed in the subsequent 30 years owing to Continental competition. In about 1882 the US annual production was around 7 million bales, of which Lancashire took 3 million. The crops of Brazil, Peru, Central America and the West Indies averaged over 300 000 bales. Turkey and Iran produced about 120 000 bales. The Egyptian crop had reached 600 000 bales but was an established and differentiated trade. India produced 2.5 million bales and China 1.425 million bales, but generally of inferior quality.

From 1890 until World War II the Liverpool market traded futures in Egyptian cotton. Futures markets in cotton also existed in Bombay and Alexandria. The whole system revolved around Liverpool. But around 1900 an export business in Lancashire began to export second-hand textile machinery to Japan, India, China and the Continent. In 1908 there were estimated to be 144 704 012 spindles in the world, of which 55 971 501 were in the UK, 31 519 766 in the USA and 11 404 944 in Germany. The peak of achievement for the Liverpool market was reached in 1911/12, when 5.23 million bales were imported into the UK - compared with just 2000 bales in 1708. This level fell to 1.631 million bales in 1941. Prophetically, the first Chairman of the International Cotton Federation, Sir Charles Macara, said in 1904: 'In any dislocation of the cotton industry its serious effects on employment generally would be widespread. But it would not end here: the retail, mercantile, banking, professional and leisured classes would all suffer severely'. He was right: the north of England has never recovered from the collapse of the English textile industry; in 1900 cotton had an 80% share of the international textile market.

World cotton production continued to increase in the twentieth century. The world gradually became less dependent on US production, through greater production in Africa and India. In 1902 the British Cotton Growers' Association was established. 1904 saw the foundation of the International Federation of Master Spinners' and Manufacturers' Association. This arbitrated in international disputes and provided from 1922 an annual review of the US crop. The original motivation for the establishment of the Federation internationally was the poor crop in the 1903/04 season and the attempt by a Mr Sully in New York to monopolise cotton supply, forcing prices to their highest ever levels. The International Congress aimed to encourage cotton production, to prevent large-scale speculation and to establish a permanent secretariat.

In 1906 the Liverpool cotton market moved indoors and the present Cotton Exchange Building was opened. The trade flourished there: cotton arrivals in the UK reached their highest level in the 1911/12 season when 5.23 million bales were imported, mostly through Liverpool. At that time Liverpool was the greatest cotton importing market in the world and Lancashire was called 'tailor to the world'; Liverpool provided an unparalleled service to spinners in being able to provide a guarantee of availability of cotton for forward delivery positions, and in holding large warehouse stocks of all growths for spot sale in Liverpool. The first part of the twentieth century was marked by population expansion in the industrialised world and a dramatic increase in textile manufacturing, especially throughout Europe. The world cotton trade grew commensurately. Expansion ceased at the time of World War I. At that time, the International Federation had 16 member countries as members, representing three-quarters of the world's total productive capacity of 129.9 million spindles. These consumed a total of 20.3 million bales of cotton annually.

After the war, the export of cotton machinery and Lancashire 'know-how' helped overseas textile industries and UK raw cotton imports began their slow decline: British mills increasingly faced competition from newer mills in other countries. Because of the continued dependence that the Lancashire industry had on the USA, evidenced in World War I, the Empire Cotton Growing Corporation was founded in 1921, later to become the Cotton Research Corporation. This quango was instrumental in developing cotton production in Africa. Cotton generally was in short supply in the early 1920s; in December 1923 quotations rose to their highest level since World War I. Starting with the record crop of 1926/27, however, prices began to fall. The 1920s also saw the introduction of tractor power for cultivation, although this did not spread until the 1930s when additional implements were developed for seedbed preparation and cultivation.

Attempts to restore the balance through voluntary reductions failed, and in 1933 the USA introduced a price support mechanism for agricultural produce which has not been withdrawn since. In the 1930s, when world trade plummeted, cotton became too expensive in many countries. In Germany and Japan, dependence on cotton was reduced through government regulations, which required that rayon staple fibre be used in place of cotton; Japan was meanwhile entering the international market for the first time and the International Federation was already concerned about the implications for world production and trade. In other countries, cotton consumption was reduced by commercial decisions in favour of rayon later in the period before World War II. US consumption reached the UK level by 1900, and doubled between 1900 and 1945, to become by far the world's largest cotton-consuming country. Cotton consumption increased in India, Russia, Japan and China; in the 1920s and 1930s the South American countries, especially Brazil, also increased their consumption substantially. Continental Europe consumed as much cotton as the USA throughout the period prior to World War II.

Cotton production increased: fresh uses were found and new outlets were sought. A new peak of production was reached just prior to World War II. However, the share of production and consumption by country was changing. The US share of world production fell from 66% in 1900 to 45% in 1939 (approximately). US production remained relatively steady, around an average of 2.75 million tons (from 2.4 to 3.6 million), but other countries, notably Russia, India and Brazil as well as many other smaller countries, were increasing their output. The Indian Department of Agriculture was established in the early 1900s, and an Indian Cotton Committee, the predecessor of the Indian Central Cotton Committee (ICCC), was established in 1914. The ICCC controlled all aspects of the Indian cotton industry until independence in 1947, establishing in 1924 the Cotton Technological Research Laboratory in Bombay. The Indian government set up the Institute of Plant Research at Indore with a cotton section. As a result of these developments, the International Federation contained only 57% of world production by World War II.

The Liverpool Cotton Market remained active until World War II. In 1941, responding to wartime exigencies, the control of purchasing and distribution of raw cotton passed into government hands. The impact of World War II on the UK textile industry was once again serious.

Cultivation from World War II to the present day, by country

After World War II cotton production increased in all producing countries. World production reached approximately 40 million bales, compared with 31.6 million average before the war. During the five years from 1946 to 1950, annual production of commercial cotton – that which reached mills rather than being processed by hand spinning, used for stuffing mattresses or padded clothing or otherwise used outside mills – averaged approximately 25.4 million bales. Trade restrictions after World War II prevented cotton from immediately returning to its original trading patterns, and may have dealt a fatal blow to Liverpool's pre-eminence in the international trade. Efforts to restore trade in cotton centred around possible ways of providing reliable credit and confidence in currencies.

By 1950 the USA was producing half the world's commercial cotton, but this percentage was slowly falling. The USA had sought even before World War II to co-ordinate producer policies, hence the formation of the International Cotton Advisory Committee (ICAC) in September 1939, although work did not really start until after World War II. The ICAC gradually became the co-ordinating body for cotton producers. The Master Spinners Federation became the International Federation of Cotton and Allied Textile Industries (IFCATI), later to become the International Textile Manufacturers' Federation (ITMF), which continued to represent the textile industry as a whole.

World cotton production in 1952 was approximately 34 million bales, compared with a pre-World War II average of 31.6 million bales. Of this, almost half – 15 million bales – was grown in the USA. US price supports were already encouraging substantial increases in production in the rest of the world. Most of the world's cotton-growing countries consumed the majority of their output domestically. Even by the 1950s, among the countries which consumed the majority of their output were the USA, the USSR, China, Korea, India and Argentina. By contrast, Pakistan, Egypt, Brazil, Turkey, Mexico, Peru, the Sudan, Uganda, the Congo and Iran exported most of their production. During the 1950s, the majority of cotton exports went to the UK, other European countries, Canada, Japan and China. In the 1950s, approximately 40% of the world output of cotton was traded internationally, making cotton one of the most important internationally traded commodities. However, immediately after World War II the overwhelming majority of cotton was produced by primitive methods. Most cotton-producing areas relied on animal power for soil preparation; the crop was planted, cultivated and harvested by hand. Yields worldwide only began to increase substantially during the 1960s.

IFCATI was concerned after World War II principally with the management of international textile production competition (see the section on 'cotton textile manufacturing' in Chapter 6) and this role

Table 1.1 World cotton production by country (thousand 500 lb bales)

	1935–39 (average)	1960–64 (average)	1969 (average)
Argentina	289	552	475
Brazil	1 956	2 235	3 500
China	2 855	5 040	6 600
Colombia	23	335	700
Egypt	1 893	2 037	2 200
Greece	77	377	450
India	5 348[a]	4 741	5 200
Iran	171	494	750
Mexico	334	2 206	2 000
Pakistan	NA	1 656	2 500
Peru	379	632	425
Spain	10	427	300
Sudan	248	675	1 000
Syria	28	656	650
Turkey	249	1 091	1 850
USA	13 149	14 795	10 036
USSR	3 430	7 370	9 500

[a] *Includes Pakistan*
Source: USDA

has continued with ITMF. Table 1.1 shows the world cotton production by country.

The USA

In the USA some form of government control of acreage or acreage allotment was in operation from 1933 until the crop year 1951 when the programme was terminated. However, intensive cultivation during the years of acreage control created greater yields per acre, which offset the area reduction. In the years from 1937 to the 1950s, about 900 000 farms went out of cotton production in the USA, either by absorption with other farms or by changing to other crops. This left about 1.1 million farms in the cotton belt, although the belt itself was redefined, expanding westward. Cotton acreage in the west increased from 10.862 million bales in 1937 to 13.95 million bales in 1952; at the same time the old southeastern former cotton belt decreased from 10.405 million bales to 7.964 million bales. The largest increase was seen in California, which between 1937 and 1952 tripled its cotton-growing area and became a major force in production and technology.

In the period to 1950, cotton rather lagged behind other US crops in its utilisation of mechanisation. By 1949 only 6% of the US crop was mechanically harvested. The trend towards mechanised planting and cultivation then quickened, however, boosted by labour shortages and the beneficial economics. By the mid-1950s more than 7000 mechanical pickers were being used in western Texas and Oklahoma, reaching 30% of the crop by the mid-1950s. The price of these machines, though high for the small farmer, was not prohibitive for co-operative harvesting or for large plantations. By the mid-1950s in the USA the only real gap in complete mechanisation for cotton farming was in chopping, or thinning, out the plants after they had come up, and this was being overcome through new methods of hill planting and through experimentation with mechanical choppers, and, more importantly, through the increased use of mechanical harvesting. By 1958 in the southwestern USA, the most technically advanced cotton-producing area in the world, somewhat less than three-quarters of the crop was manually harvested. The degree of mechanisation decreased from west to east along the cotton belt, due to farm size, field slope and competition for labour. Parts of the Mexican crop grown contiguous to the USA were as mechanised as in the USA. Yields ranged from 157.1 lb/acre in 1930 to 298.7 lb/acre in 1944. They averaged 500 lb/acre in the period 1963–68, reaching a record 527 lb/acre in 1965, falling back to 450 lb/acre in the late 1960s and rising again to over 500 lb/acre by the early 1970s. By the end of 1965, US stocks had risen to 14 million bales; 65% of the US cotton crop was being picked and stripped by machine.

The effect of the relative increase in production in the rest of the world is that the US percentage of world cotton production fell: from 60% in 1929–30, through 41% in 1939–40, 50% in 1949–50 and down rapidly thereafter to 20% by the late 1960s.

Central and South America

In Mexico, more than 2.3 million acres were planted in 1951 producing a crop of nearly 1.4 million bales. The acreage more than doubled in the six years after World War II, and production increased proportionately. Of the Mexican crop, 90% was produced in four main regions. This considerable increase was caused by the use of irrigation water provided by dams, canals and wells. The Laguna region, one of the most fertile and productive regions of the country, has always been noted for the quality of its cotton. By 1975 the crop was sown on a dramatically smaller area of 580 000 acres, with a yield of 745 lb/acre, producing a crop of 900 000 bales. By comparison in 1982 the crop was sown on only 519 000 acres, with a yield of 879 lb/acre, producing a total crop of 950 000 bales.

During the early 1950s there were about 4.6 million acres planted in Brazil; annual production was about 1.5 million bales, and exports averaged 550 000 bales. In Peru in 1951 more than 427 000 acres were planted and 400 000 bales picked; the country ranked second after Egypt in the production of extra-long staple cottons. In Argentina, the acreage planted increased steadily after World War II, reaching 1.3 million acres in 1951, producing 600 000 bales. 80% of the crop was grown in the Chaco region which was climatically favourable.

The UK

After 1954, with the gradual deregulation of the cotton trade, activity picked up considerably. In February 1963 the Liverpool Cotton Association became a limited company, with full membership being open to individuals and to limited companies actively engaged in raw cotton trading in the UK. Associate membership was made available to those engaged in cotton manufacturing and in allied industries in the UK and also to overseas organisations.

As the actual volume of cotton imported through Liverpool has shrunk, Liverpool cotton since World War II has changed from an important fact to a dominant idea. The volume and proportion of trading under Liverpool Rules and therefore ultimately subject to Liverpool arbitration has risen very significantly and is now estimated to constitute some 60% of all international cotton trade. Over 400 companies in 58 countries are registered with the Association and make use of its services. There is a relatively continuous turnover of membership.

Cotton Outlook, the world's leading cotton analysis company, has its home in Liverpool; trading firms based in the UK make their home there also; and the Liverpool Cotton Association continues to exercise huge importance. The cotton that previously was shipped to Liverpool for resale to the Lancashire industry is now traded through Members and Associate Members worldwide for shipment between various countries.

India

When in 1947 India was partitioned into the two countries of India and Pakistan, approximately 75% of the subcontinent's cotton acreage,

65% of the production, and 98% of the commercial consumption of cotton remained in India. However, most of the Pakistani cotton fields were irrigated: not so in India. In the mid-1950s, India had about 15 million acres of cotton and produced approximately 3 million bales annually compared with 25 million acres in 1914. Despite partition and the demand for more food crops, India remained the fourth largest cotton-producing country, albeit with very low yields. By 1975 India had an acreage of 18.1 million, a yield - still comparatively very low - some one-third of the US average and production of 5.7 million bales. Despite government concern over the low level of yields, they failed to rise over the succeeding decade, being only 148 lb/acre in 1982 compared with 151 in 1975, and with a planted acreage up to 2.087 million acres, the crop was 6.3 million bales. India's production only began to rise significantly during the 1980s.

Pakistan

West Pakistan's (until Bangladesh's separation) provinces of Punjab and Sind are rich agricultural areas: 94% of Pakistani cotton was produced in the west. Demands for more food acreage restricted cotton production. But the government aimed to produce 2 million bales through the extension of irrigation and high per-acre production planning. In the mid-1950s, just over 3 million acres were planted, 90% under irrigation and yielding an average of 175 lb/acre. Production was controlled by the Pakistani Central Cotton Committee, which also controlled work in improving cultivation, processing and marketing - rigorous legislation on cultivation methods was in force, but mechanisation by the 1950s was still negligible in Pakistan. The effect of mechanisation, which began to be introduced in the 1960s, was dramatic. Yields rose to 241 lb/acre by 1975 and continued to climb, reaching 333 lb/acre by 1982. Acreage in Pakistan rose from 4.7 million acres to 5.338 million acres. As a result of the combination of acreage increases and yield improvements, the total Pakistani production rose from 2.36 million bales in 1975 to 3.7 million bales in 1982. In comparison with India, production in 1975 was 41.4% and in 1982, 58.7%. The gap continued to narrow thereafter.

Egypt

Cotton became the foundation on which the Egyptian economy was based. Hand-picking itself required an immense amount of labour which was in many cases the limiting factor in the growth of production. Thus cotton was essentially a low wage crop, and the best results were obtained when hand labour was used throughout, for example in Egypt. In the 1950s nearly one-third of the cultivated land was cotton, and three-quarters of Egypt's export revenues came from cotton. Cotton was grown on irrigated land throughout the Nile valley, and cotton acreage was restricted by the government. In 1951 more than 1.5 million bales were grown on approximately 2 million acres of land. Cotton was produced largely by hand, mainly because of low labour costs. Labour requirements in 1950s Egypt totalled approximately 1500 h/500 lb of cotton, compared with 15 in the high plains of Texas. By 1975 Egypt produced on 1.4 million acres with a yield of 617 lb/acre a total of 2.205 million bales. This compared with 1.087 million acres, 883 lb/acre and 2 million bales by 1982. These trends were favourable but not impressive. Egypt's relative importance as a cotton producer has declined with the decrease in importance of extra-long staple cotton.

North Africa

The Sudan produced 225 000 bales plus in the 1950s on nearly 600 000 acres. Yields were increasing during the 1950s, averaging about 400 lb/acre. Cotton is indigenous to the Ethiopian highlands, having been produced there for textile fibre since ancient times. By the mid-1950s, however, annual production did not exceed a few thousand bales, a steady decline having occurred since the Italian occupation (1936–41), during which time cotton districts had been established, roads had been built and gins constructed. About 3000 bales were produced in its peak year for export to Italy. In Eritrea, production has always been small, with only a few thousand acres planted. In Somaliland, under Italian occupation, about 11 000 acres were planted annually, mostly on plantation concessions, but commercial production declined after 1940 and by 1950 amounted to only a few thousand bales. Production has not revived since.

Sub-Saharan Africa

Sterling's first devaluation in 1949 encouraged the saving of dollars by producing cotton in Africa. Although cotton lost out to some extent to coffee in Uganda, taken together, Uganda, Kenya and Tanzania produced approximately 350 000 bales annually in the mid-1950s, most of it being shipped to the UK and India. The Uganda crop was significant in world trade as it was the only source outside the USA of fine-fibred long staple American upland cotton. By the 1960s this output had increased to over 450 000 bales, although this was still less than 5% of world production. In the Congo, the total area under production in the mid-1950s was about 790 000 acres: production was from small plots of about an acre, using just knives and hoes, managing a yield of about 150 lb/acre. The crop was maintained at an annual level of approximately 210 000 bales by strict adherence to local laws. During the 1950s cotton constituted approximately 10% of the Congo's export revenues.

The former French territories in Africa, Nigeria and the coastal countries produced annually in the 1950s a total of about 150 000 bales. This was the main source of income for more than half of the 4 million inhabitants of French Equatorial Africa in the 1950s, forming over half of export revenue. Cotton was still a garden crop, with an average of one acre per family. No soil rebuilding was practised, but by law cotton could be planted on one plot only on a four-year rotation.

The annual production in Southern Africa was about 330 000 bales; about 90% was produced in the Portuguese colonies of Mozambique and Angola, which together planted 800 000 acres, supplying all of Portugal's needs.

Since World War II cotton production has been steadily expanding in the savanna lands of francophone Africa except for Central Africa itself; the area under cultivation has doubled since the early 1960s and is now 1 million ha. Yields have also increased dramatically, because of 'green revolution' higher-yielding varieties, from 400 kg/ha in 1970 to 1000 kg/ha plus in the early 1990s. The ginning percentage – the percentage of fibres removed from the seed cotton – is also increasing and Africa holds the world record with more than 40%. The effect has been for cotton production to increase more than 36-fold in francophone West Africa, 13-fold in Madagascar and nearly 3-fold in Central Africa.

By contrast, production in the English-speaking African countries has experienced continual decline. In 1960, 90% of cotton produced in sub-Saharan Africa came from the anglophone countries, but during the 1970s production fell in the Sudan, Uganda, Mozambique, Tanzania and Nigeria, which, apart from Egypt, were the region's major producers. Only Zimbabwe succeeded in increasing its production.

The disparity has partly to do with relative investment. In 1976 the Cotton Research Corporation (CRC) was closed down after 55 years. Cost escalation, the effect of the 1974 oil crisis on UK government funding, the emphasis on food technology and a hard backward look at the decline of the UK textile industry all contributed to this unfortunate decision. Productivity faltered in many countries where CRC staff had been based, and fibre quality failed to increase to meet rising demands from the marketplace. The catalogue of subsequent problems included a failure to maintain correctly organised production and also to distribute seed regularly for sowing. Varietal purity suffered. Infrastructures were not maintained and vehicles, depots and ginneries fell into disrepair.

There were also marketing difficulties. Farmers found themselves waiting too long for revenues, credit was hard to come by, and many small-scale farmers found it impossible to raise the money necessary to buy fertilisers and pesticides, so vital for cotton production. So food crops became more attractive: maize in Nigeria and rice in Tanzania, for instance. The high substitutability of cotton-suitable land was an important factor in this process. Cotton price falls, especially relative to food crops, combined with the rising cost of inputs, made cotton production less profitable. In addition, political and social upheaval damaged the entire agricultural infrastructure in many countries: war in Mozambique, war and famine in the Sudan and a repressive regime in Uganda, for instance. Governments appeared indifferent in many anglophone countries, and producers appeared to be indifferent to a crop which lacked adequate research, training, marketing and other support and which demanded high cost inputs. More recently the governments of these states have initiated rescue packages, but the effects have yet to be seen.

Africa as a whole produced 478 000 tonnes of cotton in 1966–70, and only 515 000 tonnes by 1982. Even 10 years later total African production amounted to 890 000 tonnes. Africa remains a very small source of cotton in production terms worldwide. However, because of particular quality requirements and the high percentage of production exported, West African cotton in particular has a higher importance in world trade – and especially into Europe – than production figures suggest.

The USSR

In the former USSR cotton was grown under government control on kolkhozy (collectives) and sovkhozy (state farms). The Ministry of Agriculture distributed seed, and its agents superintended cultivation and harvesting. The machinery was owned by the farms and held at

Machine Tractor Stations. The crop was ginned at government-owned gins and bought by the government at a fixed price. The majority of the crop came from Turkistan where it was grown under irrigation.

It was estimated that 3 million acres were planted in 1940; total Soviet production and acreage obviously fell dramatically during World War II and fewer than 2.3 million bales were harvested in 1949 from less than 3 million acres. This compares with an estimated 1.5 million bales even during World War I for the Russian crop of 1916. Plantings began to increase by 1950, up to 5.9 million acres, but production information became unreliable. It was understood that Soviet production experienced a rapid growth from 1950 onwards, from 3.215 million bales to reach a level according to the US Department of Agriculture (USDA) of 12.1 million bales in 1975 from an acreage of 7.22 million, yield 804 lb/acre and then to the ICAC estimate for 1983 of 12.57 million bales, comprising 10.37 million bales of medium-long staple, 1.76 million bales of long staple and 0.44 million bales of extra-long from a yield up some 5% only from 1975. However in retrospect these production figures may have been overestimates.

China

In 1947, under the UN Relief and Rehabilitation Administration (UNRRA), large quantities of improved and certified seed were allocated to China, and a majority of that and subsequent years' planting was of improved seed. By 1948 Chinese cotton production, immediately prior to the Communist takeover, was estimated to be about 2 million bales, a total probably quite similar to output throughout the inter-war years. Production and acreage was increasing, reaching 3 million bales from more than 10 million acres by 1951. Production was estimated by the USDA to have reached 11 million bales by 1975, rising to 14.5 million bales by 1982. Acreage was 12 million in 1975 and 13.35 million in 1982, illustrating the substantial yield growth from 440 lb/acre in 1975 to 522 lb/acre in 1982.

CHAPTER

2

The growing of cotton

Planting and growing

The climate best suited to cotton growing may be described as subtropical. However, cotton is a very adaptable plant with many different types: some types can be grown where no other crop of comparable value can be grown. Commercial cotton growing usually demands about 200 days between killing frosts and a minimum summer average temperature of 25 °C although cotton is grown in very dry, humid, semi-desert and desert climates as well as subtropical, equatorial and Mediterranean climates. However, more than half the world's cotton crop is grown in temperate zones.

Rain-fed cotton currently accounts for about 35% of production. Cotton demands ample water (about 600 mm annually), which can be obtained from either rainfall or irrigation, but good drainage is necessary to avoid waterlogging. About 100 mm of annual rainfall is sufficient to eliminate any need for irrigation, but yields are also affected by the distribution of rainfall over the year as well as the annual total. All cotton with less than 400 mm rainfall is irrigated. Rainfall/water supply is the main determinant of the crop: for rain-fed cotton the need is for adequate but not excessive rain throughout the growing season, falling off during ripening and sowing. The period of growth of the plant from sowing to picking is about five to nine months, but in many cases the picking season is very much drawn out. The plant grows adequately in a wide variety of types of soil; the principal need is a sufficient root range. Consequently neither particularly sandy, acid, or salty soil represents insuperable difficulties. In the USA, untimely rain can damage the crop, as happened during the summer of 1992; the rain stains or 'tinges' the ripe cotton and lowers its value. Of course, where irrigation is not possible the crop is liable to serious drought damage during the growing season.

A further 65% comes from irrigated cotton, including extra-long staple. Dry, warm climates are the best for irrigated cotton. The use of irrigation increased dramatically in the twentieth century: canal capacity and pumping stations have increased capacity and now are operated throughout the year. Irrigated cotton is obviously more expensive to grow, but 'it usually results in higher yields, more regular harvests, a higher quality of fibre and requires a shorter ripening period' (Morris, 1988, p. 12). A majority of the highest yields are now produced under genuine desert conditions with irrigation: for example Australia, Egypt, Israel, the Sudan, Syria, the San Joaquin valley of California and India, where cotton was at first a monsoon crop but where irrigation now provides the best growing conditions.

Almost all cotton is now grown as an annual crop from direct seeding, although perennial cotton is still grown in a few places. The sequence of production is as follows:

1 residue disposal;
2 preplant tillage;
3 seedbed preparation, including (a) fertilisation and (b) application of a broadcast soil-incorporated preplanting herbicide;
4 planting, including the application of fungicides, systemic herbicides, and pre-emergence herbicides;
5 post-emergence weed control, chemical and mechanical;
6 insect control;
7 harvesting and hauling.

In addition whatever irrigation is necessary must be carried out.

In the USA, planting begins as early as February in the lower Rio Grande valley of Texas and as late as June in the northern areas of the cotton belt, e.g. Texas–Oklahoma Plains. Normally planting takes between two and three weeks. Land preparation starts in the autumn, just after the harvest. Stalks from the old crop are chopped by rotary or flail-shredding machines and the residue is ploughed under. Some stalks are left standing in some areas to control wind erosion but in most areas the importance of pest control demands complete shredding. The important priorities are to control potentially damaging weeds and to provide a suitable seedbed. Annual weeds are usually buried, but they can be allowed to germinate and then be destroyed. Perennial weeds are a more difficult problem. In the spring, the soil is tilled to a depth of several centimetres, then smoothed and laid off in rows. The crop needs a fairly warm, moist, well-formed, well-drained, clean and firm seedbed to ensure that the crop is sown to approximately the same depth, with firm soil below the loose surface. The type of seedbed preparation carried out depends on the rainfall expected and whether the land is irrigated. Fertiliser, a key production input in many areas, is introduced at this stage. Final preparation is usually left until as late as possible to facilitate drainage, especially on sloping land. Some cotton is grown on ridges, especially in Africa. Both run-off and erosion are problems under these conditions. Federal law in the USA will require all C/D steeper slope classification farmlands to incorporate a conservation plan by 1995. Cover crops will need to be planted for cotton. Initial data reported by Cotton Incorporated (*Annual Report 1992*) suggests that yields are thereby increased in comparison with conventional tillage methods. However, weed control methods still need to be improved. Narrow-row plantings are increasing under conditions of crop rotation, giving faster canopy coverage and an earlier maturing crop with similar or even better yields. Preplant herbicides are introduced at this stage in 90% of plantings in the US Cotton Belt. Planting usually takes place when the soil temperature has reached 15 °C at planting depth.

Planting in the USA is usually done with a six or eight row tractor-drawn precision planter. The planter opens furrows, drops in the correct amount of seed at determined depths and intervals, and covers them over with soil. A wide variety of implements is used in land preparation and sowing: ploughs, ridges and rotary cultivators, each of which has a benefit for a specific type of soil and climate. In low rainfall areas the crop is planted at a depth of about 38–50 mm, and in rainfed areas 25 mm in sandy and loam soils, and 38–75 mm in clay soils. The aim is to produce uniform distribution of seeds, which is desirable for chemical weed control for maximum yields and to reduce seed cotton losses when harvesting.

In some developing countries, the crop is still sown by hand or ox hoe. Hand cultivation may encourage a family to plant more than it can weed; and insufficient oxen may be available for ploughing in semi-arid areas. Mechanisation, almost invariably, brings higher yields and more cost-effective production under almost all circumstances, except where labour is plentiful and cheap or where the mechanisation is not maintained. The careful balance between increased mechanisation and rising rural unemployment is an extremely delicate one to strike, and no developing country would claim to have achieved it yet. Munro (1987) summarises the advantages and disadvantages of mechanisation as follows:

1 Reduced labour requirements
2 Quicker operation, therefore more land can be prepared in time and
3 Maximum advantage can be taken of suitable weather
4 Dry season cultivation is possible
5 Deeper cultivation can be undertaken to break up pan.

Against this, the following disadvantages should be recognised:

1 Dependence on imported oil
2 Technical skills are necessary
3 Efficient and regular maintenance must be organised
4 Repair and replacement services are necessary
5 Tractors may not be available when conditions are ideal
6 Small scattered holdings mean excessive road travel and
7 Tractors may be under utilised during the remainder of the year.

Over time these disadvantages have become less important as the relative price of oil has fallen, the level of education in developing countries has gradually increased, and the cost of providing mechanisation has decreased relative even to quite depressed raw cotton prices.

Fertilisers and pesticides are systematically put down at the same time as the planting. This saves labour and energy and prevents the soil from becoming more compacted than it needs to be.

The young cotton seedlings emerge from the soil within a week or two after planting. Exact timing depends on temperatures and moisture conditions, and can range from as little as four days after planting, up to three weeks under very unfavourable conditions. Squares or flower buds form about 35–45 days later, hence the reference to the 'squaring' period. The first bloom usually occurs in the first node of a fruiting branch low on the stalk. Shedding of squares is common, attributable to internal and external stresses on the plant. The blossoms appear in another three weeks, turning, over the next three days, pink and then dark red (or with yellow flowers, in the case of Pima cotton) before falling off and leaving a tiny ovary attached to the plant. This is what ripens and becomes the cotton boll.

Frost is a threat to the cotton crop. In the USA, for example, the date of the first killing frost in autumn is often the determining factor in the final yield of the crop - although frost itself is also the worst enemy of certain insect pests, for example the boll weevil. Cotton often matures several weeks before the first frost. Defoliants are applied to make the plant shed leaves earlier, and desiccants to make the leaves dry sooner on the ground. This allows more cotton to be gathered at first picking, in advance of any adverse autumn weather. Droughts in summer also reduce the insect threat.

Cotton requires intensive cultivation at certain periods. After sowing, it must be chopped - thinned out - and the ground carefully cultivated - hoed - to keep it clear of weeds and to prevent unnecessary evaporation of the moisture of the soil. There is usually a period before final harvesting when the crop is 'made' or 'laid by' in US terminology.

Harvesting usually starts around mid-July in southern Texas, about mid-August in the mid-belt and around the beginning of September in the northern areas. As a result, harvesting in the northern areas is not usually completed until the end of the year, or even later if it is delayed by unfavourable weather. Only a negligible percentage of the US harvest is harvested manually.

Chemical defoliation (including aerial spraying) and desiccation check plant growth and accelerate boll opening, thereby reducing insect damage and preventing the late season build-up of insect pests that would otherwise remain for next year. Defoliants are used for cotton that is to be picked with spindle harvesters and desiccants in that to be stripper-harvested.

The cotton cycle

Cotton is a rotation crop. Its cultivation cycle is 175–225 days and in many tropical countries both cotton and other (usually food) crops are produced on the same land in one year. After pollination the multiple fruit grows, expanding to become a capsule about the size of a walnut. After another 25–45 days the dry walls of the capsule pop open and four or five fibrous pods well out, each of which encases several seeds; the fluffy mass expands beyond the walls or 'burs' of the capsule. Each boll contains between three and five locks of cotton, each of which contains 5–11 seeds per lock, densely covered with lint and fuzz in commercial varieties. These cling to the mother plant until they are picked off or are blown off by wind. Chemical defoliation is started when 60–70% of the cotton bolls have opened. Every ripe cotton seed is embedded in a thick covering of short fluff and long hairs. These form the raw material to be spun.

Harvesting is vital because the crop must be raised before weather damages or ruins the quality and yield; for example, it must be completed in California before the winter fog and rainy season starts. No mechanical picker can harvest long and extra-long staples without causing damage to the fibre.

The ripe bolls are harvested largely by machines, although a small percentage is picked by hand - the Zimbabwe crop is an example - although this in itself should not necessarily be regarded as conclusive evidence of technical backwardness. A hand picker only collects 80-100 kg of cotton daily, but the cotton is all clean and of consistent ripeness. By working the same field several times, a maximum yield per surface area is achieved. By comparison, the rotating spindles on a picking machine mesh with each stalk and remove the cotton fruit, including some leaves and stems, and can harvest more than 155 kg/h. In the USA, the Commonwealth of Independent States (CIS) (depending on the state of machine repair) and Australia cotton is for the most part harvested mechanically: either with spindle-type pickers or tractor-mounted strippers. The picker has vertical drums with revolving spindles which engage and remove the cotton from the open bolls. The US National Cotton Council (NCC) reports that a two-row picker can harvest about 2200 lb of seed cotton - or about two bales of lint cotton per hour; a four-row picker may harvest up to 8 bales/h. Over 70% of US cotton is thus harvested.

The cheaper machine stripper, better suited to the compact wind-resistant cotton grown in Texas and Oklahoma, has rollers or mechanical levers removing the entire boll from the plant - 90% of the crop there is thus harvested. It is less suited to the tall plants and humid conditions of other regions. A two-row stripper can harvest a bale an hour and 15-20 acres/day. A third type of harvester is the gleaner or ground retriever, used to harvest about 1% of the US crop.

Harvested cotton is stored in the fields, in the larger US farms, in large covered bays containing 10-12 bales, until the whole crop is completed and ready for ginning. The seed cotton is then transferred from the harvester to trailers or modules in the field and delivered to the gins either by the grower or by the gin. More than 55% of harvested seed cotton is now stored in modules, formed by tightly compressing cotton on the ground or on a pallet resting on the ground. Modules improve pre-ginning storage efficiency and are now a source of averaging grades ('module sampling'). The increased attention to storage is because adverse weather can damage any crop backed up waiting to enter a ginnery. The module is taken to the ginnery on a truck or trailer. The harvesting process in the USA can be contracted out or may be carried out by ginneries but this is comparatively rare elsewhere in the world.

The picked cotton then goes to the ginnery, where a cotton gin separates the long fibres, or lint, from the seeds. Presses then pack the raw cotton into tight bales for transport to the spinning mill. There are many different types of bales, especially in the USA. 'Gin universal density' (or gin UD) bales average 500 lb. Bales also vary in weight internationally.

Pests, diseases and weeds of the cotton plant

Cotton production before 1945 was largely dependent on the level of insect attack, so production was also geographically confined to those areas where it was not prevalent. Insect pests were second only to climate in determining where cotton could be successfully grown. Cotton is indeed particularly susceptible to insect attack. It has lush foliage, succulent leaves, large flowers - all with nectaries - and its heavy fruitage encourages insects to feed on the plant or to reside within it. Hundreds of species of insect attack cotton, among them some of the most destructive known to agriculture. In Nigeria early sowing controlled red bollworm (*Diparopsis watersi*) although the emphasis on food production confined its adoption except on research stations. Most African countries had legally enforceable closed seasons to prevent the carryover of red and pink bollworm (*Pectinophora gossypiella*) which has few or no other host plants. In Egypt, it was made law to collect the egg masses of the cotton leafworm. Trap crops, paper bags and poison baits were tried but without much success. In the 1940s it was shown that jassid resistance was related to the hairiness of the cotton leaf, and this knowledge spread over Africa. However, despite these measures, losses from pests continued to mount, causing damage estimated at more than $100 million annually in the 25 years after 1929, with the worst year being 1950, with losses of $900 million.

About seven major diseases substantially affect the cotton plant, most of them attacking the plant in its seedling stage. In some African countries and in Brazil, during the first half of the century up to 50% of the entire crop could be lost to disease. In India losses ranged from 7% to 20% and even in the USA, where control efforts were always intensive, losses reached 15% regularly.

From 1892 onwards, when the first taxonomy of cotton diseases was published, new diseases have constantly been identified. The diseases can be categorised into ***fungoid***, ***bacterial***, ***nematodal*** or ***physiological***. (These may be caused by poor soil conditions.) Fungi enter the plant through the soil; the fungus may live saprophytically for many years. The most widespread diseases currently are verticillium wilt, fusarium wilt and bacterial blight. Disease losses currently amount to approximately 1.5 million bales of the US crop annually. The increase in cotton production since 1945 was in part permitted by the new insecticides, in particular DDT and BHC, which are synthetic organic insecticides with strong contact action and long persistence. DDT was first used in the USA in 1944, and since then a wide range of synthetic organic products has appeared. The world output of pesticides has increased and world expenditure on pesticides has risen from $5 billion in 1974 to many times that in 1993.

There are over a thousand active ingredients registered in the USA, and over 100 000 tradenames. The effect of the new insecticides has been to allow cotton production to expand into areas hitherto considered uneconomic because of pest problems, as well as to increase yields. It does, however, continue to be important that farmers display adequate 'pest realisation', both in identifying that there is a problem and in knowing what to do about the specific problem. Insect losses are estimated to amount to 950 000 bales annually in the USA.

Pests

The most destructive cotton pests are, in order of importance, the boll weevil, the bollworm (red and pink), the red spider, lygus bugs, thrips, the cutworm and the flea hopper. Different insects cause damage in different parts of the world: in China for instance the 1992 crop was very badly affected by the boll weevil.

Cotton boll weevil

The boll weevil (*Anthonomus grandis*) was the best-known cotton pest, which spread from Mexico (first recorded in 1843) to Cuba (1871) and on into the US cotton belt, crossing the Rio Grande near Brownsville, Texas, in 1892 and spreading on into 85% of the US cotton belt by 1922 and then into Venezuela and Colombia; a related species is found in Peru and Ecuador, but neither has been found elsewhere. To control it, early-cropping varieties were used, which shifted the industry from the deep south to the west. In the later twentieth century, after a period of increasing dependence on synthetic insecticides, integrated pest management programmes have again emphasised the importance of cultural practices and are now dealing with whole groups of farmers. These programmes have re-emphasised the importance of planting early-maturing varieties, preparing seedbeds correctly, early planting, seed treatment, soil improvements, sometimes chemical defoliation of plants and the cleaning up of favoured hibernation sites, with special emphasis on the destruction of cotton stalks quickly after the harvest. Progress in eradicating the boll weevil entirely in North Carolina and Virginia has been made but the problem is not completely solved yet.

Other beetles

Other beetle pests include cotton stem borers (*Sphenoptera gossypii*, *S. neglecta*, *Eutinobothrus (Gaterocercodes) gossypii* and *E. braziliensis*), stem-girdling beetles, leaf-eating beetles and flea beetles. These are found in both Africa and America.

Termites

Both the mound-building termites and harvester termites can pose problems for the cotton crop. Mound-builders have caused damage in Tanzania and Kenya, and the construction of the mounds can have a serious deleterious effect on soil fertility. Harvester beetles (*Hodotermes mossambicus*) attack growing plants for the material they want, including cotton in the drier parts of Tanzania and Malawi.

Bollworms

Bollworms (*Heliothis*) attack many crops, including cotton, tobacco, corn and tomatoes. They are hard to control because insecticides fail to reach them once they are inside the boll. Cultural practices similar to those used against the boll weevil are helpful in controlling bollworms. The pink bollworm is a general pest in all cotton-growing countries apart from the USA and Mexico. It is one of the most serious insect causes of loss. The bollworm heavily damaged the 1992 Chinese crop, for example. First found in Mexico in 1911, it moved quickly to

Texas. Vigorous inspection, quarantine and control methods were immediately enforced and remain in effect. The chief control method is to heat cottonseed to 65 °C for 30 seconds or more as part of continuous ginning or to fumigate it. The lint cotton is treated by passing it between steel rollers designed to crush the worm, by fumigation of baled cotton and by high compression of cotton bales. Rigid inspection at gins and mandatory burning of gin trash in specially built incinerators also form part of the control programme.

Insecticides

Chemical insecticides remain an important part of the campaign against insects. During seasonal periods of high infestation, they are applied either by air - which enables up to 1500 acres/day to be covered - or by tractor-mounted rigs which can spray several rows at a time. Constant efforts are made to scout out cotton land to determine how much pesticide to use. There are six main types of insecticide - the actual availability of types varies as technology improves, brand names change and environmental banning changes. Different products are available in different countries, depending on, for example, toxicity to humans.

1 Inorganic, e.g. Paris green, lead arsenate.
2 Natural, e.g. nicotine, derris, pyrethum.
3 Organochlorine, e.g. DDT, dieldrin, endosulphan, toxaphene.
4 Organophosphorus, e.g. monocrotophos, parathion, malathion, dimethoate.
5 Carbamate, e.g. carbaryl, aldicarb, carbofuran.
6 Pyrethroid, e.g. permethrin, cypermethrin, fenvalerate, delatmethrin.

Insecticides used to be applied in many countries, for example Uganda, on a fixed schedule - 14 day intervals beginning 35 days from seeding. Although this has the advantage of routine and packaging, it is limited in effect when confronted with a variation in pest attacks both in severity and timing. Consequently a 'scouting' process was developed to identify what flexible schedule should be adopted. The cotton plant can tolerate a certain amount of damage; it normally produces more leaves than it needs, and can compensate for the loss of a percentage of the buds and bolls produced. Ideas on pest control prevalent since the 1980s admit a certain amount of damage and then intervene only when a certain pre-agreed level is reached. This aims to reduce the number of sprays, allowing beneficial insects to develop, and also hopefully increases the time taken for the pests to develop resistance to the insecticide.

Herbicides

Successful defence practices have been devised against almost all known cotton diseases, using a combination of methods in most cases. Rotation of crops is helpful in eliminating nematode infestation, fusarium wilt, root rot, *Ascochyta* blight, anthracnose and bacterial blight. Ethylene dibromide, used as a soil fumigant, gives almost total protection against fusarium in the presence of nematodes, especially when used with wilt-resistant varieties. Verticillium wilt is widespread throughout the US cotton belt. Only selection of seed for wilt tolerance, avoiding excessive irrigation and excessive application of organic matter and crop rotation are known to work in protection against wilt tolerance. *Ascochyta* blight can be fought with cutting and burning of old stalks and by crop rotation. Fumigation of planting seed is of use in controlling seed-borne diseases, and can prevent attack by soil-borne fungi on cotton seedlings.

Prevention

Attention is currently focused on disease prevention rather than cure. The development by cottonseed breeders of varieties with seedling vigour and resistance to fungi and other organisms was very important. In the San Joaquin Valley in California there has been considerable success in growing a variety resistant to the diseases found in certain soils.

Integrated pest management

A combination of factors led to a rethink. The cost of insecticides, the advice from entomologists to growers, and the danger to the environment were the three main factors. DDT and other chemicals have been banned for use on field crops in the USA, and subsequently in many other countries.

Peru introduced a better system, limiting the use of organic insecticides, in a programme known as '*control integrado*' or integrated pest management (IPM). The original suggestion for the integration of biological and chemical methods of pest control was made as early as 1954; the system was introduced experimentally in 1957 and Peru passed legislation making it compulsory in 1961–62.

IPM involves combining insect control measures such as close season, restricted planting dates and the destruction of alternative host plants. In Peru, bollworm egg parasites of *Trichogramma* sp. were bred locally for release into the cotton fields; the free use of calcium and lead arsenate was permitted, but official permission was required before any organic insecticide could be used, and this was given sparingly and only after investigation. The number of applications of such insecticides was reduced drastically, and the cost fell to under 25%, while yields actually rose or at least were maintained. IPM proved applicable to other countries less suitable than Peru, where reinfestation from neighbouring areas was not great. Part of IPM are the 'cultural methods' of crop control. Most important is the destruction of the crop residue straight after the harvest, burying or burning the stalks and damaged bolls. This controls pests and diseases which need the cotton plant to survive, such as boll weevil, leaf miner *Bucculatrix* and bacterial blight of cotton, and it has some effect against other pests too. Supplementing this is often a legal close season, which establishes a date by which cotton residue must be destroyed, and a date before which no cotton must be sown. The aim is to break the reproductive cycle of the pest and to reduce the number of diaporse forms. To complete an effective plan, loose seed cotton in fields and elsewhere must be disposed of; stores need cleaning out and fumigating annually at least. Next, the sowing time is arranged so the peak pest population coincides with a time when the crop is not vulnerable.

Legislation was introduced creating controlled communities and weevil-free zones in the USA, and there have also been voluntary schemes of IPM.

IPM takes many forms, usually depending on the life history and habits of the insects concerned. A thorough study of the insect pest is needed before suitable measures are planned, as measures effective against one may actually encourage another - leaf hairiness, for example, which deters jassids but encourages whitefly.

Problems with the technology

Three key problems remain, in order of seriousness. First, *pesticides damage the environment*. It was well known by the 1960s that the intensive use of these insecticides carried dangers. First, there was the destruction of beneficial parasites and predators. Second, there was the development of stains of the pest species resistant to the insecticide. Up until the 1970s, although in gradually decreasing amounts, farmers responded to these problems by increasing insecticide use, adding in new chemicals and raising doses.

Second, new research suggests that *pesticide use can become uneconomic*. This was an especial problem in the period until 1960, when the only insecticides generally available in the developing countries were inorganic chemicals such as sulphur, calcium arsenate, mineral oils and fumigants, and plant origin substances such as derris, nicotine and pyrethum. Even with the new insecticides, however, their use depended on the farmer being able to buy them. Governments provided loans to small-scale farmers to enable them to buy the insecticides.

Third, *some pests defy control*.

These factors suggest that biological control of insects (see the section on 'Cotton biotechnology' in Chapter 3) may eventually become the most important method of insect control. Already significant progress has been made in breeding resistant plants.

Weeds

Cotton's susceptibility to weeds is explained by its slow initial growth. Weeds cost US farmers an estimated 700 000 bales annually, according to the NCC, despite the persistent and expensive use of both chemical and mechanical measures to reduce them. In Tanzania in the 1960s, yields were reduced by up to 50% because of delays associated with weeding and thinning. The early period of growth is the most critical, up to two months after sowing.

The worst weed in the USA is Johnson grass, but there are many others, ranging from the nightshades of California to Bermuda grass in the mid-South, and teaweed in the Carolinas. Nut grass is another difficult weed. Annual weeds established before land preparation are usually buried, but if time and weather permit, the seeds are allowed to germinate before being destroyed: traditionally, mechanical cultivation has been used: cultivators of different types have been used, employing toothing devices or small V-shaped plough points to uproot weed seedlings and loosen the soil. In Africa and elsewhere, traditional tools are still used for hand weeding - hoes and machetes, and old type cultivators are used for inter-row cultivation with oxen and tractors.

The response to weeds has also been to use weedkiller: most US cotton acreage is treated twice or more with herbicides. Selective herbicides were discovered in 1942 which was a tremendous advance given the susceptibility of most cotton to 2,4-D. Herbicides developed have three names, chemical, common and trade, all agreed internationally. Pre-plant herbicides are put down before planting; pre-emergence chemicals are put down in a 25–35 cm band as the seed is planted. The aim is to kill weed seedlings and germinating weed seeds but to leave the cotton unharmed. Once the cotton has emerged, the use of sprays requires careful control; it is all too easy to damage the cotton plants. The 15–20 cm wide spray is aimed at the base of the plants. In the 1980s recirculating sprayers were developed, spraying from a tractor to a level above that of the cotton; a funnel on the opposite side catches any unused spray and returns it to the sprayer for rescue. This both cuts costs and reduces potentially toxic and damaging overuse. The two strategies are usually used in conjunction: 'lay-by', by which time the plants are large enough to shade the ground and themselves reduce weed growth. However, some tall and climbing weeds can interfere with picking and other field operations.

Ginning, baling and transportation

Before cotton can be spun it must be ginned, baled and transported, probably shipped, taking the original farmers' crop and transforming it into a world market commodity. The methods of picking, ginning, baling and transporting the crop still vary enormously internationally. Gin quality performance in large measure determines the grade of each bale and has a direct bearing on the spinning qualities of the fibre. India was the first country in which a mechanical device for separating lint from seed has been recorded. The earliest '*churcka*' gin consisted of two rollers through which lint, but not seeds, could pass. These were developed into the power-driven roller gins that are still used for ELS cotton. Upland however, has linters and the longer lint hairs are attached more firmly to the seed coat, hence the development of the saw gin.

The seed cotton arrives at the ginning mill in a loosely bundled mass, with varying degrees of dirt, leaves and bits of stalk depending on the efficiency of the picker. The first priority is for the cotton to be dried. Driers condition the seed cotton for smoother and more continuous operation of the gin by removing the excess moisture and fluffing the partly opened locks. Dried cotton surrenders more of its foreign matter and the ginned lint is smoother. It is then dried and cleaned. Lint cleaning at the gin is still a somewhat contentious issue, despite having been introduced 40 years ago. The requirements for speed often result in excessive lint damage, to the detriment of the spinning operation.

The dried cotton is then fed in turn into the bale-breaker, the opener and the scutcher. These machines break up the bales, beat and suck out the burrs, sticks, grass, dirt, stems and sand and open out the fibres into their original fluffy state. These cleaning machines increase the lint grade and thus the value of the cotton, reducing the manufacturing waste for mills. The types of cleaning equipment used depends on investment policies; stripper-harvesting demands extra cleaning equipment and therefore ginning charges are higher. The cotton fibres then emerge from the scutcher as a continuous fleecy sheet, or lap.

The volume of seed cotton needed to produce one bale of lint, a ratio known as the 'ginning out-turn' varies substantially from year to year, depending on the method of harvesting and the area of growth. The correct setting of the gin is crucial. The 'ginning out-turn' is vital to the profitability of the gin. Machine-picked seed cotton produces the least waste at this stage. The separated cotton seeds are covered in downy fibres, or linters, which are too short for normal spinning but have many alternative uses (see the section on 'The use of cotton' in Chapter 5).

At a gin, the seed cotton is drawn through an air pipe, usually of some 25–30 cm width, connected with a separator fitted with a screen through which the convecting air, laden with fine trash, is exhausted. The seed cotton is then dropped through a vacuum wheel into an air line, and blown through a drying unit by heated air to reduce its moisture content and to enable trash removal and sticks in later cleaning processes. Cylinder cleaners and burr-extractors remove immature locks of cotton and other extraneous material. All these operations facilitate processing and improve fibre quality. In the extracting process, the seed cotton locks are taken as they pass beneath a stripper or beater: burrs, sticks, stems and other foreign parts are removed. The achievement of the pre-ginning cleaning is that the majority of foreign matter is removed from rough hand-picked cotton and mechanically picked cotton and 90% from snapped and machine-stripped cotton. In mechanised cleaning processes great care must be taken to avoid damage to the lint.

From these cleaners the cotton is fed to the gin stands at a controlled rate. Ginning separates the lint from the seed by mechanical means, producing spinnable cotton and undamaged seed. The seeds are removed in the cotton gin by revolving circular saws (in a saw gin) which pull the lint through closely spaced ribs that prevent the seed from passing through. Leaves and dirt are removed at the same time so that only the long fibres, or 'lint', remain. The lint is taken from the saw teeth through either air blasts or rotating brushes. Seed cotton weighing 100 kg normally yields approximately 35 kg of fibre, 55 kg of seed and 2 kg of waste material. In the USA, almost all the crop, American upland, is ginned by saw gins with an average output of 1000 kg/h. Well-equipped gins can produce as many as 35 bales/h depending on the number of gin stands, and the whole process takes just minutes, from the lowering of the suction pipe onto the raw cotton to the emergence of the finished bale from the press, and one individual can oversee several gin stands.

Power for the gins can come from any suitable source, usually electricity but also gas or steam. In the USA there are new roller gins in operation, for example in the Tornillo area of Texas, but these are only used for ELS and other specialised types of cotton, a relatively small percentage of the crop. Roller gins work through opposing rollers which pull the fibres from the seed. In some cotton-growing countries, for example India, the Sudan and Egypt, there is still use of the McCarthy type and other roller gins. In some parts of Russia, China and India cotton is grown which, because of the in-growing character of the lint, must be plucked bodily from the plant and the lint extracted later. In the past the better preparation of cotton processed through saw gins, however, usually resulted in a small premium for such cotton above that for roller gin cotton of equivalent grade. Currently, however, roller ginning of, for example, C32 cotton attracts a premium. Gins worldwide are becoming larger and more centralised, increasing transport requirements and the need for technical expertise; the justification for the investment is there, however, in much reduced overall ginning costs. The size of the ginnery depends on the dispersion of the crop – clearly, the more widely dispersed the crop, the less economic pressure for concentration will be.

The separated cotton is then cleaned again, whilst the cottonseed is taken to storage. Lint cleaners improve quality but can reduce bale weight by 22 kg or more. The quantity of foreign matter removed depends on the harvesting method, number of cleaners and initial trash content, so such cleaning may not be profitable.

From the saws the cotton is blown to condensers which drop it into the press box, while the seed from which the lint has been cut is carried by conveyor to a storage bin. Growers are accustomed to selling their seed cotton to the ginner at prevailing prices. Gins are regularly inspected, and care in disposal of gin trash is usually legislated for. The press box serves to press the lint into bale shape through the use of mechanical trampers, emerging as a unit of approximately 180, 230 or 330 kg weight, depending on the country of origin. Traditionally in the USA cotton was compressed at the gin into 'gin-flat' bales with a density of 12–13 'standard density' (23 lb/ft^3) for domestic shipments or into 'high-density' bales (32 lb/ft^3) for overseas shipments. This compression reduces transport and warehouse costs. The bales wrapped in polypropylene, polyethylene, cotton or jute bags and tied with steel bands for shipping may be compressed for shipping to almost twice their original density. The cotton bales are stored in warehouses or in the open until they are transported to mill buyers domestically or to shippers, but there are now only a few compresses left in operation. Gins also provide sampling services, both mechanical and automatic. Ginners in some cases buy a sizeable portion of the crop and most cottonseed is bought by ginners, either for their own needs or for resale to oil mills. A large proportion of cottonseed is burned, for example in the Ivory Coast.

Ginning charges, obviously, vary very widely between countries and different seeds, harvesting methods and what services are provided; many gins are farmer co-operatives and therefore part of the ginning charge may be rebated to the producer according to various different schedules. They are also charged on different bases: per bale, per kg, etc. US ginning costs in the 1991/92 season were approximately as given in Table 2.1.

Research in ginning currently surrounds the best structure facilities and aeration for cotton storage. Surveys show an increase in gin storage of cottonseed in the early 1990s in the USA. There were approximately 1500 gins in the USA in the early 1990s, with a tendency towards larger, newer gins becoming very marked. The average volume processed per gin in the USA is now upward of 10 000 bales, compared with 7096 in the 1989/90 season (Table 2.2).

In the USA ginning capacity has increased in its ability to cope with the crop in under four months. Cotton warehouses provide receiving, compressing, storing and outhandling services. Warehouse receipts are widely regarded as evidence of their quality and representation and are used as proof of ownership. Storage itself is a complex and skilful activity: in the USA alone there were 344 warehouses with a total capacity of over 18.9 million bales in 1992/93 (2.5 million in the Southeast, 6.3 million in the Delta, 7 million in the Southwest and 3.1 million in the West). Worldwide storage capacity is in excess of 40 million bales. The warehouses are either owned independently, as part of warehouse chains, or as co-operatives or government-owned, as in India and Pakistan. Charges for using

Table 2.1 Average charges for saw-ginned Upland cotton, average charges for selected marketing services, and related information, by US state, 1990/91 season (except for italic figures, 1991/92 season)

Item	Unit	USA	AL	AZ	AR	CA	GA	LA	MS	MO	NM	NC	OK	SC	TN	TX
Bales ginned[a]	Thousand	15 038	375	954	1 063	2 727	400	1 179	1 799	300	91	262	363	137	485	4 903
Active gins[b]	No.	1 533	72	90	122	138	59	80	192	48	26	39	63	40	70	494
Average volume per gin	No.	9 810	5 208	10 600	8 713	19 761	6 780	14 737	9 370	6 250	3 500	6 718	5 762	3 425	6 929	9 925
Ginning and wrapping charges:																
Total charge per 480 lb net-weight bale[c]	$	*42.61*	*35.10*	*41.88*	*36.20*	*45.54*	*41.04*	*36.54*	*36.39*	*38.95*	*57.33*	*49.06*	*50.47*	*46.90*	*34.19*	*48.93*
		43.68	34.78	41.95	37.63	46.32	41.59	36.84	38.20	40.61	56.26	47.81	50.46	46.59	34.06	48.47
Method of harvesting																
Machine-picked	%	73	100	99	100	100	100	98	100	100	90	100	13	99	99	26
Machine-stripped	%	27	0	0	0	0	0	0	0	0	8	0	87	0	0	74
Machine-scrapped	%	[d]	[d]	1	0	0	[d]	2	[d]	0	2	0	0	1	1	[d]
Weight of seed cotton per 480 lb (217 kg) net-weight bale:																
Machine-picked	lb	1 468	[e]	1 473	1 466	1 496	[e]	1 430	[e]	1 492	1 517	[e]	1 550	[e]	1 436	1 461
Machine-stripped	lb	2 187	0	0	0	0	0	0	0	0	1 890	0	2 132	0	0	2 195
Machine-scrapped	lb	1 854	0	1 950	0	0	0	[f]	0	0	2 033	0	0	0	1 575	1 925
Cotton ginned from: Trailers	%	43	49	32	67	25	52	64	54	55	84	97	46	96	87	30
Modules	%	57	51	68	33	75	48	36	46	45	16	3	54	4	13	70
Charges for warehousing and related services:[g]																
Charge per bale for receiving	$	2.60	3.04	0	2.88	1.50	2.85	3.46	3.11	1.98	1.88	2.90	2.06	2.43	3.12	2.59
Charge per bale per month for insured storage	$	1.77	1.69	1.97	1.89	1.85	1.60	2.02	1.96	1.84	1.64	1.34	1.53	1.47	1.90	1.55
Charge per bale for compressing to universal density	$	7.62	6.85	5.75	7.75	6.25	0	7.70	8.50	7.75	7.25	0	7.50	0	8.25	8.35
Charge per bale for outhandling	$	5.62	5.30	4.82	7.93	5.11	4.71	7.85	8.16	7.81	4.33	2.48	3.67	3.57	8.04	4.19

[a] Based on report of June 1 1991 by Bureau of the Census, and includes both American Pima and upland cotton. Excluded are 26 000 bales ginned in Florida and Kansas. [b] Based on Bureau of Census information; excludes two active gins in Florida and one in Kansas. [c] Includes bagging and ties, drying of seed cotton, lint cleaning, and insurance, but does not reflect any patronage dividends, rebates, transportation to warehouses, industry organization dues or cotton classing fees. [d] Less than 0.5%. [e] Seed cotton usually not weighed. [f] No data available. [g] Based on published tariffs.

Source: USDA

Table 2.2 US cotton ginnings

Crop year	Running bales ginned (1 000 bales)	Average bale weight (lb)	Average ginning charge ($/ bale)	Number of active gins
1950	9 910	484	11.19	7 572
1951	15 076	481	12.04	7 653
1952	14 955	485	12.44	7 367
1953	16 317	484	12.69	7 141
1954	13 619	482	12.83	7 070
1955	14 542	485	12.75	6 929
1956	13 151	485	13.56	6 662
1957	10 880	483	14.51	6 349
1958	11 435	482	14.75	5 798
1959	14 515	481	15.02	5 630
1960	14 265	479	15.42	5 395
1961	14 325	479	16.83	5 224
1962	14 864	479	17.08	5 132
1963	15 290	480	16.80	5 058
1964	15 149	480	16.78	4 989
1965	14 933	480	17.30	4 870
1966	9 562	479	18.30	4 608
1967	7 439	480	18.60	4 203
1968	10 917	480	18.64	4 218
1969	9 937	483	19.15	3 942
1970	10 112	484	19.40	3 759
1971	10 229	492	20.33	3 623
1972	13 269	496	21.01	3 517
1973	12 611	494	23.74	3 285
1974	11 328	489	29.38	3 219
1975	8 151	489	32.13	2 856
1976	10 347	491	32.87	2 771
1977	14 018	493	34.73	2 689
1978	10 549	494	36.16	2 461
1979	14 262	492	39.31	2 336
1980	10 826	493	43.77	2 254
1981	15 150	495	42.90	2 189
1982	11 526	497	43.46	2 000
1983	7 504	496	45.87	1 852
1984	12 545	496	45.64	1 860
1985	12 988	496	44.86	1 774
1986	9 438	494	44.91	1 662
1987	14 358	493	45.82	1 653
1988	14 985	493	45.14	1 645
1989	11 884	492	44.26	1 584
1990	15 064	493	43.68	1 535

Source: USDA and US Dept. of Commerce.

them vary both from year to year and also greatly from country to country.

The main methods of moving cotton domestically within countries are, of course, road and rail. Trucks have now replaced rail worldwide as the main method of transportation, with over 80% of the crop in the USA being moved by truck. In Europe, sea and river transportation is very important with road traffic being used only for much shorter journeys. The road alternative has been picked for the usual freight reasons: competitive truck rates, more flexible scheduling, shorter delivery periods of truck transportation and the efficiency of containerisation, especially for exports. Rail does allow for the 'transit privilege', however, allowing intermediate stops to consolidate lots whilst charging only for the direct journey.

CHAPTER

3

The genetics of cotton

Cotton agronomy

In addition to climate, the method of tilling and production generally affects yields, as does pest control, which is required because cotton-growing is monoculture. Yields vary enormously. In Uganda, for example, the yield in 1991/92 was 140 kg/ha compared with 1633 kg/ha in Israel and 1720 kg/ha in Australia. Experience in the past 30 years suggests that yields can be dramatically increased. Morris (1988) cites the fact that the average yield for developing countries rose from 228 kg/ha during the 1960/61 season to 445 kg/ha in 1987/88, although he notes that 'it must be remembered that yields in the developing countries were at a much lower level initially than those of developed country producers' (p. 13). It is therefore argued that cotton production could be increased dramatically without using more land, just by the employment of more efficient cultivation methods, as shown in the USA, see Tables 3.1 and 3.2.

Following World War II, great advances were made in cotton agronomy. The aim, predictably, was to breed desirable varieties – high-yielding, disease-resistant, early-maturing or suitable for temperate climates – and thereby to increase per hectare profit. The new varieties and strains developed met the needs of agricultural

Table 3.1 US cotton yield per planted acre (lb)

Crop and area	1982/83	1983/84	1984/85	1985/86	1986/87	1987/88	1988/89	1989/90	1990/91	1991/92[b]	10-year average	5-year average
Upland southeast	736	406	716	727	468	564	486	591	526	710	593	576
Alabama	769	401	694	793	503	569	468	560	474	619	585	538
Florida	588	584	847	637	689	635	497	546	623	682	633	596
Georgia	692	448	771	696	395	649	507	619	548	804	613	625
North Carolina	690	344	594	638	638	490	507	604	628	678	581	581
South Carolina	767	369	785	697	354	424	463	616	449	773	570	545
Virginia	0	0	480	480	960	240	480	432	595	749	442	499
Mid-south	733	541	691	675	563	781	657	646	659	747	669	698
Arkansas	625	485	625	726	590	779	721	670	674	744	664	718
Louisiana	690	608	780	557	557	775	619	646	697	773	670	702
Mississippi	845	620	758	757	560	821	712	711	722	867	738	767
Missouri	636	324	547	644	529	792	600	603	608	612	589	643
Tennessee	626	329	476	592	559	692	524	491	453	542	528	540
Southwest	225	281	321	375	251	469	437	291	436	351	344	397
Kansas	0	0	0	0	480	480	480	48	168	288	194	293
Oklahoma	238	218	207	370	252	415	316	224	483	273	299	342
Texas	223	286	330	375	251	473	447	296	433	356	347	401
West	1 056	1 016	1 026	1 122	1 090	1 252	1 024	1 204	1 150	1 189	1 113	1 164
Arizona	1 116	1 196	1 225	1 237	1 296	1 405	1 186	1 298	1 112	1 187	1 226	1 238
California	1 069	986	992	1 124	1 078	1 248	1 004	1 216	1 193	1 237	1 115	1 180
New Mexico	474	600	542	487	472	647	636	630	661	522	567	619
Other upland[a]	480	0	0	0	0	0	0	0	0	0	48	0
Total upland	505	469	557	601	460	677	587	541	600	592	559	599
Extra long staple	667	722	782	886	883	990	846	881	744	774	818	847
Arizona	754	752	828	918	960	1 124	904	935	745	860	878	913
California	0	0	0	0	0	960	853	1 072	1 072	1 040	500	999
Mississippi	0	0	0	0	0	0	0	300	591	369	126	252
New Mexico	480	698	576	660	742	651	634	705	609	552	631	630
Texas	552	698	720	840	757	765	760	755	648	440	694	674
All cotton	506	471	559	603	465	681	591	553	603	595	563	605

[a] Other includes Illinois, Kentucky and Nevada. [b] = preliminary.
Source: NASS, USDA.

Table 3.2 US cotton yield per harvested acre (lb)

Crop and area	1982/83	1983/84	1984/85	1985/86	1986/87	1987/88	1988/89	1989/90	1990/91	1991/92[b]	10-year average	5-year average
Upland southeast	749	415	722	741	493	571	515	603	531	717	606	587
Alabama	775	409	699	795	506	572	486	571	476	626	592	546
Florida	627	608	847	693	707	646	566	557	640	696	659	621
Georgia	714	467	784	725	455	662	564	631	555	813	637	645
North Carolina	699	350	600	646	646	495	515	615	631	683	588	588
South Carolina	783	369	785	708	370	428	473	626	452	777	577	551
Virginia	640	360	528	443	554	373	510	498	562	728	520	534
Mid-south	747	564	701	689	577	788	688	664	672	766	686	716
Arkansas	657	535	632	767	602	786	742	687	692	759	686	733
Louisiana	702	623	786	565	567	782	705	672	715	825	694	740
Mississippi	853	640	767	764	571	829	736	732	728	878	750	781
Missouri	648	377	554	653	588	796	607	618	641	621	610	657
Tennessee	638	337	498	600	567	700	529	497	461	551	538	548
Southwest	297	315	365	402	347	500	462	357	478	423	394	444
Kansas	0	240	388	320	336	480	373	240	280	320	298	339
Oklahoma	254	232	233	380	288	431	334	244	496	316	321	364
Texas	301	322	376	404	353	506	472	367	477	430	401	450
West	1 068	1 036	1 036	1 139	1 110	1 264	1 038	1 220	1 165	1 202	1 128	1 178
Arizona	1 118	1 225	1 227	1 241	1 301	1 410	1 190	1 303	1 119	1 190	1 232	1 242
California	1 077	996	999	1 132	1 088	1 259	1 015	1 228	1 204	1 250	1 125	1 191
New Mexico	551	715	605	631	595	689	710	698	735	554	648	677
Other upland[a]	343	0	0	0	0	0	0	0	0	0	34	0
Total upland	589	506	599	628	547	702	615	602	632	653	607	641
Extra long staple	672	725	786	891	890	1 000	848	893	758	796	826	859
Arizona	760	768	841	927	965	1 126	904	936	751	885	886	920
California	0	0	0	0	0	1 173	853	1 078	1 080	1 040	522	1 045
Mississippi	NA	NA	NA	NA	NA	NA	NA	436	591	600	163	325
New Mexico	511	683	595	687	718	642	634	707	609	558	634	630
Texas	561	689	744	868	751	787	769	794	682	463	711	699
All cotton	590	508	600	630	552	706	619	614	634	656	611	646

[a] Other includes Illinois, Kentucky and Nevada. [b] Preliminary.
Source: NASS, USDA.

mechanisation, responded better to chemical fertilisers (although this factor has progressively become less important) and were more disease-resistant. One important aim has been to reduce barky grades, through the introduction of the multistage stick extractor which removes trash from stripper-harvested cotton more efficiently than do traditional machines; this is now in commercial production.

Stickiness of cotton lint is a major problem, in part because of the failure to establish effective control of whitefly. In 1991, an ITMF survey reported that 27% of cotton was sticky. This was a deterioration of the situation from 1989 when the same survey reported 21%. The Sudan, Cameroon, Chad, India, Mali, USA (Pima) and Brazil (North) reported over 50% stickiness. In 1991 only six origins reported no stickiness, compared with 12 in 1989. Clearly either the industry is becoming more aware of the problem or it is genuinely getting worse. Sticky cotton results from insect deposits - the whitefly is the main culprit, explaining most of the problems, for example, in the Sudan - on raw cotton as well as plant sugars and other naturally occurring residues and there is a prevalent belief amongst purchasers that the industry does insufficient to prevent it - it can pose serious problems for textile production. In the USA, however, the industry is now putting massive resources into solving these problems. It was adjudged sufficiently serious in the USA to launch a Sticky Cotton Action Task Force in 1990. Compounds were tested in the early 1990s for application during harvesting; and, for example, tests have begun into the chemistry of sugars to develop a field-applicable enzyme system that will make them ineffective in cotton. Prykhanja of the Sudan has currently extensive ongoing research into this area. It also has been noted that rain-fed cottons have a lower sugar level. In general, available methods of detecting stickiness in raw cotton are inadequately slow and unreliable, and work is under way to improve them. These are research programmes that are currently being worked on. Samples are collected each week from USDA Classing Offices, and are evaluated for the content of extractable sugars, stickiness, neps, short fibre and fibre maturity, with the objective not only of obtaining accurate quality information on the US crop, but also analysing the relative measurements of different instrument systems.

Cotton is effectively a self-pollinating crop, although some control of outcrossing is required if the genetic lines are to be kept. The aim of a breeding programme is to select and propagate the better material from a heterogenous population. Programmes usually proceed from primary selection and introduction to secondary selection and hybridisation. Selection, the key to breeding, is 'the act of choosing a plant which is different from its neighbour, and harvesting the seed of that plant separately to produce progeny in the next generation; the progeny may breed true to type or may segregate' (Munro, 1987, p. 232).

The best breeding method is hybridisation. In several generations of breeding, successful new breeds have been produced. Once the strain or variety is established and accepted by the growers, the breeder's responsibility becomes to maintain and improve its standard of excellence, as the growers adopting it for planting and the manufacturers who use it need a continuous dependable supply of seed and lint. In the USA, this breeding is a major commercial business. In the 1980s there was developed in Texas an early, fast-fruiting cotton type, which allowed fewer inputs and less weather risk but was lower yielding; this has proved very popular in Texas and Georgia. Varieties established by seed breeding are registered jointly by the American Society of Agronomy and the US Department of Agriculture. Seed has become increasingly standardised, through the establishment of one-variety communities, gins and plantations.

The annual Cotton Spinner–Breeder conference and the Beltwide Conference in January/February in the USA foster the establishment of better varieties of cotton, resulting in a greater return over area to the farmer and better spinning results for the manufacturer. Cotton Incorporated, the US industry's jointly owned development company, also plays a major role in crop improvement: they reported (Annual Report 1991) that

> breeding programmes are continuing to address the need for increased strength in the mid-south and south Texas. New cultivars have been released in the past year (1991) which significantly increase strength without yield reduction. The hybrid breeding programme in New Mexico, after more than a decade of research, is scheduled to release a series of parent lines for hybrids with unique fibre and agronomic qualities. A newly released cotton variety, 'Georgia King', has come first in several of the regional variety tests in both yield and quality.

Cotton Incorporated stressed the importance of continued government support for the state breeding programme.

The cultivation of GL7, the new variety of glandless cotton, is expanding rapidly throughout the world. The key to this variety's success is that it produces seed without gossypol, whilst its ginning percentage is 46%, higher than that of any other variety commercially produced. Gossypol present in ordinary cottonseed and throughout the plant is toxic if ingested by man or other monogastric animals, and only ruminants can safely eat it. Scientists have been producing gossypol-free varieties for decades, but apart from a few trials, they have not been grown commercially. The main reasons have been their weakness to pest attacks and that they produce fewer fibres. But the high ginning percentage of GL7 is different, and in 1991 a significant area was planted with it: 8500 ha in Côte d'Ivoire, 7000 ha in Burkina Faso and 3500 ha in Benin. Undoubtedly, the proportion of cotton acreage sown with glandless cotton will rise: then it will be the producers' task to maximise the gain from these protein-rich seeds (see the section on 'The uses of cotton' in Chapter 5).

The attention to step-by-step technical innovations in the 1990s contrasts with the belief held in the 1980s that there was great value in, for example, the GOSSYM–COMAX cotton crop simulation model which was 'designed to aid farmers in making their on-farm management decisions on irrigation, fertilisation, defoliation and other production inputs and practices' (USDA, 1986, p. 40).

Cotton biotechnology

Since the 1980s, the use of biotechnology has enabled substantial increases in cotton productivity in many countries. 'Biotechnology' is a general word encompassing breeding, biological control, tissue culture and molecular genetics. Biotechnical solutions have been preferred to chemical ones because of the absence of adverse environmental effects. There remains a fear that genetic engineering generally, including perhaps that of plants, may result in problems in generations to come, and there are already many regulatory controls over this kind of research, and it can be expected that there will be many more; in general, studies by the US government have suggested that scientists are well aware of any potential environmental disbenefits from genetically engineered improvements to any plants, including cotton. There are also fears that the expense of biotechnical applications will reduce still further the number of family farms, but in practice this is unlikely to be an important factor, and if biotechnical remedies for disease and insects eventually become very cheap, it may even reverse the trend. Biotechnology may also result in the genetic engineering of very specific crops with high local and quality factors which will best bȩ grown by small-scale producers.

Cotton Incorporated in the USA believes that 'biotechnology offers great potential for many improvements in the cotton genome' (*Annual Report for 1991*, p. 25). It is already 'playing a role in extensive programs supporting the marketing of cottonseed products' (*Annual Report for 1992*, p. 27). The most obvious application of biotechnical research in cotton production is to pest management strategies, which has been the primary focus of biotechnology companies. They have pressed ahead in such areas as herbicide resistance and insect resistance received through the *Bt* toxin and other proteins. But there is also interest in the biotechnical control of pests, improving the quality of fibre and stress tolerance of the cotton itself. In the future it can be expected that biotechnology will be applied to cottonseed protein and oil quality.

In controlling diseases, weeds and insects, the basic use of biotechnology is in using selected micro-organisms to use as biocontrol agents, controlling the production of pathogens, weeds and insects. Much of this work is suitable for a number of crops, for example cotton, groundnuts, tobacco and coffee. In controlling seedling disease, there are already two commercially available bacteria preparations; some fungi are also available. A number of research programmes have currently been discontinued owing to lack of funds in a period of low commodity prices. Some fundamental research is still being undertaken by the USDA and at the Hebrew University in Israel. The biocontrol of nematode pests in cotton has not yet been achieved: companies active in the field hope to be able to engineer host resistance through gene implantation. Using micro-organisms to control specific insect pests is a longer-established discipline. Initially during the 1970s spray techniques were used, but more recently the objective has been to ensure the genetic engineering of the bacterial genes into the crops themselves. Problems such as the low persistence of the spray after application and the increased resistance of older larvae make timing of applications critical. In some cases, chemicals are economic competitors to viral agents, and only a combination of regulatory restraint and increased insect resistance to chemical agents suffices to raise producer interest in the viral alternative, which usually takes longer to have an effect. As yet, weed problems in cotton cannot be addressed adequately with mycoherbicides, but in the future it is hoped that some weed species will be amenable to control in this way.

Amongst the major crops, cotton has been extremely important in biotechnical research into ovule and cell culture, and the regeneration of plants from cells. These developments have allowed the genetic engineering of cotton, and have permitted the use of cell and organ culture in order to make basic studies of cotton physiology and genetics. A number of US projects are involved with the introgession of more exotic germplasms into upland cotton, in particular in relation to pest resistance. Cell culture research aims to generate variations which will be toxin-resistant, and better suited to certain different temperatures. The regeneration of cotton plants from callous tissue has now been established: however, there are significant variations between the different cotton genotypes in terms of their regenerative ability. This is currently a major factor influencing the research and general strategy of the biotechnology companies working with cotton.

So far as genetic engineering of cotton is concerned, a number of underlying principles appear to dominate research strategy according to *Stewart (1991)*. First, that the seed of the modified crop will have sufficient value added that it can be sold at a substantially higher price than the unmodified seed. Secondly, that the market volume for the seed is large. Thirdly, that the market share of the particular pesticide could be increased by engineering genetic resistance to the particular chemical; to this can increasingly be added the general biotechnical objective of improving environmental control by replacing chemical applications with biological organisms. Fourthly, are general questions of government regulations and funding; and finally are such factors as the availability of well-characterised genes with economic potential, methods for genetically transforming particular species, and of regeneration.

Both resistance to broad-spectrum herbicides and insect resistance based on the toxins produced by the bacterium *Bacillus thuringiensis* have been initially important strategies in the development of genetic engineering for cotton. Few commercial strains of cotton appear easily manipulable; others demand particle accelerators or the direct transformation of meristems or germ-cell lines. Herbicide resistance has been the most important achievement of genetic engineering in cotton to date. This resistance has been achieved through the selection in tissue culture of cells that can grow in the presence of the herbicide. There may be a mutation that causes overproduction of the enzyme system that falters in the presence of the herbicide. In the USA, four major companies were in 1991 developing herbicide-resistant cotton, Calgene, Dupont, Monsanto and Phytogen, and it was expected that other companies would become involved. One of the principal herbicides to which resistance is being imparted is glyphosate; another is bromoxynil; a third is the sylfonylurea group, marketed by Dupont. The company hopes to produce and market a stronger, more effective version of the herbicide and the modified cotton gene at the same time. Finally efforts are underway to impart 2,4-D resistance into cotton. 'A cultivar successfully transformed by USDA ... is now awaiting government approval for release to breeders' (Cotton Incorporated, *Annual Report for 1991* p. 25).

The use of biocontrol agents to reduce insect infestation will diminish the chemical load on the environment. However, they must be effectively applied to the crop. The cheapest method, and the most effective, is to impart the resistance genetically. Insect resistance is the next most actively researched area after herbicides. The economic calculation undertaken is to work out the difference in cost of production of the plant with and without resistance. The interest of the industry in cotton is explained by the very high level of pesticides applied to the cotton crop to control insects and pathogens. The average cost of controlling cotton insect pests in the USA was estimated to be $60/acre in 1991, and will probably have increased since then. Companies are aiming to increase the level of *Bt* toxin in plants, but the problems appear to be more connected with the limited effect that the implantation has rather than the technical difficulties of doing so. A number of US companies are seeking to improve the effectiveness of their programmes for the implantation of *Bt* genes: Agracetus, Agrigenetics, Calgene and Phytogen, as well as groups at Arizona and New Mexico Universities. Other peptides, less well-known than *Bt* toxins, also have possible applications to insect control.

All the companies are concerned to keep details of exactly which genes are under secret study; however, general impressions of which categories are under study are known: protease inhibitors, lectins, arachnid (spider) venoms, and genes from the target insect itself, for example neuropeptides. Research is now centred on how to identify, remove and then utilise these genes, although in most cases it has not been made public. Research effort allocated to pathogen resistance is less substantial at present, because of the low value-added potential for pathogen-resistant seed. Techniques for virus resistance have been successful in other crops, but the limited scale of the problem in the USA itself has discouraged research. Problems in other countries such as Mexico must be the spur.

There are two other areas in which biotechnology may have an important role: environmental stress and hybrids. Stress resistance has not yet been genetically engineered into cotton, but some work on drought and flooding stress is underway in the USA. Recent research indicates that the genes of selected enzymes from cold-tolerant plants may be placed into cotton to increase resistance to cold, and there is also a project to improve photosynthesis underway at the University of Arizona. Cotton Incorporated is sponsoring work in this area, reporting that

> Cotton tissue cultured and selected for cold tolerance have been successfully grown into plants and have produced seed. The high temperature tolerance has not been adversely affected in these plants. Only a few degrees of additional cold tolerance can translate into many additional bales of premium micronaire cotton throughout the cotton belt (*Annual Report, op cit*, p. 7).

However, in the long term this is surely a dangerous line of research for existing cotton-producing areas. In Belgium, work is being carried out to improve the effectiveness of hybrid breeding using genetic engineering. The aim of the project is to allow producers to buy new seed every year. Finally work is under way to develop a cotton plant that produces gossypol glands in all parts but the seeds. There is also a promotional programme in the USA creating a demand for gossypol-free seeds as feed for hogs and catfish.

Biotechnology may be able to improve fibre quality, too. Agracetus is one company investigating this area through the possible development of fibre-specific genes. A number of US universities are also working on this, and different programmes are concentrating on the effects on fibre strength at different times in the development process.

As yet, molecular mapping of the genome of cotton has not been achieved but there are no theoretical objections to its achievement in the next decade or so; the work is underway under USDA auspices.

Currently there is an international problem over technology transfer. The biotechnology companies work for profit, and guard their recipes jealously through the courts. The companies themselves have ploughed many millions of dollars into developmental research on molecular genetics and in many cases have failed to make any profits from their investments. Currently, this profit motive is the main spur to biotechnical research, which is concentrating on adding high value to cottonseed, or increasing proprietary chemical sales. The target market is therefore high-input producers, mainly in the developed world.

The UN has neither the resources nor the legal expertise to unravel the patents required for the general application of top-class biotechnical research throughout the world's cotton-growing countries. The patents themselves attempt to cover future developments and restrict distribution. The result is likely to be that biotechnological innovations remain at the research and development stage, and concentrated in the USA, for some considerable time. The long term effects on global yields and available acreage, however, will be significant.

CHAPTER

4

World cotton production

Types of cotton

Cotton is a member of the sub-kingdom Spermatophyta, seed-bearing plants. It is an Angiospermae because the boll encloses the seeds. It is in the order of Malvales (which includes other fibre crops such as jute, kenaf, baobab, and cocoa), family Malvaeae and genus *Gossypium* (named by Linnaeus in 1753). There are over 40 known species and a multitude of varieties, of which four, principally, are cultivated. Yet despite cotton's enormous economic importance, it was only in 1926 that the Empire Cotton Growing Corporation opened its research station in Trinidad to pursue research which led in 1947 to research worker Hutchinson's fundamental report. Opinion still differs as to the exact number of cotton species but probably the most reliable source is Fryxell (1979).

Work begun in the 1920s showed that the basic chromosome number is n=13, with species having either n=13 or n=26. The n=13 species fall into five major groups (following Beasley, 1940, 1942) between which, with one exception, there is little chromosome homology. These are *old world cultivated (A), wild species of Africa (B), wild Australian (C), wild American (D) and wild Arabian (E)*. The cottons with n=26 chromosomes carry a complement which consists of a set of 13 homologous with that of the wild American species. Crosses between species belonging to different main groups are usually difficult and sometimes impossible; when they do occur they are sterile.

There are eight species groups or sections of *Gossypium*, which differ in genetic architecture, or between which cytological or genetic differences exist. Of the eight sections, six are lintless: Sturtiana, Erioxyla, Klotzshiana, Thurberana, Anomala and Stocksiana. These are all n=13 varieties, not cultivated, and wild-growing in tropical and subtropical areas. The two linted species groups are Herbacea and Hirsuta. Herbacea species are n=13: *G. arboretum* and *G. herbaceum.*

In the Hirsuta section, n=26, varieties occur in both the wild and cultivated state in America. There are two cultivated species of tetraploid Hirsuta: firstly *G. barbadense*. This species is now grown in tropical South America (especially Peru where the Tanguis variety is grown), Central America, Antilles, Egypt, Sudan, South Carolina and Georgia. Fine-linted *G. barbadense* was brought to the West Indies in the eighteenth century, and from there in 1786 to South Carolina, and in the nineteenth century it became Sea Island cotton. Now the limited quantities of *G. barbadense* grown in the USA are in west El Paso Texas, New Mexico, Arizona and also in California. There are no defined races of *G. barbadense*, but brasiliense, or kidney cotton, is distinguishable. The fibre length of this cotton is usually 35–45 mm.

The second tetraploid Hirsuta species is *G. hirsutum* L. From this variety, producing high-medium and medium staple cotton, and known as upland cotton, the New York Cotton Exchange estimates that 99% of US cotton and 88% of the world's commercial cotton crop is grown – although there are actually more than 80 varieties grown in the USA. Indian Cambodia cotton also belongs to this species. The crucial distinction in cultivated cotton is between New World tetraploid cotton and Old World diploid cotton. It has now been accepted that the Asian species differentiated before they were domesticated. The fibre length of upland cotton is usually 22–32 mm.

In the Herbacea section, *G. arboretum* has become widely distributed geographically, with the result that there are a number of genetically distinct races, although there is gene exchange – and these include the races of indicum, burmanicum, cernuum, bengalese and soudanese. *Gossypium herbaceum* was first developed as an annual in Iran and surrounding countries, spreading into Pakistan, India and China. There are perennial forms in Africa. There are five races, separated geographically from one another: persicum, kulijanum, acerifolium, wightianum and africanum. Prentice (1972) believed that the cultivated *G. herbaceum* cottons were domesticated

Table 4.1 A key to the cultivated species of cotton

A	Bracteoles entire or dentate; teeth usually less than three times as long as broad.	
B	Bracteoles flaring widely from flower, usually broader than long; upper margin usually 6–8 teeth; capsule rounded or with prominent shoulders.	*G. herbaceum*
BB	Bracteoles closely investing flower, longer than broad, entire or with 3–5 teeth set near apex; capsule tapering.	*G. arboreum*
AA	Bracteoles deeply lacinate; teeth usually more than three times as long as broad.	
B	Leaves deeply lacinate for two-thirds length into 3–5 lobes; petals usually bright yellow with basal reddish spot; anthers compactly arranged; pollen deep yellow; capsule coarsely pitted with black oil glands.	*G. barbadense*
BB	Leaves less deeply lacinate for half length or less, or rarely five lobes; petals usually pale yellow or cream without basal reddish spot; anthers loosely arranged; pollen pale yellow or cream; capsule surface smooth.	*G. hirsutum*

Source: Purseglove J W (1968) *Tropical Crops, Dicotyledons*, Longman.

from the Arabian and Baluchistan race acerifolium.

The New World cotton arose from a cross between two diploid species, and there is an academic dispute about how the linted diploid not originally indigenous to the New World reached it. Some have suggested continental drift, land bridges and routes via Antarctica and the Bering Straits; others suggest that man himself brought the species across. Another suggestion was that the cotton seeds floated across the Atlantic, as they can survive aquatically for up to a year. Similarly there is no exact clarity as to the origin of the Peruvian *G. barbadense* cottons. However, these upland cottons have spread from a small area in Mexico to become most of the world's cotton crop. It has been suggested that seed was taken from Mexico to the USA about 1700, where types capable of fruiting with a shorter day length were selected. There are three principal cultivated varieties: Marie-galante, Punctatum and Latifolium.

Table 4.1 gives a key to the cultivated species of cotton. The main basis of distinction between individual cotton types is on the basis of staple lengths. Staple length is still almost universally measured in inches and 32nds, for example in the quotes published by Cotton Outlook in Liverpool, and is internationally traded thus.

Extra-long staple cotton

Extra-long staple (over 35 mm) is a specialised product grown only in a few countries from varieties of *G. barbendense* or crosses between *G. barbendense* and *G. hirsutum*. There are numerous technical problems associated with growing cotton of this staple length, especially in Africa, but these have in large measure been overcome. In the USA, Peru, Israel and Morocco the extra-long staple (ELS) variety is called Pima cotton – breeding work is now centred in Tempe Arizona. There are other varieties of extra-long staple cotton, such as Del Cerro.

The Ishan cotton of Nigeria, a perennial *G. barbadense*, was the source of the perennial cotton introduced into Egypt in 1820 by Jumel; when crossed with Sea Island cotton about 1850 it developed into modern Egyptian cotton. In 1900 Egyptian cotton was planted, experimentally, in the USA. The first plantings, in South Carolina, were unsuccessful, but in the Salt River Valley of Arizona, in Texas and in the Imperial Valley of California a number of hybrids were made of different varieties of Egyptian, Sea Island and American upland cottons. By 1911 the first commercial crop of 30 bales of a hybrid labelled Yuma was sold at a high premium. Yuma was selected from Mit Afifi, an Egyptian cotton. It was also grown in Peru and in French North Africa, and a new Egyptian variety was grown from its seed. In 1914 a strain of Egyptian Sakellaridis was given to the US field station in Sacaton, Arizona. It was crossed with Pima, resulting in $S \times P$, an extra-long staple cotton deemed a strategic material for military purposes. After World War II further American-Egyptian strains were developed. In 1952 a total of 88 000 bales of the extra-long staple were produced, the main types being Pima 32 and Amsak.

The extensive variation in production between countries over time is accounted for by the wide price differential between ELS and medium staple cotton – between 20% and 250%. ELS cotton can be spun into finer, stronger yarns than medium cotton, although the gap is reducing every year, thereby reducing the demand for ELS cotton. Countries as varied as Israel, Morocco, the Democratic Republic of Yemen and Zimbabwe at different times produce large quantities of ELS cotton, attempting to exploit favourable price differentials.

The longest and finest cotton in the world, it is argued, is still Sea Island Cotton, although Sea Island is now a really insignificant product. Demand for ELS cotton fluctuates substantially and although production is no more than 1000 tonnes of lint annually there has been a continued overproduction problem and premiums have shrunk dramatically this century. Its markets are mainly high-value products such as sewing thread and expensive clothing items.

Long staple cotton

Long staple cotton (28–35 mm) needs regular water to develop and mature the long, fine fibre, to make it free from neps, irregularities in the spun yarn caused by tangles of immature fibres. The majority is therefore grown under irrigation or in equatorial regions where the dry season is not extensive. Long staple has been grown in Egypt for decades, and in India, as well as in Peru (Tanguis was rebred from 1943 onwards).

Annually, considerable quantities of US cotton seed are exported for planting throughout the cotton growing world. The seeds of *G. hirsutum* bear a heavy coat of lint hairs, ranging from 18 to 30 mm in length. Underneath the lint, closely attached to the seed, is a layer of fuzz, also a valuable commodity, known as linters. Much of the 'long staple' is in fact just normal upland varieties grown under favourable conditions. Examples are Acala 1517 from California, BP52 and BPA from Uganda, and MCU 5 and MCU 9 from India. During the 1960s many established upland cotton breeders in the USA grew high-quality versions of established varieties, but this proved relatively unprofitable as the overall quality of high-medium and medium staple cotton has increased.

In 1988/89 global output of long and extra-long staple cotton increased in all major producers, except Egypt, the world total rising by about 10% to 1.1 million tonnes. Since that time production has continued at levels more than ample to satisfy world demand.

Short staple

Most short staple cotton is produced in India and Pakistan mainly from Old World diploid cotton. It is known together as 'desi' cotton. This is a declining part of cotton production; it now forms rather less than 5% of Pakistani production. There is some production in China, although there too there is a trend towards longer staple lengths and production of *G. hirsutum*. There is a very small amount of short staple production in Central and South America.

Each new variety of cotton possesses similar fibre characteristics to its predecessor, but is improved in respect of yield, ginning percentage, disease resistance or vegetative characteristics. The ICAC publishes a full list of all the cotton varieties grown in each country, and the list is constantly changing owing to new additions – at an increasing rate.

Wild cotton

Wild species include one with 26 chromosome pairs, *G. tomentosum* and approximately 20 with 13 pairs. These diploid species occur in the Indian subcontinent, Australia and Africa and Arabia and throughout the Americas.

Overview of world cotton production

World cotton production has increased dramatically in the past 200 years. In 1834, for example, production is estimated to have been fewer than 2 million bales of 480 lb each. By 1900, the level of production had probably reached 18 million bales. During the period from 1900 to 1925, the USA was the largest producer of cotton in the world, followed by India (which included what is now Pakistan and Bangladesh). The USA usually supplied more than 50% and India about 20% of the world's cotton. By the 1924/25 season production was already estimated to be about 26 million bales, 93% produced by five countries: USA, India, China, Egypt and Brazil. Forty years later, world production was twice as high and 90% was concentrated in 15 countries. Cotton is a tropical plant, adapted to temperate zones and is cultivated in most tropical and subtropical zones, i.e. between 47 °N and 36 °S. Cotton is difficult to grow at latitudes further north than 40 °N, and is confined to small belts in Russia, Hungary and the former Yugoslavia. It thrives best in high-temperature climates with considerable sunshine, abundant but not excessive rainfall and high soil fertility. More than 80 countries within the zone of 47 °N and 35 °S produce cotton for commercial purposes. More than 70% of the world's cotton crop is grown in five countries: the USA, China, India, Pakistan and Uzbekistan.

World production levelled off in 1974/75 after slowing down from 5% growth in 1972/73 to 1.5% in 1973/74, despite larger acreage. Production averaged 11.4 million tonnes in the late 1960s, reaching 13.6 million tonnes by 1974. There was a dramatic fall in 1975, because of smaller plantings stemming from lower prices. In 1976 there was something of a recovery, by 5%, but at nearly 12.5 million tonnes it was still substantially less than in the early 1970s. The increase was accounted for by increased plantings, especially in the western hemisphere, because of favourable price prospects and government action to return profitability to cotton in comparison with food crops. This increase was repeated in 1977/78, with more area planted and higher yields. A record crop of 13.9 million tonnes ensued, from some 33 million hectares. Developing countries, including China, produced 7.6 million tonnes by 1977, compared with 7 million tonnes in the late 1960s, whilst the developed countries as a whole, including the centrally planned economies, produced 6.3 million tonnes in 1977 compared with 4.4 million tonnes in the late 1960s. The biggest increase came almost exclusively from the USA. The following year world production fell back by about 6% to 13 million tonnes, which was caused by low prices at planting time and low yields caused by poor weather. 1979/80 saw a record world production of 14.3 million tonnes: yields increased dramatically, more than compensating for some decline in the area under cotton. Throughout this period, the volatile size of the crop most affected world production totals. World production almost reached the same level in 1980/81, and achieved a new record of 15.4 million tonnes in 1981/82. There was no certainty of a continual rise in production amongst traders, however - the following year saw their scepticism justified as US production curtailment and poor weather took world output back down to 14.7 million tonnes, a level which was not quite maintained the next year.

It was the 1984/85 growing season that really changed the world production outlook, and, in retrospect, probably changed the fundamentals of the world cotton market. Suddenly and coincidentally, yields reached their highest ever levels, acreage increased considerably, and the world faced a 30% rise in production to over 19.2 million tonnes (ICAC). These increases came most notably from developing countries, including China. Production slipped back to 17.4 million (ICAC) tonnes and then to a low level of 15.3 million tonnes (ICAC) in 1986/87, responding to the low international prices that had themselves been caused by the high output level of 1984/85. Production recovered in 1987/88, fell the following year but in 1988/89 production rose by 3.5% to almost 18.3 million tonnes (UN Food and Agriculture Organization, FAO), the second highest level recorded. This trend was reversed in 1989/90 when production fell by more than 5% to 17.4 million tonnes; there was a revival to 19 million tonnes in 1990/91 and a new record of 20.9 million tonnes was reached in 1991/92. The New York Cotton Exchange's (NYCE) 1992 summary of world production growth is telling: 'In 1980, world cotton production was 65m bales. In 1991 world cotton production was 95.5m bales, a 47% increase' (p. 4). See Tables 4.2-4.4.

In 1991/92 the world cotton crop reached 21 million tonnes, over 9% greater than the record set in 1984/85. The majority of the increase

Table 4.2 Cotton production (million tonnes)

	1966–70	1971	1972	1973	1974	1975	1976	1977	1978	1979	1980	1981
Developed												
USA	2.098	2.281	2.983	2.821	2.537	2.085	2.303	3.133	2.364	3.185	2.422	3.405
Western Europe	0.170	0.163	0.199	0.158	0.177							
Others	0.078	0.099	0.090	0.104	0.123	0.277	0.280	0.359	0.372	0.367	0.409	0.452
Developing												
Latin America	1.566	1.656	1.702	1.643	1.705	1.268	1.599	1.764	1.727	1.620	1.650	1.602
Near East	1.430	1.620	1.671	1.620	1.655	1.320	1.417	1.563	1.398	1.361	1.341	1.396
Far East	1.611	2.005	1.909	1.902	1.875	1.714	3.843	3.849	4.021	4.408	4.829	5.227
Africa	0.478	0.529	0.539	0.502	0.532	0.481	0.535	0.441	0.475	0.512	0.477	0.448
USSR/Central Asia	2.069	2.385	2.471	2.602	2.84	2.526	2.631[1]	2.783[1]	2.701[1]	2.830[1]	3.095[1]	2.840[1]
China	1.864	2.081	1.778	2.147	2.147	2.385	2.342	2.049	2.168	2.212	2.710	2.970
India												
Pakistan												
Other CPEs	0.025	0.020	0.023	0.023	0.023	0.018						
Total	11.3489	12.839	13.365	13.522	13.611	11.797	12.608	13.892	13.058	14.283	14.223	15.570

	1982	1983	1984	1985	1986	1987	1988	1989	1990	1001	1992	1993	1994
Developed													
USA	2.605	1.692	2.826	2.928	2.119	3.214	3.36	2.66	3.38	3.84	3.53	3.70	3.57
Western Europe													
Others	0.374	0.435	0.579	0.645	0.616	0.669	0.78						
Developing													
Latin America	1.283	1.402	1.979	1.660	1.289	1.942	1.80	0.67[5]	0.72[5]	0.67[5]	0.42[5]	0.41[5]	0.5[5]
Near East	1.455	1.465	1.493	1.410	1.377	1.284	1.38						
Far East	5.879	6.468	9.063	7.325	6.616[2]	7.308[2]	7.44						
Africa	0.492	0.558	0.633	0.629	0.730	0.786	0.85	0.77					
USSR/Central Asia	2.617[1]	2.689[1]	2.567[1]	2.711	2.673	2.479	2.77[3]	2.65[4]	2.64[4]	2.47[4]	2.07[4]	2.27[4]	2.33[4]
China	3.598	4.638	6.249	4.152	3.540	4.246	4.15	3.79	4.51	5.68	4.51	4.20	4.76
India								2.31	1.99	2.05	2.33	2.38	2.38
Pakistan								1.46	1.64	2.18	1.56	1.95	1.97
Other CPEs													
Total	14.705	14.709	19.140	17.308	15.42	17.682	18.38	17.40	19.00	20.80	17.90	18.40	19.30

Source: USDA and ICAC

1 Including Eastern Europe 2 Asia and the Pacific 3 Excluding Eastern Europe 4 Central Asia 5 Brazil only

Table 4.3 World cotton production (thousand 480lb bales)

	1987/88	1988/89	1989/90	1990/91	1991/92 Estimate	1992/93 Estimate
World total	81 109	84 423	79 866	86 982	95 947	83 150
China	19 500	19 100	17 400	20 700	26 100	20 800
USA	14 760	15 411	12 196	15 505	17 614	16 260
FSU — 12	11 491	12 686	12 203	11 909	11 250	9 600
India	7 140	8 276	10 599	9 135	9 430	9 800
Pakistan	6 744	6 551	6 687	7 522	10 000	7 200
Brazil	3 514	3 376	3 030	3 215	3 445	2 600

came from China, the USA and Pakistan. Forecasts of the 1992/93 season showed continued falls in production estimates for both the USA and other major world producers. In August 1992 the ICAC forecast that the 1992/93 season would see production and consumption coming more into line, with production at 20.2 million tonnes, a 3% fall. At the same time the USDA revised its forecast to 93 million bales, with a marginally larger US crop offsetting a reduced crop in the former Soviet Union. By December 1992 world production for 1992/93 was estimated by the ICAC to be only 18.9 million tonnes and the USDA forecast it to be 86 million bales, down 10.5% from the record crop of the year before. Out-turn was estimated in May 1993 by the USDA at only 83.2 million bales, down 13.3% from the 1991/92 production level (see Fig. 4.1). In July 1993 Cotton Outlook estimated world cotton production at 86.35 million bales, compared with 82.62 million bales in 1992/93. By the end of 1993 world cotton production

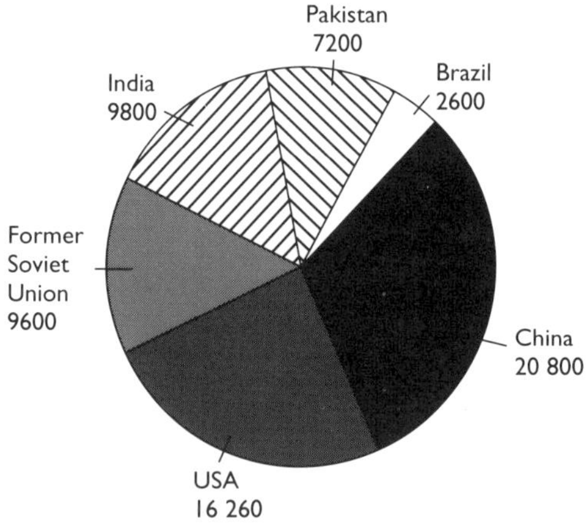

4.1 World cotton production: 1992/93 estimate (thousand 480 lb bales) (source: USDA/FAS).

Table 4.4 World cotton production and yield (thousand 480 lb bales)

Marketing year	1000 ha	Yield (kg/ha)	Production
1960/61	32 120	305	45 066
1961/62	32 413	299	44 467
1962/63	31 780	322	46 958
1963/64	32 882	336	50 761
1964/65	33 551	349	53 770
1965/66	33 326	372	56 886
1966/67	31 210	365	52 272
1967/68	30 964	362	51 524
1968/69	31 921	388	56 872
1969/70	32 487	367	54 699
1970/71	31 801	377	55 072
1971/72	33 043	390	59 114
1972/73	33 538	401	61 776
1973/74	32 849	415	62 638
1974/75	33 530	413	63 656
1975/76	29 889	393	53 942
1976/77	30 568	401	56 366
1977/78	33 568	414	63 863
1978/79	32 901	395	59 654
1979/80	32 211	443	65 472
1980/81	32 055	432	63 607
1981/82	32 978	453	68 630
1982/83	31 482	461	66 679
1983/84	30 883	464	65 825
1984/85	33 609	574	88 646
1985/86	31 457	556	80 289
1986/87	29 214	526	70 602
1987/88	30 793	573	81 109
1988/89	33 676	546	84 423
1989/90	31 482	552	79 866
1990/91	33 033	573	86 982
1991/92	34 748	601	95 947
1992/93	32 485	557	83 150

Source: USDA/FAS

was estimated by the ICAC to be 17.9 million tonnes for 1993/94, about the same as in the previous season. These figures show the enormous variation which can exist in cotton production from year to year, as well as the considerable disparity both between forecasting organisations and also between the same organisation's forecasts over time.

A total of 33 million hectares was allocated to cotton growing in 1992. Of the 82 countries in which cotton was grown in 1988 - since the break-up of the Soviet Union, several more - 73 were developing, of which 25 were classified by the UN as being least developed. Regionally, cotton was then produced in 34 African developing countries, of which 18 were least developed, 21 in Asia - of which 6 are least developed - and 18 in Latin America. World recession will no doubt have increased these numbers.

The USA, Pakistan and India are the big producers whose increase since 1983/84 now appear to mean that it is almost impossible that the world will be short of cotton in the period to 2000. The four largest producing countries accounted for half of world production in the early 1980s, but two-thirds in the early 1990s. What appears to make the current situation different is that in many cases there are no obvious alternative crops in many countries to which farmers can consistently turn to make better money. World crops of the order of 20 million tonnes must now be considered not unusual - almost twice the amount of cotton grown in the late 1960s.

It is widely recognised that in many countries production is still falling way below its potential, especially in Africa and India. There is no one international agency responsible for co-ordinating technical assistance to national cotton industries - in the 1970s the Cotton Research Corporation provided a great deal of assistance to English-speaking African countries, and the same role since the 1980s for the Francophone countries is being provided by the Institut de Recherches du Coton et des Textiles (IRCT).

However, cotton faces competition as a crop in some areas. In the Southeast of the USA, the main competitors for cotton are soyabeans and corn. Soyabeans account for over 10 times as much acreage there and corn over 5 times as much. Net cash returns are roughly equivalent for all three crops and therefore cotton acreage is very sensitive to government programmes and general price evolutions and expectations. Peanuts and tobacco have also been high cash earners but their acreage has been controlled by allotments and therefore the effect on cotton acreage has been small. In the US Delta, it is soyabeans again that have been the main competitor. In Texas, Oklahoma and New Mexico, it is grain wheat and sorghum which are the competitors. Wheat, barley and alfalfa are the main competitors in the West, accounting for 20, 17 and 10% by comparison with cotton's 32% of planted acreage in the mid-1980s. Yields from competitor crops to cotton tend to increase further north. Cotton is also more sensitive to competition from other crops in some of the non-irrigated parts of the central zone of the Cotton Belt than in other areas (Table 4.5).

In other countries, such as those in the Commonwealth of Independent States (CIS), cotton is still more or less a monoculture - despite government concern - although the area allocated for fruits and vegetables is increasing - and there is therefore little competition for production. In some developing countries, there is a real choice to be made between cash crops, such as cotton, and food crops, such as rice. In more sophisticated farming strategies, it becomes clear that the most productive land for rice growing is usually the least productive for cotton and vice versa.

Cotton is a very labour-intensive crop in developing countries, where about 200 million people depend for some or all of their incomes on cotton production and a further 50 million on processing. This amounts to almost 7% of the working population in these countries and makes cotton the most important cash export crop of the developing world. For a large number of rural communities in developing countries, growing cotton for cash sale is the only alternative to subsistence farming.

In terms of land use, however, cotton is sufficiently intensive that it takes only a small percentage of total available land. In global terms, cotton accounts for about 2.5% of land utilisation; for developing countries, about 3%, and in Asia, 4%. In Africa less than 1% of utilisable land is used for cotton.

Cotton production by country

Africa

In French West Africa (the African franc zone), a typical increase over 1960–80 was seen in Chad, where production rose from 160 000 bales to 240 000 bales, and yields from 128 to 194 kg/ha. Ethiopia scored significant success with its cotton production programme between 1949 and 1980. Although the statistics are relatively unreliable, and yields fell from 269 down to 156 kg/ha, because the area sown

Table 4.5 Crop values and harvested acres. Cotton vs. competing crops: USA

Crop year	Cotton and cottonseed	Soya beans	Grain sorghum	Corn	Wheat	Rice
Value of production ($000)						
1975/76	2 355 942	7 617 984	1 775 023	14 789 368	7 534 851	1 071 924
1976/77	3 680 365	8 786 979	1 450 085	13 471 796	5 851 433	811 358
1977/78	4 002 973	10 352 409	1 433 991	12 944 467	4 742 686	941 248
1978/79	3 530 604	12 449 679	1 464 187	16 280 688	5 280 634	1 087 000
1979/80	5 088 995	14 250 214	1 879 823	19 904 013	8 070 378	1 383 993
1980/81	4 561 189	13 559 595	1 696 008	20 571 244	9 277 608	1 873 007
1981/82	4 624 444	12 071 341	2 087 223	20 405 973	10 224 661	1 654 413
1982/83	3 777 844	12 462 779	2 090 774	22 039 409	9 813 015	1 246 608
1983/84	2 989 626	12 774 974	1 382 097	13 534 930	8 532 790	874 004
1984/85	4 115 387	10 748 050	2 054 805	20 084 595	8 755 033	1 119 373
1985/86	3 993 752	10 571 324	2 538 322	21 029 100	7 647 801	893 377
1986/87	2 738 885	9 326 189	1 315 833	12 387 438	4 861 364	528 256
1987/88	5 029 720	11 391 000	1 179 444	14 107 705	5 497 712	971 167
1988/89	4 908 743	11 487 742	1 337 424	12 661 362	6 683 999	1 091 817
1989/90	4 370 571	10 916 145	1 287 738	17 896 810	7 542 464	1 134 039
1990/91	5 798 139	11 042 010	1 220 499	18 191 643	7 184 427	1 047 242
1991/92[a]	5 803 753	11 078 442	1 347 366	18 063 205	5 800 704	1 166 077

Harvested acres (thousand)						
1975/76	8 796	53 579	15 355	67 505	69 391	2 818
1976/77	10 914	49 358	14 723	71 300	70 771	2 480
1977/78	13 275	57 830	13 797	71 614	66 686	2 249
1978/79	12 400	63 663	13 410	71 930	56 495	2 970
1979/80	12 831	70 343	12 901	72 400	62 254	2 869
1980/81	13 215	67 813	12 513	72 961	71 125	3 312
1981/82	13 841	66 163	13 677	74 524	80 642	3 792
1982/83	9 734	69 442	14 137	72 719	77 937	3 262
1983/84	7 348	62 525	10 001	51 483	61 390	2 169
1984/85	10 379	66 113	15 355	71 915	66 928	2 802
1985/86	10 229	61 584	16 782	75 224	64 734	2 492
1986/87	8 468	58 292	13 859	69 159	60 723	2 360
1987/88	10 030	56 977	10 604	59 208	55 960	2 333
1988/89	11 948	57 373	9 042	58 250	53 189	2 900
1989/90	9 538	59 538	11 103	64 703	62 189	2 687
1990/91	11 732	56 512	9 089	66 952	68 283	2 823
1991/92[a]	12 842	57 951	9 820	68 842	57 693	2 750

[a] Preliminary
Source: NASS, USDA.

increased so substantially, over a hundredfold, production was able to increase from a negligible amount to an estimated 100 000 bales - a rate of increase that was to be dashed by the civil war which resulted in no increase in production after 1983/84 according to the ICAC.

USDA figures profile the development of production since World War II. They show that, in some cases, African yields made little improvement between 1947 and 1980. Angola, for instance, produced 24 000 bales of cotton in 1947 and 20 000 bales in 1980, from an area of 40 000 and 24 000 ha respectively - a yield improvement of just 40% in 33 years. In French Equatorial Africa before independence, the country produced 114 000 bales with 243 000 ha in 1947. The successor Central African Republic produced only 30% more cotton in 1980 than in 1960, although yields had improved by over 80% in the same period. Production reached 240 000 bales in 1980.

In West Africa Benin's yields have always been relatively high, but fell from 1965 (when the USDA started reporting them) to 1980. The area sown to cotton in Benin increased dramatically, however, from 9000 to 50 000 ha, so production was able to rise from 12 000 bales to 65 000 bales. These were, however, very small amounts. The Cameroon produced virtually nothing after World War II, and apart from two years of exceptionally high yields in 1968 and 1969, production remained low until steady yield growth caused a rise in production after 1978.

Amongst the French overseas territories that produced cotton, French West Africa produced 14 000 bales in 1947 and 90 000 bales in 1964 after a rise in yields of over 60%. Subsequent states such as Mali, on the evidence of the USDA, had lower yields and production: Mali produced only 37 000 bales in 1965, and 190 000 by 1980 after a dramatic rise in yields from 143 to 487 kg/ha. Upper Volta produced 12 000 bales in 1965 with a dismal yield of only 50 kg/ha, rising to 90 000 bales and a very respectable 245 kg/ha by 1980, with most of the improvement occurring during the 1970s in both cases. Niger produced only 9000 bales in 1965 with a yield of 118 kg/ha; by 1980 Niger was still lagging, with production of 14 000 bales and a yield of 169 kg/ha. The Ivory Coast produced 17 000 bales in 1965, but by 1980 had increased production dramatically to 300 000 bales. Area sown had increased from 12 000 to 145 000 ha, and already high and therefore internationally competitive yields had risen from 315 to 450 kg/ha. Senegal's production in 1965 was negligible, but by 1980 it was producing 50 000 bales from a yield estimated at 218 kg/ha.

What has happened generally in Africa since 1980? Anglophone production percentage has now fallen to 55% of sub-Saharan African production, with no sign of the decline relenting. In the early 1990s the continuing low prices caused by the widespread availability of CIS and Chinese cotton, the rising cost of production inputs, demanding quality requirements aided by the spread of high-volume instrument (HVI) equipment (see later in the chapter for details of this), and the high fixed costs of international marketing are imposing a crisis on African cotton producers.

In 1974/75, drought led to falls in production in Uganda, whilst there was also a shift to food crops which led to a fall in production in Egypt and the Sudan. Small improvements generally occurred in Africa in the following year but this momentum was only carried forward after a delay of one year. Output rose in 1979/80 despite setbacks in planting in a few countries due to bad weather and inadequate input supplies. Sudan in particular experienced difficulties with irrigation supply, labour and other production inputs such as fertiliser. Drought in Africa resulted in the crop staying roughly level in 1981/82 despite increased acreage, and small increases were recorded the following year. The Sudan, Tanzania and Zimbabwe harvested larger crops in 1983/84 and output continued to rise through 1986/87. The increase of 1987/88 was mainly accounted for by Zimbabwe and the Côte d'Ivoire. In 1988/89 output rose in Africa, largely because of favourable weather in the major producing countries in West Africa, which more than counterbalanced the continued decline in ginning in parts of East Africa. The following year, output declined because of dry weather at planting time combined with low farm prices and insufficient ginning capacity. Production recovered modestly in 1990/91. African harvested area remains similar in southern Africa to the levels of the early 1980s, but has increased by almost 70% in West Africa. Production similarly has increased by 23% and 83% respectively. ICAC has recorded that the West African subtotal of production increased from 303 000 tonnes in 1983/84 to 556 000 tonnes in 1992/93, whilst the southern African total, including South Africa itself, increased from 287 000 to 354 000 tonnes, most of this increase arising from Tanzania and Nigeria.

After their successes of the 1980s, cotton has become entrenched as a major part of the economies of French West Africa - countries where costs can be kept down to permit competition internationally. In countries such as Mali, Burkina Faso and Chad, cotton accounted in the early 1990s for between 40% and 60% of the export trade and 5-7% of GDP. In the Côte d'Ivoire, Cameroon and Senegal, cotton is not so important nationally but it is vital for whole regions of all three countries. In the savanna zones cotton is now the only crop which brings in regular cash income, and is regarded as a mainspring of development.

The cotton companies in francophone countries were originally established by the Compagnie Française pour le Developpement des Textiles (CFDT), but were handed over to the host countries in the 1970s. All cotton production activities, from actual production through harvesting and processing to marketing and sales are controlled by these companies. Farmers tend to be relatively well informed and trained; fertiliser, seed and pesticides are available on credit and supply is constant. There is ample technical support throughout the season, and above all there are government support programmes to ensure a preset guaranteed price for the cotton crop, paid on delivery: even small-scale farmers have confidence. The French institute for research on cotton and other textiles, the IRCT, also suggests new seed varieties, improving spraying schedules and analysis of fertiliser requirements. In 1987/88, 77% of the cotton-producing land in the francophone countries had fertiliser applications of 145 kg/ha, and 79% received three applications of fertiliser. IRCT must take much of the credit for the production improvements since the 1960s.

Cotton generates cash for West Africa, and the income gives farmers sufficient money to buy livestock and equipment for animals that can be used for power generation. Ploughing with even this mechanisation can be speeded up fivefold, and it is most advanced in the cotton zones: in Mali draught power is now used on 90% of all cotton-growing holdings. The cotton companies have contributed significantly to the development of draught power through animal training units, education and supplying equipment - which has also included the supply and maintenance of some small tractors. This is part of the general extension of the work of the cotton companies beyond cotton alone, into food crops. Cotton growing also benefits food crops because food crops thrive on the fertiliser residue, so the cotton-growing areas boast the most productive farms, and also more developed infrastructure.

Only some 10% of southern African fibre is processed locally, with the rest exported. African fibre is frequently able to command a market premium for its high quality.

The USA

Cotton is grown as an annual crop in the USA, as in the majority of the world's crop-producing areas, by comparison with the perennial growth in some frost-free areas of South and Central America. Seventeen states grow cotton in the USA: California, Arizona, Texas, Arkansas, Missouri, New Mexico, Oklahoma, Alabama, Louisiana,

Mississippi, Tennessee, Georgia, Kansas, South Carolina, North Carolina, Florida and Virginia. Most cotton production in the USA is spread over the South and Southwest of the country in the area usually referred to as the cotton belt. Prior to 1932 this term referred to the area growing cotton without irrigation. Thereafter so much irrigation was employed that the term referred to the entire area along the southern US border from the Atlantic and the Gulf of Mexico to the Pacific. The northern limit is usually said to be along parallel 37°N lat., although it extends above that line in Virginia, the Mississippi valley and California, and it falls below that parallel in the Appalachian mountains in North Carolina and Tennessee and in the high plains region of Texas and Oklahoma. The southern limit extends along the entire border with Mexico. The boundaries of the belt are set by temperature and frost-free periods to the north and national boundaries elsewhere. The limit is usually said to be 200 frost-free days; and yields tend to decrease northwards in the belt - mainly because of the risk of late spring and early summer frosts. There are four regions, each with distinctive production characteristics and with changing relative attractiveness for cotton production:

1 The southeastern region, which extends from the Atlantic westward to the Mississippi lowland region and which includes North Carolina, Georgia and Alabama and small parts of Virginia and Florida. There is a humid climate (120 cm or more of annual rainfall), red and yellow earth soils, hilly topography, small farms, intensive culture and high per-acre yields. About 10% of the US crops is produced here.

2 The Mississippi valley region includes Mississippi, Tennessee, Missouri, Arkansas and Louisiana. It also has a humid climate (110–130 cm of annual rainfall), much alluvial soil, but a smooth topography. Yields are high and the usual staple length is about 1 1/32 in. A little under 30% of the US crop is now produced in this region.

3 The Mexican Gulf southwest region includes most of Texas and Oklahoma. In the east this region is humid, the soils are red and yellow earths; in the centre and the west, the climate is milder. Most cotton is produced on the mildly humid black prairie-earth and black-earth soils in east central and western parts of the region. The rainfall varies from 95 mm annually in the prairie-earth section to 45 mm on the western high plains. The topography is flat to rolling and soils are fertile. The typical staple is between 26/32 and 1 in. Yields are not as high as in other parts of the belt but per-person

production is high. The region usually produces about 30% of the US crop.

4 The strip of land along the southern border from the Gulf of Mexico to the Pacific, including south and west Texas and the southern parts of Arizona, New Mexico and California. The area is semi-arid and there is not enough rainfall for cotton production; the soil is rich in minerals but lacks organic matter. Cotton is rotated in production with alfalfa. The yields are variable, averaging about 2.5 bales/acre in California and Arizona to about 1 bale/acre in West Texas. The crop is highly standardised and the main variety grown is Acala. Three strains predominate: Acala 442 in California, Acala 44 in Arizona and Acala 1517 in Texas. Acala's typical staple length varies from 1 1/32 to 1 5/32 in. Most ELS cotton grown in the USA is grown in this region in Texas, Arizona and New Mexico, mostly Pima 32 with a typical staple of 1 13/36 in. The region as a whole usually produces somewhat under 35% of the total US crop. In the United States in 1910, the USDA's cotton publication listed nearly 600 varieties of American upland cotton; by the mid-1950s throughout the redefined cotton belt, six varieties of American upland were grown on approximately 90% of the acreage. They were Deltapine, Acala, Coker, Empire, Stoneville and Rowden, with staple lengths averaging between 25 and 30 mm. About 70 000 acres are planted with ELS cotton annually, with an output in the region of 120 000 bales.

US production has shifted westwards because of the greater fertility of the soils; relativities have changed because of soil erosion in the Southeast and Texas Blacklands, for example. About 35% of the US crop is irrigated, although irrigation costs vary substantially between areas: pumping costs in Arizona, for example, are four times as great as in California. Government programmes continue to maintain production in some areas that would not otherwise produce it. Production has shifted from hilly to flat land generally. Other factors that have encouraged production to shift are cotton disease problems, off-farm employment opportunities and increased profitability from livestock farming. Along with this westwards shift has been an increase in farm size, from an average 24 acres in 1949 to over 250 acres in the 1990s. Many fewer farmers now farm cotton: 30 000 compared with 200 000 in 1970, although family-owned operations still constitute three-quarters of production.

The US crop acreage increased from fewer than 8 million acres at the end of the Civil War through 44 million acres in the mid-1920s with production of 18 million bales in 1926. Planted area then declined to about 10 million acres in 1966. Cotton support programmes from the early 1930s onwards have supported prices, adjusting acreage and therefore production to perceived market needs, hence the variety of acreage allotments, marketing quotas and parity price supports that existed during 1933–65 resulting in peak stocks of 17 million bales at the end of the 1965 crop year. After the Cotton–Wheat Act of 1964 and the Food and Agricultural Act of 1965, programmes became more market-oriented, with an emphasis on price support based on world price levels and direct payments when prices were low. In the late 1960s the US crop averaged 9.65 million bales, and steady increases were recorded in the early 1970s. The Agricultural Act of 1970, the Agricultural and Consumer Act 1973 and the Food and Agricultural Act 1977 were all efforts to establish price and income safety nets for producers while not giving them *carte blanche* to produce unwanted cotton and other commodities. In 1974/75 the US crop continued to fall, following drought in Texas and excessive rains in the central belt. However, average staple length increased significantly, a policy aim at the time. Plantings and yields increased the following years so output rose, but there was a savage decline in 1975 due to unattractive prices at planting time and low yields resulting from adverse growing conditions. It was this production shortfall in the USA which accounted for much of the decline in world production that year. The following year production was up again, from 8.27 million bales to 10.56 million bales and again in 1979/80 to 14.7 million bales. Drought hit the USA in the 1980/81 season and output fell to 11.02 million bales as yields fell to their lowest levels for 20 years. The Agriculture and Food Act of 1981, another attempt to establish safety nets, therefore conflicted with adverse market conditions. The measure of the turnaround possible in just one season is provided by what happened the following year. With ideal growing conditions, as well as excellent weather at harvest time, yields recovered tremendously and output rose by nearly 40% to 15.6 million bales, the highest in nearly 30 years.

Output fell again the next year and in 1984 the Agricultural Programs Adjustment Act did just that – adjusted the programme. The FAO pointed out that 'The impact of the government's programmes of area curtailment, comprising voluntary land diversion and incentive payments in kind (PIK), was compounded by adverse weather in major growing areas'. Production fell further, to 7.8 million bales, which was the lowest level since 1967/68, in response. The PIK programme was promptly discontinued as opinion believed cotton output had fallen too far. Yields were at record levels and acreage

Table 4.6 US cotton area and production

	1991	1992 est.	1993 proj.
Base (000 ha)	5 898	6 024	6 070
Programme (%)	5.0	10.0	7.5
Participation (%)	84	87	88
Required diverted (000 ha)	249	521	401
Permitted plantings (000 ha)	4 726	4 690	4 941
Participation planting (%)	91	93	93
Actual participation planting (000 ha)	4 302	4 362	4 595
Non-participant base (000 ha)	924	813	728
Non-participant planting (%)	132	112	95
Non-participant planting (000 ha)	1 222	909	692
Triple base devoted to cotton (000 ha)	61	54	40
Total planting, upland (000 ha)	5 585	5 324	5 328
Total planting, Pima (000 ha)	100	107	91
Total planting (000 ha)	5 687	5 431	5 419
Abandonment (%)	8	16	8
Harvested (000 ha)	5 245	4 538	4 985
Average yield (kg/ha)	731	777	723
Production (000 tons)	3 835	3 528	3 604

planted had been increased, so output was able to rise. But the US government then intervened again, with farm programmes that required farmers to set aside one quarter of their base cotton acreage in order to benefit from assistance.

The Food Security Act 1985 was of great importance in setting policy over 1985–90 and was of great importance. The Act was developed when the market was in recession. There were acreage limitations, non-recourse loans and target prices, but also discretionary authority for the Secretary of Agriculture to administer annual commodity programmes. The Act specified target price minima through to 1990 and set loan rates as an average of previous market prices with provisions for loans to be repaid at below the loan rate if market competitiveness might be hampered by the formula rate (Table 4.6).

That and disadvantageous weather pushed output down again and had a significant effect on prices. ELS cotton was brought within the Act too, specifying that the ELS cotton loan rate be equal to 85% of the average of the middle three average annual prices of the past five years. The following year, 1987/88, record yields resulted in an output of 14.7 million bales (all cotton), putting the USA back as the second largest cotton producer in the world. In 1988/89 the USA produced 15.15 million bales (FAO), having a second year of record yields. In 1989/90 production fell by over 20% because of area reduction programmes, reduced plantings due to unfavourable prices and lower yields because untimely rains and low temperatures delayed poll maturation in many areas. These prices increased, however, and in 1991/92 cotton production increased in the USA by more than a quarter to 17.6 million bales, the largest crop in 54 years. Table 4.7 shows costs and returns for the years 1975–90.

The reasons for the high 1991/92 US crop were put down to seed quality, good weather encouraging high square retention (and hence better, more sustained growth and boll retention), few pests (in particular, low levels of pink bollworm and verticillium wilt), uninterrupted spraying operations, continued good weather for boll maturation and little boll rot. The crop has longer staple length, a higher micronaire, more length uniformity and is stronger. USDA figures released on 12 August 1992 suggested that there would be 16.5 million bales of production, 9% below 1991/92, reflecting mostly the abandoned acreage in Texas; this estimate stayed relatively constant throughout early 1993, being 16.3 million bales in May.

The USDA can make announcements with changes in production levels that are so substantial – not just for the USA but internationally – and their effect on price levels internationally so considerable, that it is very understandable that traders' attention should be so fixed on estimates produced by them throughout the year. The final forecast for the year, released in December 1992, was 16.3 million bales, which was 1.4 million bales less than the year before. By December 1992 the cotton harvest was 81% complete, 3% ahead of the previous year and at the five-year average. At the end of October 1992 the USDA announced its preliminary decisions about the upland programme for 1993/94. The target price and loan rate remained unaffected; the acreage reduction programme (ARP) fell from 10% in 1992/93 to 7.5%. On 1 March, the Secretary of Agriculture announced that producers who enrol in the USDA's ELS cotton programme may request 50% of estimated deficiency payments during sign-up. US production and yield are shown in Table 4.8. Figure 4.2 shows a sample request for certification form.

By November 1993 US cotton production was estimated to be 17 million bales, with estimates falling steadily throughout the year. This would be the fourth largest cotton crop in US history. Several factors contributed to the lower estimates: a combination of weather and pest pressure, with the largest declines seen in Memphis. Speculation continues to centre around these crop estimates.

Table 4.7 US cotton production economic costs and returns 1975–90 ($/planted acre)

Item	1975	1976	1977	1978	1979	1980	1981	1982	1983	1984	1985	1986	1987	1988	1989	1990
Gross value of production (excluding direct Government payments):																
Cotton	211.67	277.75	262.49	226.06	316.28	274.16	286.69	319.22	312.23	318.77	330.89	225.95	446.48	314.46	362.87	404.44
Cottonseed	40.73	48.04	51.59	36.27	49.74	39.62	38.86	35.16	64.82	46.48	32.97	29.73	47.13	58.58	48.57	59.20
Total gross value of production	252.40	325.79	314.09	262.33	366.02	313.78	325.55	354.38	377.05	365.25	363.86	255.68	493.61	373.04	411.44	463.64
Economic full ownership costs:																
Variable cash expenses	141.08	143.71	160.73	159.90	189.24	200.34	216.25	235.80	224.92	237.27	245.50	217.84	256.14	255.92	260.04	271.51
General farm overhead	14.23	15.33	16.37	12.53	13.28	9.61	20.89	23.88	23.23	24.28	21.16	21.29	21.59	23.46	22.29	25.80
Taxes and insurance	6.14	6.57	7.64	8.27	9.28	11.50	9.95	9.33	9.30	10.06	10.55	12.66	14.13	14.58	15.00	15.14
Capital replacement	31.25	39.59	41.10	34.98	42.74	46.23	38.21	42.58	44.88	44.06	45.57	50.35	55.16	59.78	62.64	64.69
Operating capital	2.63	2.25	2.82	3.70	5.29	6.80	8.83	7.76	5.80	6.67	5.38	5.09	6.13	7.22	8.92	8.35
Other non-land capital	8.23	9.76	9.04	7.19	8.63	9.13	6.23	6.65	6.36	7.88	7.66	8.76	9.39	11.83	15.07	17.16
Land	25.86	33.42	32.10	40.77	49.31	43.24	50.55	58.00	62.84	60.50	58.63	51.44	83.85	71.00	73.36	82.38
Unpaid labour	5.11	6.01	6.97	9.48	10.77	11.33	8.82	9.53	9.51	9.46	10.05	20.65	21.36	21.66	22.71	23.46
Subtotal	234.53	256.64	276.78	276.83	328.54	338.18	359.73	393.52	386.84	400.18	404.50	388.08	467.75	465.45	480.03	508.49
Residual returns to management and risk	17.87	69.15	37.31	−14.49	37.47	−24.40	−34.18	−39.14	−9.79	−34.93	−40.64	−132.40	25.86	−92.41	−68.59	−44.85
Harvest period price (dollars/lb)	0.51	0.64	0.52	0.58	0.63	0.75	0.55	0.58	0.66	0.57	0.55	0.49	0.65	0.53	0.66	0.67
Yield (lb/planted acre)	416.00	435.00	505.00	390.00	502.40	367.19	523.88	550.73	470.40	554.65	598.35	462.97	684.66	593.33	549.80	603.64

Note: Survey base changed in 1986.

Table 4.8 US cotton production and yield

Marketing year	1000 ha	Yield (kg/ha)	Production (thousand 480 lb bales)
1960/61	6 195	500	14 237
1961/62	6 327	492	14 283
1962/63	6 301	512	14 827
1963/64	5 752	579	15 294
1964/65	5 689	580	15 144
1965/66	5 510	591	14 951
1966/67	3 866	538	9 555
1967/68	3 236	501	7 443
1968/69	4 112	579	10 925
1969/70	4 474	486	9 990
1970/71	4 514	492	10 192
1971/72	4 643	491	10 477
1972/73	5 255	568	13 704
1973/74	4 844	583	12 974
1974/75	5 086	494	11 540
1975/76	3 560	508	8 302
1976/77	4 417	522	10 581
1977/78	5 372	583	14 389
1978/79	5 018	471	10 856
1979/80	5 193	613	14 629
1980/81	5 348	453	11 122
1981/82	5 601	608	15 646
1982/83	3 939	661	11 963
1983/84	2 973	569	7 771
1984/85	4 200	673	12 982
1985/86	4 140	706	13 432
1986/87	3 427	618	9 731
1987/88	4 061	791	14 760
1988/89	4 835	694	15 411
1989/90	3 860	688	12 196
1990/91	4 748	711	15 505
1991/92	5 245	731	17 614
1992/93	4 514	784	16 260

Source: USDA/FAS.

In the Autumn of 1993, it was expected that the USDA would reduce upland cotton acreage in 1994 near the maximum of 25% of the Crop Acreage base. Planting restrictions were expected to be significantly above the 7.5% in force for the 1993 season.

There may be a continuation in the geographical shift of production away from California, because participation in the government programme is lowest in California as the farms are bigger. Moreover, the Acala varieties grown there usually command a

U.S. Department of Agriculture Agricultural Marketing Service Cotton Division **Request for certification**	Check services desired ☐ Combination *(Original, Review and Micronaire)* ☐ Original classification	☐ Review classification ☐ Micronaire	Lot No. ME 1773 No. B/C 25
To: Marketing Services Office *(City and State)*		Date of request 3/17/87	

Withdrawn or transferred		Tag Nos	Original class: Grade, staple, and remarks	Review class: Grade, staple, and remarks	Mike	Withdrawn or transferred		Tag nos.	Original class: Grade, staple, and remarks	Review class: Grade, staple, and remarks	Mike
NT	1	285350		4136	50		26				
	2	351		4136	45		27				
	3	352		4135	45		28				
	4	353		5036	49		29				
	5	354		4135	47		30				
NT	6	355		5236	43		31				
	7	356		5134	47		32				
	8	357		4134	44		33				
	9	358		4135	42		34				
	10	359		5035	40		35				
	11	360		5136	45		36				
	12	361		5133	49		37				
NT	13	362		4235	48		38				
	14	363		4135	46		39				
	15	364		5036	45		40				
	16	365		4134	43		41				
	17	366		3134	42		42				
	18	367		4035	46		43				
	19	368		5035	47		44				
NT	20	369		5234	46		45				
	21	370		5137	45		46				
	22	371		5134	48		47				
NT	23	372		5234	45		48				
	24	373		5736	44		49				
	25	285374		5036	47		50				

Original Class		**Review and Micronaire**	
Recorder	Classer	Recorder	Classer
Marketing Services Office *(City and State)* *(Area Director)*		☑ Review ☑ Micronaire **Grading Section** **Memphis, Tennessee** Paul R. Dickson	
Date issued		Date issued 3/27/87	Date reported 3/30/87
Date inspected 3/20/87		Location of cotton *	
For use by Marketing Services Office only		No. tenderable 20	No. untenderable 5
Owner and agent A. B. Cotton Co. Box 333 Memphis, TN 38101		Received class certificates for above bales *(Date)* A. B. Cotton Co. 3/31/87 By BKS	

Summary tenderable for grade, staple and micronaire

Grade	Staple 33	Staple 34	Staple 35 & +	Total
11				
21				
30				
31		1		1
40			1	1
41		2	5	7
50			5	5
51	1	2	3	6
12				
22				
32				
Total	1	5	14	20

Form CN-331 (8-79) (Edition of 4-79 is obsolete

** Use separate sheet for each warehouse or compress*

4.2 Sample USDA form.

market premium, which makes the loan less attractive to farmers. However, the Californian drought is over, so in the short to medium term there may be a revival in output, and the Delta and Texas may well step up production, especially if the ARP falls further. US cotton planted acreage is shown in Table 4.9.

Pakistan

When Pakistan gained independence in 1947, production was estimated by the USDA at 925 000 bales. Yields were always better than obtainable in India, and in 1947 they were 159 kg/ha (USDA). These yields actually increased generally during the 1950s, and reached 232 kg/ha by 1960. The area sown to cotton had hardly increased, with production therefore rising to 14 million bales (Indian and Pakistani bales are 170 kg; USDA and ICAC figures are usually in bales of 480 lb). The 1960s saw a further rapid rise in yields to 311 kg/ha in 1970, and the area sown increased, with the result that production rose to 2.5 million bales.

During the mid-1970s, cotton production did reduce somewhat due to a shift towards food crops and there was a particularly low crop of 1.921 million bales in 1975/76. By 1979/80 production had recovered - 3.1 million bales in 1980 - and these gains continued into the 1980s, assisted by incentive schemes. Pakistan had four record crops in succession in the mid-1980s, illustrating the success of high-yielding, early maturing seeds, effective use of fertilisers, and better irrigation. Production was 0.494 million tonnes in 1983/84 (ICAC), rising to 1.468 million tonnes in 1987/88. In mid-summer, flood damage reduced average yields and hence output for the 1988/89 season fell, but not greatly. In 1989/90 output was maintained at 1.45 million tonnes. The following season price levels stimulated output, which rose to 1.64 million tonnes, aided by more irrigation water, despite the fact that the rains did cause some damage to crops. Production overall therefore showed a considerable rise from 1983/84, of almost fourfold. Yields trebled over the same period. Pakistan is justly proud of its cotton production record since the early 1980s, and produced close to 2 million tonnes with an average yield of 761 kg/ha in 1991/92 (ICAC, 1992). Most recently, in 1992/93, production has fallen 10% down to 7.1 million bales (USDA), as a result of the heavy rains and floods in the Sind and Punjab, causing a decline in area and yields in the cotton-growing areas of Multan and the Vehari district. The higher than usual humidity has caused an outbreak of leaf curl virus, also cutting yields. However, Pakistan responded to the problems well and production is likely to revive.

Table 4.9 US cotton planted acreage (thousand acres)

Crop and area	1982/83	1983/84	1984/85	1985/86	1986/87	1987/88	1988/89	1989/90	1990/91	1991/92[b]	10-year average	5-year average
Upland southeast	634	481	703	823	761	833	1 047	854	1 133	1 584	885	1 090
Alabama	287	219	309	330	315	335	390	328	380	415	331	370
Florida	16	13	17	25	20	30	33	26	37	50	26	35
Georgia	163	120	175	255	225	250	350	265	355	430	259	330
North Carolina	71	60	97	88	82	96	126	112	201	460	139	199
South Carolina	97	69	104	124	118	120	145	120	155	211	126	150
Virginia	0	0	1	1	1	2	3	3	5	18	3	6
Mid-south	2 429	1 755	2 669	2 647	2 608	2 820	3 440	2 984	3 583	4 072	2 901	3 380
Arkansas	410	320	470	465	490	555	695	610	770	1 000	579	726
Louisiana	605	420	650	640	580	605	735	645	810	875	657	734
Mississippi	1 000	687	1 045	1 050	1 020	1 020	1 230	1 050	1 230	1 245	1 058	1 155
Missouri	154	108	164	152	178	200	245	214	248	332	200	248
Tennessee	260	220	340	340	340	440	535	465	525	620	409	517
South-west	6 280	4 320	5 776	5 371	5 251	5 101	6 061	5 022	5 882	6 842	5 591	5 782
Kansas	0	0	1	1	1	1	1	2	2	2	1	2
Oklahoma	480	320	425	370	400	400	460	370	380	440	405	410
Texas	5 800	4 000	5 350	5 000	4 850	4 700	5 600	4 650	5 500	6 400	5 185	5 370
West	1 930	1 307	1 917	1 760	1 313	1 506	1 777	1 351	1 519	1 399	1 578	1 510
Arizona	471	291	430	360	250	290	350	240	350	360	339	318
California	1 380	960	1 410	1 330	1 000	1 150	1 350	1 050	1 100	970	1 170	1 124
New Mexico	79	56	77	70	63	66	77	61	69	69	69	68
Other upland[a]	1	0	0	0	0	0	0	0	0	0	0	0
Total upland	11 275	7 863	11 065	10 601	9 933	10 259	12 325	10 211	12 117	13 897	10 954	11 762
ELS	71	63	80	84	112	138	190	377	231	247	160	237
Arizona	42	30	51	57	74	91	128	245	125	106	95	139
California	0	0	0	0	0	1	2	18	26	60	11	21
Mississippi	NA	NA	NA	NA	NA	NA	NA	2	1	1	0	1
New Mexico	10	11	10	8	11	14	18	30	19	20	15	20
Texas	20	22	20	20	26	32	42	82	60	60	38	55
All cotton	11 345	7 926	11 145	10 685	10 045	10 397	12 515	10 587	12 348	14 144	11 114	11 998

[a] Other includes Illinois, Kentucky and Nevada. [b] Preliminary
Source: NASS, USDA.

India

In 1947, the year of independence, India produced 2 625 000 bales (USDA) with a yield of 129 kg/ha. Stocks were maintained at a relatively high level of 3 630 000 bales. Given the vagaries of the statistics in this period, it seems reasonable to conclude that yields actually fell more or less continuously until 1952, to a minimum of approximately 100 kg/ha, and showing little sustained or material improvement until the end of the decade. Production increased because the area sown to cotton did, reaching 4.63 million bales in 1960. Yields oscillated after 1960 but keeping an average above 120 kg/ha, but the 1960s saw no sustained rise in production. After a rise in the early 1970s, with peak production of 6 million bales in 1974, production fell, recovering to 6.1 million bales by 1980. Production then rose steadily, only to fall again in 1982/83. Incentive schemes were introduced in the 1984/85 season, but severe drought reduced India's output in the mid-1980s. When this threat receded, India experienced in 1989/90 a dramatic increase in yields, of 25%, and output of lint rose by 20% to 2.2 million tonnes (FAO). There was also a use of improved seeds, more irrigation and management, combined with much better weather and few pests. In 1988/89 output increased 10%, with very satisfactory prices in the growing season encouraging production in more non-irrigated land. Indian production has benefited from the increase in yields since 1989/90, and production has almost doubled since 1983/84, from 7.9 million bales in 1984 (USDA)/1.3 million tonnes (ICAC) to 2.3 million tonnes in 1992/93 (ICAC), some 13% of world production. India has the largest land area under cotton, estimated at 7.6 million ha, 23% of the world's cotton area, and about 1.5 × greater than the area currently cultivated in China and the USA. Output was expected in late 1993 to equal the previous season in 1993/94.

However, some Indian yields remain amongst the lowest in the world, for example 471 kg/ha in the Punjab, Haryana and Rajasthan. This is mainly because of state monopoly procurement systems which set limits on returns to farmers, although cotton in northern India is irrigated and is therefore better yielding than elsewhere, where the crop is dependent on monsoon rains for the moisture necessary to make a cotton crop. In the last 25 years, unfavourable monsoons have outnumbered favourable ones. Water storage and irrigation facilities remain limited, but progress is being made. In 1947 only 2% of the cotton area in India was irrigated – by 1993 the figure was 31%. Overall, however, Indian yields are not expected to reach world average 1981/82 levels until 2000.

The former Soviet Union

What used to be described by the cotton trade as 'Russian' cotton was never really that - it is grown in the Central Asian countries of Kazakhstan, Transcaucasia in Russia, Uzbekistan and Azerbaijan. Most Central Asian cotton is medium staple. The cotton is grown quite far north, between 37°N and 42°N, being quite frost-resistant.

Statements about former Soviet cotton output must be caveated with the unreliability of Soviet statistics, but agricultural figures - especially where there were sizeable exports - were more difficult to falsify than industrial ones. Unfortunately, information is not yet much more reliable from the successor states. However, from the information provided to the FAO, Soviet cotton output reached its fifth consecutive record in 1974/75, and again in 1975/76, rising 5% in response to favourable weather conditions and more irrigation. Another 10% was recorded in 1977/78, but output fell slightly in 1978/79 before recovering the following year. In 1980/81 output was recorded as 3.1 million tonnes, 9% above the previous year. Unfavourable weather knocked back production by 4% in 1981/82, and both heavy rains and cold weather adversely affected harvesting in 1982/83. Production did not recover in 1983/84 as expected, officially explained by 'low ginning extraction rates'. The following year the explanation for the further fall was 'late sowings, significant resowings and hot summer weather in the growing areas'. The temptation must be to conclude that a combination of the system breaking down and a readjustment of official figures was under way, but it is very difficult to establish exactly how the balance lies between the two. Some economic reforms were introduced in the following year, so that although bad weather caused low yields, output decline was constrained by more efficient ginning, reflecting a new procurement price structure which paid for seed cotton on a quality differential basis. This was a one-off improvement, however, and output fell again in 1987/88 - there was even snow during harvesting. There was a recovery in 1988/89: better weather led to an increase in the crop of 12% to almost 2.8 million tonnes. Some decline in output occurred in 1989/90: snow and frost again, so that one-third of the cotton area had to be replanted. Production again recovered slightly in the following year.

Currently, cotton is produced (in order of volume of production) in Uzbekistan, Turkemenia, Tadzhikistan, Azerbaidzhan, Kazakhstan and Kirgizia. According to the ICAC, Uzbekistani production was 1.522 million tonnes in 1991/92, falling to 1 462 million tonnes in 1992/93. These may well prove underestimates of the falls involved, as it is difficult to believe that the USSR succeeded in producing almost 2.8 million tonnes in 1987/88.

The separate republics, despite all the economic disasters and the collapse of the central procurement mechanism that has occurred since then, achieved about 2.5 million tonnes. In 1993 the ICAC quotes a figure of 2.27 million tonnes for Central Asian production, an increase of 0.2 million tonnes on the previous year. Production for 1994 was estimated to be 2.33 million tonnes: a slow revival of production levels. In terms of country breakdown, Tadzhikistan was estimated to have produced about 400 000 tonnes with again a similar amount projected for 1992/93, Armenia less than 200 000 tonnes and falling because of the war with Azerbaijan, Kazakhstan a very rough estimate of 100 000 tonnes, and Beloruss 22 000 tonnes. Estimates of production in the Central Asian countries tend to be falling continually in 1992/93, down to 9.4 million bales (USDA) in December 1992 - 16% down on the previous season - as plantings are reduced and there is cool, rainy weather in Uzbekistan and Turkmenistan. In Uzbekistan, the government has already announced plans to diversify away from cotton to grain, vegetable and forage crops: the intention is to maintain output of cotton - its main source of hard currency - for the foreseeable future by raising yields. Civil war and economic disruption are the main causes for the declines in production in Tadzhikistan.

The USDA also, in mid-1993, published updated figures on production, area, yield and stock figures for the countries of the former Soviet Union; Table 4.10.

China

Since the mid-1980s, China has been the world's largest cotton-producing country. Its crop in 1991/92 was the largest since 1984/85, although production has fallen for the 1992/93 crop year in response to pest infestation. There are two main cotton-growing areas in China, a northern area which lies chiefly in the basin of the Hwang-Ho river (which includes cotton growing as far north as 40°) and a southern which lies in the Yangtze river basin. The majority of Chinese cotton is short staple, some being only 11 mm or less. Some of it is of a harsh, rough character, which is mixed with wool to make blankets. Chinese cotton has only recently begun to compete internationally.

Little was known about Chinese cotton production in the mid-1970s, beyond that, for example, in 1975/76 production declined slightly, or that in 1976/77 the crop declined considerably due to poor

Table 4.10 The former Soviet Union's cotton production and yield (thousand 480 lb bales)

	1985/86			1986/87			1987/88			1988/89		
	Area (ha)	Yield (kg/ha)	Production	Area (ha)	Yield (kg/ha)	Production	Area (ha)	Yield (kg/ha)	Production	Area (ha)	Yield (kg/ha)	Production
Armenia	0	0	0	0	0	0	0	0	0	0	0	0
Azerbaijan	297	869	1 185	310	845	1 203	303	743	1 033	298	681	932
Byelarus	0	0	0	0	0	0	0	0	0	0	0	0
Estonia	0	0	0	0	0	0	0	0	0	0	0	0
Georgia	0	0	0	0	0	0	0	0	0	0	0	0
Kazakhstan	130	762	455	128	844	496	127	756	441	129	767	455
Kyrghyzstan	29	621	83	30	700	96	31	645	92	32	750	110
Latvia	0	0	0	0	0	0	0	0	0	0	919	0
Lithuania	0	0	0	0	0	0	0	0	0	0	0	0
Moldova	0	0	0	0	0	0	0	0	0	0	0	0
Russia	0	0	0	0	0	0	0	0	0	0	0	0
Tajikistan	313	930	1 337	310	945	1 346	324	852	1 268	319	922	1 350
Turkmenistan	562	690	1 781	653	542	1 626	630	603	1 745	640	641	1 883
Ukraine	0	0	0	0	0	0	0	0	0	0	0	0
Uzbekistan	1 985	871	7 936	2 044	794	7 450	2 112	713	6 912	2 014	860	7 956
Totals	3 316	839	12 777	3 475	765	12 217	3 527	709	11 491	3 432	805	12 686
USDA estimate 1/	3 316	839	12 777	3 475	765	12 217	3 527	709	11 491	3 432	805	12 686

	1989/90			1990/91			1991/92			1992/93		
	Area (ha)	Yield (kg/ha)	Production	Area (ha)	Yield (kg/ha)	Production	Area (ha)	Yield (kg/ha)	Production	Area (ha)	Yield (kg/ha)	Production
Armenia	0	0	0	0	0	0	0	0	0	0	0	0
Azerbaijan	280	686	882	264	682	827	245	722	813	230	491	519
Byelarus	0	0	0	0	0	0	0	0	0	0	0	0
Estonia	0	0	0	0	0	0	0	0	0	0	0	0
Georgia	0	0	0	0	0	0	0	0	0	0	0	0
Kazakhstan	119	832	455	120	850	468	117	778	418	110	700	354
Kyrghyzstan	27	852	106	30	833	115	26	731	87	25	600	69
Latvia	0	0	0	0	0	0	0	0	0	0	0	0
Lithuania	0	0	0	0	0	0	0	0	0	0	0	0
Moldova	0	0	0	0	0	0	0	0	0	0	0	0
Russia	0	0	0	0	0	0	0	0	0	0	0	0
Tajikistan	309	939	1 332	304	842	1 176	298	856	1 171	250	748	859
Turkmenistan	634	626	1 823	623	701	2 007	604	710	1 971	570	679	1 778
Ukraine	0	0	0	0	0	0	0	0	0	0	0	0
Uzbekistan	1 969	841	7 605	1 830	870	7 316	1 720	859	6 790	1 665	787	6 021
Totals	3 338	796	12 203	3 171	818	11 909	3 010	814	11 250	2 850	733	9 600
USDA estimates 1/	3 338	796	12 203	3 171	818	11 909	3 010	814	11 250	2 850	733	9 600

1/ USDA Import and Export data show only the external trade of the former USSR — not the trade among the republics.
Source: FAS/USDA

weather conditions. The FAO and ICAC made estimates of Chinese production, which suggested that China produced on average 1.9 million tonnes of cotton annually in the late 1960s, and had slightly increased that figure to 2.15 million tonnes by 1974. These were years of exceptional political turbulence in China, not conducive to effective crop management or investment. China's potential was not being used effectively. This is illustrated by the fact that comparing the late 1960s with 1991, the USA produced 68% more cotton, but China increased production by 265%. There was a slight revival of production the following season, due both to increased planting and higher yields, accelerating through 1979 to an output in 1980 of 2.7 million tonnes. The use of high-yielding varieties and a larger sown area helped boost the following year's output to another record of 2.97 million tonnes and records continued to be broken throughout the 1980s. In 1982/83 China's output at 3.6 million tonnes was 38% greater than that of the USA making China for the first time the world's largest cotton producer. Not content to rest on these new laurels, however, Chinese output was increased still further throughout the remainder of the 1980s. Although the once-off gains from changes in varieties planted could not be repeated, farmers were offered attractive guaranteed procurement prices. By 1984/85 Chinese output at 6.5 million tonnes accounted for almost one-third of world production.

However, since then the picture has been more mixed. Conscious of lower likely export prices, China changed its producer price policies in 1984/85, curtailing the area under cotton to two-thirds of the record level reached two years before. Output fell to 4.15 million tonnes. The following year, the weather was poor, too, and output fell drastically to 3.5 million tonnes, more that 45% less than the record. Once again government policy changed: the area sown to cotton was increased, yields were at record levels, and production recovered to 4.2 million tonnes. Then output was reduced to 4.1 million tonnes because of poor weather which lowered yields to 1982 levels; and also because of domestic price policies which made competitive crops – especially food crops – more attractive. This policy continued into the 1990s, with output declining to 3.8 million tonnes, caused by 'unattractive prices and unsatisfactory procurement practices compounded by adverse weather' (FAO, *Annual Review (Cotton)* 1990–91). In 1990 the government responded again by raising the procurement price, and output bounded back up to 4.5 million tonnes in 1990 and almost 5.7 million tonnes in 1991/92 (ICAC figures) although output for 1992/93 fell to 4.51 million tonnes; this relative failure of the 1992 season crop, announced to market surprise by the USDA via its agricultural attaché in Beijing in November 1992, down to some 21 million bales, caused a brief revival in New York future prices when it was announced during trading hours. Estimates are that drought and an insect infestation had affected the crop, not only in Shandong, but also in Hebei and Henan, although not in the south and in Xinjiang.

China's farmers are being urged to grow more cotton following the drought and insect infestation in the major cotton-producing provinces which reduced 1992/93 production prospects from 26.1 million bales to an estimated 20.8 million bales. The government has promised to purchase at a 'good' price in order to ensure that returns on cotton growing are adequately met. Whether China's cotton farmers believe the government will follow through with its promise will probably determine the outcome of cotton production in the near future. The ICAC puts Chinese production at 4.2 million tonnes for 1993 and 4.76 million tonnes for 1994 as these measures begin to take effect.

A centrally planned economy, as China alone amongst major cotton producers is, is inevitably slow to respond to world market changes, and is often too late in doing so. Such countercyclical policies are almost inevitably unprofitable and disruptive. There can be little sense in constant changes to the procurement price in such an unmanageable way, bearing little on a world price, which is itself changed by the Chinese government decision. It is a vivid example of the classic pricing problem of the central plan: as one Soviet economist is supposed to have said, you will need at least one free market left under Communism - so as to know how to set all the others. The consequence of these changes is a muddled investment policy and a failure to maintain the rate of advance of the mid-1980s. Nevertheless, there is ample cotton for domestic needs and the ability to supply the world market if the price becomes sufficiently attractive.

Argentina

Heavy rains, followed by extremely high temperatures, poor planting conditions, lack of farmer credit, delayed disaster assistance payments (from last year) and the lack of incentive to plant caused by falling cotton prices have together caused lower production in 1992/93, which the USDA estimated at 0.9 million bales down 22% from even the flood-damaged crop of the previous year.

Australia

Production is expected to fall from 502 000 tonnes in 1991/92 to 325 000 tonnes in 1992/93, down 30% from the previous year (USDA: 1.6 million bales). Poor planting conditions, especially cool weather in New South Wales, recognised in mid-1992, and an acute water shortage causing lack of irrigation supplies, are the culprits. The planted area is expected to fall 20%; however, this is probably against the general trend towards more cotton production.

Brazil

Two widely separated areas produce the bulk of the Brazilian crop. The oldest cotton-producing areas are in the northeast, where short staple cotton is grown in the upper coastal area adjacent to the lowlands, and fine-fibred long-staple tree cotton is grown on the uplands. The newer cotton area is in southern Brazil and includes São Paulo, Rio de Janeiro, Parana, Minas Gerais, Matto Grosso, Goiaz and Santa Catarina. Expansion in these areas originally came from land formerly planted to coffee, and newly cleared former rain forest land.

Smaller crops in 1974/75 were the result of shifts to more profitable food crops; this trend reversed the following year but fell back in 1976/77 and the year afterwards. Crops increased in 1980/81 and unevenly thereafter. Brazil reached a new production record in 1985 as farmers diverted land from soyabeans but output fell back the following year before recovering in 1987/88.

Production fell in 1988/89 as in most of Latin America, reflecting lower international prices at the time of planting, but a recovery was under way by 1990/91. Brazilian yields show a gradual increase from 240 kg/ha in 1983/84 to almost 400 kg/ha, and there is reason to suppose that further improvements will be possible. Output has not significantly increased, however, because the area sown to cotton has declined by a third over the same period, resulting in a fall of output from the record of almost 4.6 million bales in 1984/85 to the current level of approximately 3.67 million bales, due to low real prices for cotton after inflation which is cutting into farmers' income. The USDA said in December 1992 that production would be 3 million bales and this was subsequently revised down to 2.6 million bales, partly reflecting the lower area sown – 1.58 million ha, and partly caused by the dry weather that occurred during late December and early January 1993. Further large-scale reductions in planted area are expected for the next season, farmers switching to soyabeans instead. The government recognised in 1992 that there is clearly an inadequate level of plantings, and an increase in production is likely. The ICAC recorded that production was still well down in 1993, to 0.41 million tonnes, but was expected to recover in 1994 to 0.5 million tonnes.

Colombia

The domestic procurement price system has been dropped, as has the fixed quantity that producers must set aside for home textile industry consumption. In 1992/93 the area harvested and production were estimated by the USDA in May 1993 to be 115 000 ha and 310 000 bales respectively.

Egypt

The Egyptian government has a policy (under International Monetary Fund, IMF, guidance) whereby farmers will be paid a proportion of the export price of their cotton, rather than a fixed, artificial low value, which was largely responsible for the decline of cotton in the country. The objective of locking cotton prices in at 66% of international market levels in 1991/92 was not achieved. Production therefore fell in 1991/92. Production rose by 15% the next crop year, because of better growing conditions and use of pesticides - showing the continued importance of both to cotton production. The production increase came from long staple cottons, with ELS production falling.

Mexico

90% of the Mexican crop is produced in four main regions. The crop is now planted almost entirely to American uplands cotton and Acala in particular.

In 1974–77, favourable weather and successful pest control, together with increased acreage, boosted production. A decline was experienced in 1981/82 and the following year government policies favoured staple food crop cultivation which reduced output despite growth in yields. The following year, favourable cotton prices caused farmers to increase production. Production see-sawed throughout the 1980s, declining drastically in 1989/90 and recovering slightly the following year before falling again, partially as a result of difficulties associated with land reform programmes. The ICAC estimated Mexican production in 1991/92 as under half of its level in 1983/84, with traditionally high yields little changed from that crop year.

Paraguay

Planting intentions are 21% down on 1991/92 although yields are expected to improve and production was expected by the USDA to reach 900 000 bales, up 29% on the previous year.

Peru

The country produces a great deal of ELS cotton, both Pima and the native Tanguis, the latter including some '*soca*' or perennial cotton, which requires replacing after the second or third crop. Production stabilised at about 270 000 bales in the early 1980s and has remained at that level, with yields if anything showing a tendency to fall.

Southern Africa

The cotton area is estimated at 400 000 ha in 1992/93, up 25% from the severely drought-affected previous season - irrigation is limited in the communal areas of Zimbabwe where cotton is grown, for example.

South Africa's production (Table 4.11), originally expected by the USDA to reach 200 000 bales in 1992/93, itself below the average production level of the 1980s, finally reached only some 90 000 bales, 55% below the previous season.

Zimbabwe's production (Table 4.12) - hand-picked, mostly rain-fed, mainly short staple Delmac and long staple Albar cotton - was optimistically forecast in 1992 by the ICAC at 400 000 bales, more than three times the previous year's production. Prices closer to world prices, supplementary payments to farmers and improvements in the quality of planting seed were cited as likely reasons. However, in practice less than adequate rain prevented a complete recovery from the drought. Zimbabwe's production was predicted by the USDA in May 1993 to be only 275 000 bales.

Table 4.11 South African production, yield, supply and utilisation (thousand 480 lb bales)

Marketing years	1000 hectares	Yield (kg/ha)	Beginning stocks	Production	Imports	Total supply	Consumption	Loss	Exports	End stocks
1983/84	99	363	86	165	180	431	301	0	0	130
1984/85	107	429	130	211	115	456	294	0	0	162
1985/86	126	373	162	216	106	484	338	0	0	146
1986/87	160	381	146	280	102	528	359	0	0	169
1987/88	205	380	169	358	80	607	358	0	0	249
1988/89	208	374	249	357	45	651	344	0	23	284
1989/90	165	364	284	276	37	597	367	0	15	215
1990/91	127	382	215	223	65	503	339	0	24	140
1991/92	78	279	140	100	140	380	310	0	20	50
1992/93[a]	120	363	50	200	160	410	320	0	0	90

[a] Forecast.

Table 4.12 Zimbabwean production, yield, supply and utilisation (thousand 480 lb bales)

Marketing years	1000 hectares	Yield (kg/ha)	Beginning stocks	Production	Imports	Total supply	Consumption	Loss	Exports	End stocks
1983/84	190	479	78	418	0	496	101	0	260	135
1984/85	231	446	135	473	0	608	111	14	332	151
1985/86	192	464	151	409	0	560	136	14	316	94
1986/87	243	357	94	398	0	492	133	14	267	78
1987/88	272	427	78	533	0	611	150	14	325	122
1988/89	248	367	122	418	0	540	141	14	250	135
1989/90	229	294	135	309	0	444	150	13	187	94
1990/91	273	269	94	337	0	431	178	10	153	93
1991/92	236	111	93	120	155	368	250	4	25	89
1992/93[a]	275	317	89	400	50	489	300	13	75	151

[a] Forecast

West Indies

The Windward and Leeward Islands grow, without irrigation, what is said to be the longest and finest cotton in the world, *Sea Island cotton*. It was developed in Georgia, South Carolina and Florida in the nineteenth century; seed was imported into the British West Indies in 1903, where the sugar industry was meeting severe competition. Cotton proved successful as a replacement or a catch crop between the last ratoon and replanting. The West Indies is the only major source of Sea Island cotton in the world. Demand for it has however been very erratic and there is always a danger of overproduction even though the total output of six islands rarely exceeds 1000 tonnes of lint. A sizeable quantity of manufactured goods claim to use Sea Island cotton, although production there is falling and constitutes an insignificant amount of production in global terms.

Extra-long staple cotton

Global production of extra-long staple reached 2.57 million bales in 1978/79, largely accounted for by Peru, then the Soviet Union and India. Production stagnated in Egypt in the 1970s and declined in the Sudan and USA. After the early 1980s, production went into decline, production shifting to long staple especially in the USA, India, Egypt and the Sudan. In 1982/83 there was a further decline in production in Egypt and especially in Peru - excess rainfall destroyed the crop there. In 1984/85 India produced almost 1.6 million bales of ELS, almost as much as the other countries combined. Global production increased again in 1986/87, with increases registered in both the USA and India, despite falls in Egypt and the Sudan which continued the following year. World production declined a further 6.5% to 4.6 million bales in 1990/91, caused by weaker prices for Pima cotton in the USA in 1989/90 and the desperate need for food production in the Sudan, although there was some recovery of production in the then Soviet Union. Production is estimated at 4.24 million bales in 1992/93, compared with 4.28 million bales the year before, with most of the decline in production being in Central Asia - both poor weather and deliberate policy being responsible.

Production is being raised in the Sudan, and Egyptian yields are improving. There is some small increase in Californian Pima production observed - California produced 270 000 bales in 1992/93, up 80% from the year before and over 50% of estimated ELS production in the USA - but in the USA overall it is declining, expected to be about 410 000 bales this season. On 1 December 1992, the US Secretary of Agriculture, Madigan, announced the 1993 price support loan for ELS and an acreage reduction percentage of 20% based on an evaluation of carry-over stocks estimated at 267 000 bales by the end of the 1992/93 season. The loan rate is based on the average of the mid three years of prices received by ELS producers over the previous five years.

The long-term trend for ELS cotton appears to be downwards, only partly offset by a continuing improvement in yield, now reaching 918 lb/acre. In the USA, the USDA characterised the situation in May 1993 as being one of reduced demand, excessive domestic supplies partly caused by the 497 000 bales output, the second largest on record, and a resultant sharp downturn in prices. By late 1993, the general opinion was that supplies of both extra-fine and medium cotton were declining, whilst supplies of cotton for coarse count yarns were rising.

The measurement of quality and type

Historically, the grading of cotton and the provision of quality guarantees based on grades is a relatively recent phenomenon. What happened in the mid-1980s was that the forwarder guaranteed the purchaser in Europe delivery of a quality that was 'fair average for the season'. This precluded suits based on quality deficiency, because there was no objective means of assessing the price effect of quality differences. Even if a 'standard' or 'fair' were available at the Liverpool Cotton Exchange, the 'of the season' proviso prevented any case. The 'standard' was usually quite aged anyway, based on several seasons ago. Usually, forwarders who provided higher than average quality were accustomed to receive premiums in Europe from purchasers. Gradually the demand for proper categorisation and guarantees based on them became unstoppable.

The classification of cotton is vital both from a technical and an economic standpoint. Differences in staple length and micronaire, fibre strength, the amount of unwanted extraneous material such as sand leaf, dirt, twigs, shell, seed coat fragments, neps, dead and immature fibres, grass and insect damage causing caked or matted fibres and immature locks, the colour of the cotton and other physical qualities greatly affect how it can be spun and hence on the value of the cotton. For example, low-grade quality of cotton of any given type may be worth less than half as much as the best of the same type.

Individual varieties are classified into categories according to their staple length. Cotton of a given staple produces fibres of about the same length within climatic limitations. Fibres may vary within a bale, however, so the classifier designates the grade according to the length adjudged for the majority of fibres. Short staples are less than 26 mm long; medium staples are up to 29 mm long, long staples are up to 38 mm long and extra-long staples are 39 mm long or more. In terms of world trade, 90% is short and medium, long is 7–8% and extra-long a mere 2–3%. Generally, the longer the staple, providing that the fibres are regular in length, the finer the fibre with the least percentage of the types of defects noted above, the higher the price of cotton. As a consequence, every bale must have its own quality.

The term 'grade' is used to categorise cotton from a quality standpoint: in terms of extraneous matter (trash content), colour and gin preparation. The grade standards are established by the national government concerned. Grade does not, therefore, refer to staple length, strength or other features of the cotton. The strength of individual fibres varies up to four times, and is measured in grams per square cm or inch.

Finally there are micronaire, which is a measure of how the fibre is filled out, and fibre maturity. The quality characteristics of cotton can very widely, both from year to year and between locations.

Cotton quality measurement has improved considerably since World War II. Cotton fibre quality itself is continuously improving and more data are constantly available to the industry. Currently measurement is now made of: strength (using the Pressley and other testers); fineness; length uniformity (obtained from the Fibrograph and the Baer sorter); and maturity – thick or thin walled fibres – as determined by the use of a microscope or a differential dye test. These four test results are expressed as follows:

1 Strength in thousands of kg per cm of fibre.
2 Fineness in μg per cm.
3 Length uniformity as a ratio.
4 Maturity as a percentage of the total fibres in the sample (one of the most complex properties to measure).

In many ways these qualities are more important than staple length or grade in determining the usefulness of the cotton for spinning. However, these measurements demand the use of technical instruments under conditions of constant temperature and humidity.

An important innovation in cotton quality assessment over the past decade has been the introduction of high-volume instrument (HVI) testing. In Bremen in February 1992 it was reported by the ITMF Committee on Test Methods at the ITMF Conference that there were now HVI systems in 47 countries, with a total of 545 systems capable of measuring at least micronaire, length uniformity, strength and colour. Difficulties still exist in the calibration of HVI equipment and the reproducibility of test results. HVI systems use a length uniformity index that is not completely accurate in determining short fibre content, and efforts are now under way to increase the specific fibre quality measurement process. Studies are in progress to evaluate how strong are individual fibres and bundles (the availability data suggest that measured bundle strength increases significantly when the brushing force used for the test is raised). The achievement of this measurement, although judged unlikely to occur in the short term, would provide the necessary economic incentive to gins to adjust their cleaning procedures to minimise short fibres and settle the argument as to whether US cotton really does have a larger percentage of short fibres than other origins: evaluations are being carried out to see whether HVI equipment can be used directly to measure short fibres rather than to use a flawed index. Work is also underway to improve the repeatability of HVI strength measurements because as yet different instruments give different readings.

In the USA classifying of cotton is carried out at a network of USDA Classing Offices throughout the cotton belt. Cotton is re-tested many times throughout its life - by merchants or the co-operative, by mills and consumers. The classing fees for cotton are set by a formula in the Uniform Cotton Classing Fees Act of 1987. In 1992 the USDA increased its fees classification for growers to $1.92/bale, up from the $1.73/bale which had held since 1989. There is a 5 cent discount to producers or agents who centralise their ledgers.

Beginning with the 1993 crop, US cotton classification has been divided by both leaf and colour. The actual classification of the cotton is very similar to the previous method, with colour grades being called by names - middling fair, strict good middling, strict middling, middling, strict low middling, low middling, strict good ordinary and good ordinary - and number coded by two digits. Leaf is classified by a one digit grade of one to seven. Gray, light gray, plus grades and average rule grades have been eliminated. A new loan chart was necessary to match these new grades, because in the past the loan values of leaf and colour have been combined. The USDA Leaf Schedule Plan is implemented for white grades 51 and better and light spot grades 42 and better while the alternative Conversion Plan - supported by the merchants and mills - was adopted for all others.

Table 4.13 Premiums and discounts 1933-crop upland cotton
1993 CCC loan values at Greenwood, MS and equivalent locations, micronaire 4.3 through 4.9 and 3.5 through 3.6

Colour/leaf	Leaf	32	33	34	35	36	37+
White SM and better 11 and 21	1–2	50.80	52.30	54.20	54.70	54.75	54.80
	3	50.70	52.20	54.10	54.60	54.65	54.75
	4	50.30	51.80	53.70	54.20	54.25	54.30
	5	49.85	51.35	53.05	53.55	53.65	53.70
	6	46.90	48.40	49.35	49.65	49.70	49.75
	7	39.80	41.30	41.30	41.30	41.30	41.30
MID 31	1–2	50.80	52.30	54.20	54.55	54.65	54.75
	3	50.70	52.20	54.10	54.55	54.65	54.75
	4	50.35	51.85	53.50	54.00	54.05	54.15
	5	49.85	51.35	53.05	53.55	53.65	53.70
	6	46.90	48.40	49.35	49.65	49.70	49.75
	7	39.80	41.30	41.30	41.30	41.30	41.30
SLM 41	1–2	50.35	51.85	53.05	53.55	53.65	53.70
	3	50.20	51.70	53.05	53.55	53.65	53.70
	4	49.85	51.35	53.05	53.55	53.65	53.70
	5	48.45	49.95	51.05	51.40	51.45	51.50
	6	46.90	48.40	49.35	49.65	49.70	49.75
	7	39.80	41.30	41.30	41.30	41.30	41.30
LM 51	1–2	47.80	49.30	50.25	50.55	50.60	50.65
	3	47.70	49.20	50.15	50.45	50.50	50.55
	4	47.30	48.80	49.75	50.05	50.10	50.15
	5	46.90	48.40	49.35	49.65	49.70	49.75
	6	40.90	42.40	42.50	42.70	42.70	42.75
	7	39.80	41.30	41.30	41.30	41.30	41.30
SGO 61	1–2	39.80	41.30	41.30	41.30	41.30	41.30
	3	39.80	41.30	41.30	41.30	41.30	41.30
	4	39.80	41.30	41.30	41.30	41.30	41.30
	5	39.80	41.30	41.30	41.30	41.30	41.30
	6	39.80	41.30	41.30	41.30	41.30	41.30
	7	37.90	39.40	39.40	39.40	39.40	39.40

Colour/leaf	Leaf	32	33	34	35	36	37+
Spotted SM and better 13 and 23	1–2	46.55	48.05	48.05	48.15	48.20	48.25
	3	44.65	46.15	46.15	46.15	46.15	46.15
	4	44.65	46.15	46.15	46.15	46.15	46.15
	5	40.60	42.10	42.10	42.10	42.10	42.10
	6	37.70	39.20	39.20	39.20	39.20	39.20
	7	36.70	38.20	38.20	38.20	38.20	38.20
MID 33	1–2	44.65	46.15	46.15	46.15	46.15	46.15
	3	44.65	46.15	46.15	46.15	46.15	46.15
	4	40.60	42.10	42.10	42.10	42.10	42.10
	5	40.60	42.10	42.10	42.10	42.10	42.10
	6	37.70	39.20	39.20	39.20	39.20	39.20
	7	36.70	38.20	38.20	38.20	38.20	38.20
SLM 43	1–2	44.65	46.15	46.15	46.15	46.15	46.15
	3	40.60	42.10	42.10	42.10	42.10	42.10
	4	40.60	42.10	42.10	42.10	42.10	42.10
	5	37.70	39.20	39.20	39.20	39.20	39.20
	6	37.70	39.20	39.20	39.20	39.20	39.20
	7	36.70	38.20	38.20	38.20	38.20	38.20
LM 53	1–2	40.60	42.10	42.10	42.10	42.10	42.10
	3	40.60	42.10	42.10	42.10	42.10	42.10
	4	37.70	39.20	39.20	39.20	39.20	39.20
	5	37.70	39.20	39.20	39.20	39.20	39.20
	6	36.70	38.20	38.20	38.20	38.20	38.20
	7	36.70	38.20	38.20	38.20	38.20	38.20
SGO 63	1–2	37.70	39.20	39.20	39.20	39.20	39.20
	3	37.70	39.20	39.20	39.20	39.20	39.20
	4	37.70	39.20	39.20	39.20	39.20	39.20
	5	36.70	38.20	38.20	38.20	38.20	38.20
	6	36.70	38.20	38.20	38.20	38.20	38.20
	7	BG	BG	BG	BG	BG	BG

GO 71	1–2	37.75	39.25	39.25	39.25	39.25	39.25
	3	37.75	39.25	39.25	39.25	39.25	39.25
	4	37.75	39.25	39.25	39.25	39.25	39.25
	5	37.75	39.25	39.25	39.25	39.25	39.25
	6	37.75	39.25	39.25	39.25	39.25	39.25
	7	37.75	39.25	39.25	39.25	39.25	39.25
Light spotted SM and better 12 and 22	1–2	50.30	51.80	53.40	53.55	53.60	53.70
	3	50.20	51.70	53.05	53.45	53.50	53.60
	4	49.80	51.30	52.90	53.05	53.10	53.60
	5	48.40	49.90	51.50	51.65	51.70	51.70
	6	43.60	45.10	45.10	45.10	45.10	45.10
	7	38.55	40.05	40.05	40.05	40.05	40.05
MID 32	1–2	49.95	51.45	53.00	53.50	53.55	53.60
	3	49.85	51.35	53.00	53.50	53.55	53.60
	4	48.45	49.95	50.80	51.15	51.25	51.35
	5	48.05	49.55	50.80	51.15	51.25	51.35
	6	43.60	45.10	45.10	45.10	45.10	45.10
	7	38.55	40.05	40.05	40.05	40.05	40.05
SLM 42	1–2	48.85	50.35	52.00	52.45	52.50	52.55
	3	48.20	49.70	50.70	50.85	50.90	50.95
	4	47.85	49.35	50.20	50.50	50.55	50.60
	5	45.05	46.55	46.55	46.80	47.05	47.05
	6	43.60	45.10	45.10	45.10	45.10	45.10
	7	38.55	40.05	40.05	40.05	40.05	40.05
LM 52	1–2	47.75	49.25	50.20	50.50	50.55	50.60
	3	47.65	49.15	50.10	50.40	50.45	50.50
	4	43.60	45.10	45.10	45.10	45.10	45.10
	5	43.60	45.10	45.10	45.10	45.10	45.10
	6	38.55	40.05	40.05	40.05	40.05	40.05
	7	38.55	40.05	40.05	40.05	40.05	40.05
SGO 62	1–2	38.55	40.05	40.05	40.05	40.05	40.05
	3	38.55	40.05	40.05	40.05	40.05	40.05
	4	38.55	40.05	40.05	40.05	40.05	40.05
	5	38.55	40.05	40.05	40.05	40.05	40.05
	6	38.55	40.05	40.05	40.05	40.05	40.05
	7	BG	BG	BG	BG	BG	BG

Tinged SM 24	1–2	37.85	39.35	39.75	39.75	39.75	39.75
	3	37.35	38.85	39.25	39.25	39.25	39.25
	4	37.35	38.85	39.25	39.25	39.25	39.25
	5	37.75	37.25	37.25	37.25	37.25	37.25
	6	33.90	35.40	35.40	35.40	35.40	35.40
	7	BG	**BG**	**BG**	**BG**	**BG**	**BG**BG
MID 34	1–2	37.35	38.85	39.25	39.25	39.25	39.25
	3	37.35	38.85	39.25	39.25	39.25	39.25
	4	35.75	37.25	37.25	37.25	37.25	37.25
	5	35.75	37.25	37.25	37.25	37.25	37.25
	6	33.90	35.40	35.40	35.40	35.40	35.40
	7	BG	BG	BG	BG	BG	BG
SLM 44	1–2	37.35	38.85	39.25	39.25	39.25	39.25
	3	35.75	37.25	37.25	37.25	37.25	37.25
	4	35.75	37.25	37.25	37.25	37.25	37.25
	5	33.90	35.40	35.40	35.40	35.40	35.40
	6	33.90	35.40	35.40	35.40	35.40	35.40
	7	BG	BG	BG	BG	BG	BG
LM 54	1–2	35.75	37.25	37.25	37.25	37.25	37.25
	3	35.75	37.25	37.25	37.25	37.25	37.25
	4	33.90	35.40	35.40	35.40	35.40	35.40
	5	33.90	35.40	35.40	35.40	35.40	35.40
	6	BG	BG	BG	BG	BG	BG
	7	BG	BG	BG	BG	BG	BG

Leaf content: Leaf level 1 and 2 are combined. Leaf level 8 is Below Grade.

Cotton classed as 'Yellow Stained' (middling and better grades) will be eligible for loan, if otherwise eligible, at a discount 200 points greater than the discount for comparable quality in the colour group 'Tinged'.

BG = Below Grade (loan ineligible)

Micronaire

Micronaire reading	32 and shorter	33 and longer
5.3 and above	−540	−375
5.0 through 5.2	−350	−245
4.3 through 4.9	0	0
3.7 through 4.2*	+5	+10
3.5 through 3.6	0	0
3.3 through 3.4	−130	−220
3.0 through 3.2	−300	−450
2.7 through 2.9	−740	−890
2.5 through 2.6	−1120	−1200
2.4 and below	−1595	−1595

* Premium applies only to white grades 11-41, leaf 1-6; 51, leaf 1-5; light spotted grades 12-32, leaf 1-5; 42, leaf 1-4; and 52, leaf 1-3.

Fibre strength	Points
Below 18.5	INELIGIBLE
18.5 - 19.4	−235
19.5 - 20.4	−170
20.5 - 21.4	−120
21.5 - 22.4	−85
22.5 - 23.4	−45
23.5 - 25.4	0
25.5 - 26.4	+5
26.5 - 27.4	+40
27.5 - 28.4	+60
28.5 - 29.4	+85
29.5 - 30.4	+110
30.5 and above	+135

Extraneous matter

CODES:	1st digit	(Type of extraneous matter)		
	0 =	Prep	4 =	Oil
	1 =	Bark	5 =	Spindle twist
	2 =	Grass	6 =	Other
	3 =	Seed coat fragments		
	2nd digit	(Level of extraneous matter)		
	1 =	Light	2 =	Heavy
DISCOUNTS:	LIGHT	−350 pts/lb		
	HEAVY	−750 pts/lb		

Remarks

75 - Other side 2 or more colour grades higher
76 - Reginned
77 - Repacked
91 - Upland cotton ginned on roller gin
92 - Pima cotton ginned on saw gin

LOAN INELIGIBLE
96 - Mixture of upland and pima
97 - Fire damaged
98 - Water damaged

Location differentials

MISSISSIPPI:	Equivalent, *except*						
		Aberdeen	(465515)	+15	Holly Springs	(479505)	+15
		Batesville	(466305)	+15	Houston	(480005)	+15
		Canton	(469005)	+15	New Albany	(489510)	+15
		Grenada	(477005)	+15	Tchula	(499501)	+15
ARKANSAS:	Equivalent, *except*						
		Bradley	(130308)	−45			
LOUISIANA:	Equivalent, *except*						
		Alexandria	(400510)	−45	Jonesville	(409209)	−45
		Belcher	(401500)	−45	Plain Dealing	(415510)	−45
		Cheneyville	(032035)	+35	Huntsville		
TENNESSEE:		Brownsville	(787005)	+15	Memphis	(794505)	+15
		Brownsville	(787010)	+15	Memphis	(794510)	+15
		Covington	(788005)	+15	Memphis	(794516)	+15
		Covington	(788020)	+15	Memphis	(794540)	+15
		Elora	(789505)	+35	Memphis	(794544)	+15
		Jackson	(792004)	+15	Memphis	(794550)	+15
		Jackson	(792025)	+15	Memphis	(794555)	+15
		Mason	(794205)	+15	Milan	(795005)	+15
MISSOURI:	Equivalent						
ALABAMA:		Albertville	(002010)	+35	Fayette	(039512)	+35
		Aliceville	(003010)	+20	Huntsville	(054560)	+35
		Decatur	(403310)	−45	Huntsville	(064010)	+35
		Eclectic	(034512)	+35			
CALIFORNIA:	−190	TEXAS:	Harlingen, Lubbock	−75	N. CAROLINA:	Gastonia	+155

Note: Numbers of Parenthesis = USDA Warehouse Codes
Source: Staplcotn

The result is that the overwhelming majority of US cotton will have somewhat narrower leaf discounts, while the narrower grade cottons will generally receive the same loan value that they have done in the past. Extraneous matter schedules have now also been separately added, although the effect of this should be insignificant, see Table 4.13.

The significance of these changes, which are eventually likely to become worldwide standards, are an increased sophistication in the evaluation of different grades of cotton leading to tighter production controls and difficult multi-layered pricing structures, making the trader's job ever more difficult in sourcing, but easier in reducing uncertainty over supply.

CHAPTER

5

World cotton consumption

The uses of cotton

Cotton fibres, though thinner than the human hair, are actually relatively strong – the finer and longer the fibre, the stronger it is. Cotton fibre can be stretched by 8–10% whether wet or dry, and it is actually more resilient when wet. Cotton has exceptional hydrophilic tendencies, which explains why it is comfortable to wear and is easy to turn into a finished fabric. It absorbs up to 20% of its own weight in water without becoming 'wet' to feel; it does not drip until it has absorbed 65%. Wet processing does still demand the removal of extremely high volume of run-off effluents, which is one reason why the textile industry conflicts with environmental considerations. But this water resistance is one of the many reasons why cotton is used in apparel.

The overwhelming bulk of cotton, except for small amounts used for hand spinning and stuffing mattresses, is consumed in spinning mills. The downy fibres which are too short for normal spinning have many uses, as raw material for making rayon, plastics and varnishes. Seed fuzz from upland cotton is a by-product of the oil-milling industry called linters and is used primarily as a source of cellulose. Bleached and sterilized linters are known better as cotton wool. The cottonseeds are crushed to produce cottonseed oil which has many human and animal nutrition uses.

US textile mills consumed 9.5 million bales in 1991/92 and the increase is steady and expected to be 10 million bales in 1993/94. On mill delivery, and using US figures as typical examples of utilisation in developed countries, a bale of raw cotton averages about 500 lb. Of this, approximately 20 lb, or 4%, is bagging and metal ties or bands (tare). More cotton is now wrapped with improved material, weighing as little as 5 lb, which makes for reduced shipping costs compared to unbagged cotton. The 480 lb of raw cotton contain 22 lb of non-lint waste such as dust and vegetable matter. An additional 38 lb of usable waste is produced in the first stages of the production of yarn, sold to an industry which uses it mainly for padding and upholstery filling. About 15 lb goes into non-woven products. On average, 405 lb is woven into yarn, 85 lb goes into knitted goods, 7 lb into sewing thread and carpet and tufting yarns account for 2 lb. Most of the yarn, 311 lb, goes into fabric. Finished cloth is the main outlet for cotton fibre, using 242 lb, more than half the original bale. Unfinished grey goods, raw unbleached fabrics, account for 19 lb and are used mainly for industrial uses. Nearly 50 lb are used to produce yarn-dyed fabrics where yarn is first dyed and then woven. The USDA published a similar analysis in diagrammatic form in 1986 (Fig. 5.1).

Most cotton denim products are made from yarn-dyed fabric and account for a large part of total cotton use. Traditionally denim is made from an indigo yarn-dyed warp and an ecru weft, giving a 50 : 50 finish.

The US National Cotton Council records the number of bales used in each class of product (Tables 5.1 and 5.2). They divide the products into the three conventional categories:

1 Apparel. A little over 50% of cotton is used for apparel, an increase since the 1970s. Cotton is used for almost every type of clothing, from coats and jackets to underwear. Most of its apparel usage is for men's and boys' clothing which is 47% cotton, taken up in jeans, underwear and shirts. Over 60% of fabric in apparel is woven. In many newly industrialising countries (NICs), almost all cotton is used for apparel.
2 Household uses. About a third of cotton is used for home furnishings. It is used for bedspreads, sheets and pillow-cases (almost 50% of the market), window shades and a host of other uses. It is the main fibre used for towels with more than 95% of the US market. Of this cotton, 95% is woven.
3 Industrial uses. These account for the remainder of the cotton and

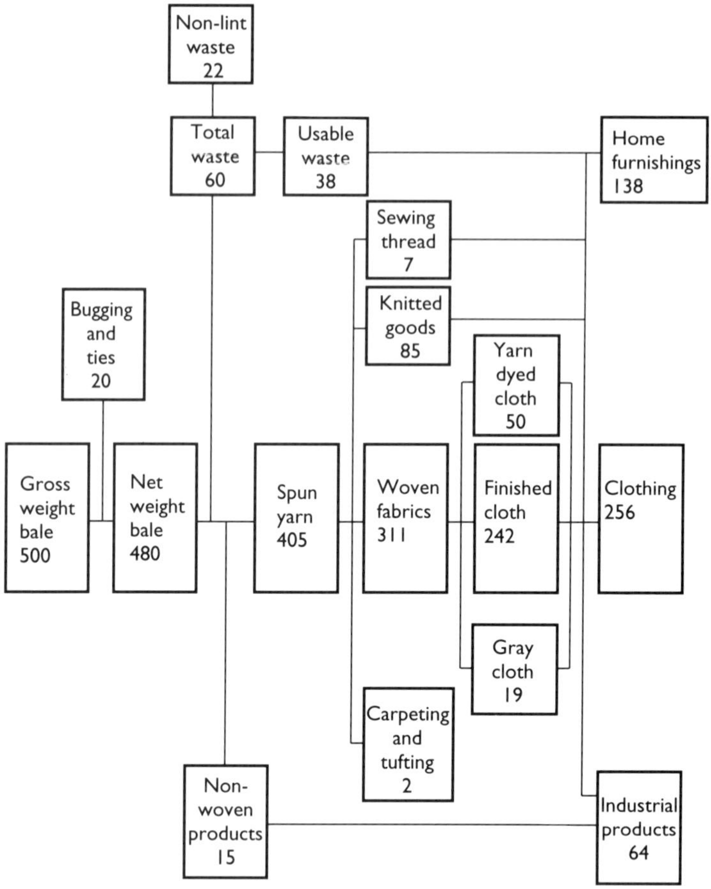

5.1 Distribution of an average bale of US cotton (lb) (*source*: USDA, 1986)

which are as diverse as tarpaulins, book bindings and zipper tapes. The major cotton users in this category are medical supplies, industrial thread, and shoes and boots. Among industrial uses, medical supplies provide the largest outlet for cotton, but cotton as a percentage of total medical supply consumption of materials has fallen to below 60%. Next in importance is industrial thread, but cotton has fallen here in absolute and relative terms. Abrasives are a category in which cotton consumption has increased both in bales and percentage terms. Wall-coverings and wiping and polishing cloths have both shown an increase in cotton consumption. Other important uses are in machinery belting and various protective uses such as tents, trailer covers and boat covers, but all of these uses have declined since the 1960s. Cotton has even been used for astronauts' space suits.

Table 5.1 Consumption of cotton and competing materials summarised by major end use (A, total materials consumed, thousand 480 lb bales[b]; B, cotton's percentage share of the total market[c]; C, cotton consumed, thousand 480 lb bales)

		Revised 1989	Revised 1990	Preliminary 1991
End uses[d]	(A)	5 483	5 122	5 394
Men's, youth's and boys'	(B)	63	67	69
apparel (4)[d]	(C)	3 449	3 422	3 723
1. M/B active apparel	(A)	90	96	105
	(B)	26	32	32
	(C)	24	30	33
2. M/B coats, jackets and	(A)	199	178	169
vests, excl. leather	(B)	30	32	30
	(C)	60	56	50
3. Gloves and mittens, work	(A)	135	137	124
type, incl. women's	(B)	79	75	74
	(C)	106	103	91
4. M/B handkerchiefs, woven	(A)	10	10	9
	(B)	53	61	59
	(C)	5	6	5
5. Hosiery, men's only, knit	(A)	317	284	314
	(B)	43	46	48
	(C)	137	130	150
6. M/N jumpsuits/overalls/	(A)	79	54	65
coveralls and ski suits/coveralls	(B)	53	57	57
	(C)	42	31	37
7. M/B pyjamas and other	(A)	126	58	48
nightwear	(B)	22	23	24
	(C)	28	13	11
8. M/B raincoats, excl. plastic	(A)	11	12	11
and rubberised	(B)	40	39	39
	(C)	4	5	4
9. M/B robes, dressing gowns	(A)	37	28	22
and smoking jackets	(B)	30	35	37
	(C)	11	10	8
Grand total, all uses	(A)	26 671	25 826	26 370
	(B)	34	34	36
	(C)	9 158	8 842	9 375

Apparel uses, total	(A)	9 869	9 348	9 716
	(B)	57	60	62
	(C)	5 577	5 566	6 039
Men's, youths' and boys' apparel	(A)	5 483	5 122	5 394
	(B)	63	67	69
	(C)	3 449	3 422	3 723
Women's, misses and juniors' apparel	(A)	3 291	3 221	3 300
	(B)	44	47	50
	(C)	1 464	1 514	1 655
Girls', children's and infants' apparel	(A)	1 095	1 006	1 022
	(B)	61	63	65
	(C)	664	630	661
Home furnishings	(A)	12 021	11 697	11 906
	(B)	24	23	23
	(C)	2 942	2 650	2 721
Industrial uses	(A)	4 781	4 780	4 748
	(B)	13	13	13
	(C)	640	626	614

Source: US National Cotton Council.

Cottonseed

With every 100 lb of fibre produced, about 162 lb of cottonseed are also produced. About 5% of the total seed crop is reserved for planting; the remainder is used as raw material for the cottonseed processing industry. About 30 million tonnes of cottonseed oil is produced annually, of which two-thirds is produced in developing countries. Asia is the largest producer, with almost 45% of the total, then America and Africa with under 10% each.

Every million tonnes of picked cotton yields about 0.6 million tonnes of seed and 0.35 million tonnes of lint, but the lint value is approximately seven times that of the seed. Seed is a by-product of the industry, but it can yield a substantial amount of edible oil and protein-rich food for livestock. Cotton seed has four major components in order of extraction: linters, hulls, oil and cake/meal. Delinting may be carried out at the ginnery or oil mill. The hulls can be used for fuel in factory boilers, and they can also be reduced to 'hull bran' and used as part of animal feed. The cracked seeds can be turned into undecorticated cotton cake, but the market for this has declined dramatically.

Table 5.2 Cotton consumption by end use, arrayed by size (thousand 480 lb bales)

Preliminary 1991

1	M/B shirts (14)	1320
2	M/B trousers & shorts (7)	1239
3	Towels & wash cloths	828
4	M/B underwear	656
5	Drapery, upholstery & slip cover fabrics (28)	650
6	WMJ slacks, dungarees & jeans	617
7	Sheets & pillowcases (35)	494
8	WMJ blouses & shirts	340
9	Retail piece goods (33)	206
10	WMJ dresses	201
11	GCI slacks, dungarees, & jeans	182
12	Blankets & blanketing	150
13	Hosiery, men's only, knit (12)	150
14	Medical supplies (44)	143
15	GCI hosiery, size 3 to 11	141
16	GCI blouses & shirts	116
17	WMJ skirts	103
18	Bedspreads	94
19	Thread, industrial	93
20	Gloves & mittens, work type, incl. women's	91
21	Comforters & quilts (27)	89
22	WMJ underwear	76
23	GCI underwear	72
24	M/B lining & pocketing	69
25	Tarpaulins, woven	67
26	GCI playgarments	62
27	WMJ shorts & other playgarments	58
28	Curtains, excl. shower	51
29	M/B coats, jackets & vests, excl. leather (7)	50
30	WMJ sweaters & jerseys, knit (16)	47
31	GCI dresses	46
32	WMJ nightwear	44
33	WMJ coats, jackets & vests	43
34	Abrasives, principally woven	42
35	Textile table cloths, napkins & place mats	42
36	Rugs & carpets (34)	40
37	Rope, cordage & twine excl. hard fibre/jute	39
38	Book bindings, excl. looseleaf binders (43)	48
39	Pads, quilted, excl. filling	37
40	M/B jumpsuits/overalls/ coveralls & ski suits	37
41	WMJ active apparel	36
42	M/B active apparel	33
43	Shoes & boots, excl. waterproof (48)	27
44	M/B sweaters, knit (16)	27
45	WMJ washable service apparel	25
46	WMJ robes, dressing gowns, & housecoats	23
47	Wiping & polishing cloths, woven	23
48	Automobile uses, excl. protective/replacement	19
49	GCI nappies (24)	19
50	Wall covering fabric (coated), woven & nonwoven	18
51	M/B washable service apparel	17
52	Belts, machinery (textile content only)	16
53	Sleeping bags, woven (50)	16
54	Pillows & cushions, excl. filling, woven	15
55	Tents & camper trailers, woven	15
56	WMJ foundation garments (22)	15
57	Boat covers & tops, woven	12
58	M/B pyjamas & other nightwear	11
59	Thread & yarn for home use	11
60	GCI coats & jackets	9
61	WMJ sheer hosiery (23)	8
62	Mattresses & box springs, excl. filling	8
63	Luggage, handbags & purses	8

64 M/B robes, dressing gowns & smoking jackets	8
65 Bags, woven	7
66 GCI skirts	7
67 Awnings, woven	6
68 Zipper tape, narrow woven	6
69 WMJ lining & pocketing	6
70 M/N handkerchiefs, woven (11)	5
71 Filter cloth, woven	5
72 WMJ suits & pant suits	5
73 Tape, pressure sensitive, excl. medical	4
74 M/B raincoats, excl. plastic & rubberized (7)	4
75 WMJ swimsuits	4
76 WMJ raincoats	4
77 M/B suits (8)	3
78 Hose (textile content only)	3
79 M/B swimsuits (7)	3
80 Tobacco cloth	2
81 Papermakers' felts, woven	2
82 GCI sweaters & jerseys	2
83 Mattress covers	2
84 WMJ handkerchiefs	1
85 Slip covers, furniture excl. plastic	1
86 Window shades	1
87 GCI nightwear	1
88 GCI swimsuits	1
89 GCI lining & pocketing	1
90 GCI suits, excl. ski & snow	1
91 GCI rainwear	0
92 GCI robes & housecoats	0

Source: US National Cotton Council.

After being separated from the lint at the gin, the seed is transported to a cottonseed crushing mill. There it is cleaned and conveyed on to delinting machines which, operating on the same principle as a gin, remove the short linter fibres. The linters go through further processing before being incorporated into a variety of different products, from mattresses to photographic film. One of their earliest uses was for fireworks and other explosives.

After removal of the linters, the seed is put through a machine using knives to loosen the hulls from the kernel. It then goes through shakers and beaters; the separated hulls are sold in bags or in bulk for livestock feed or industrial products.

The kernel is then ready for the extraction of the oil, which is the most valuable and important by-product of the seed: by the mid-1980s about 5 million tonnes of cottonseed produced 0.6 million tonnes of oil in the USA annually. Cottonseed oil can be extracted from the seed either mechanically in a screw press or chemically with a solvent (*N*-hexane). There is still not sufficient capacity to process all the cottonseed oil which could be produced, but despite this it still ranks fifth as a source of edible oil, with 8% of the total, after soyabean, sunflowerseed, rapeseed and groundnut. Most of the older mills use the screw press, yielding oil and cottonseed cake with about 5% oil. The economics of oil extraction mean that a minimum of 75% of the available oil must be extracted; solvent extraction, leaving less than 1% of the oil in the meal, is being used in new plants for treating low-oil seed. After refining, cottonseed oil has a low fat content, is light in colour and has no odour or taste. It has gained ground in competition with other vegetable and plant oils as an edible oil used in cooking oil, salad oil, shortening and margarine, although because cottonseed has only 16–18% protein content, it does tend to be used for oil rather than as feed. Limited quantities also go into soaps, insecticides, pharmaceuticals, cosmetics, textile finishes, glue for water-resistant non-abrasive plywood and other products. Currently in West Africa all cottonseed is ginned at the place of production and most is pressed for oil. Wide differences exist internationally between the effectiveness of these processing and extracting plants, such that those in, for example, West Africa and Egypt are not internationally competitive, and in many cases are very inefficient indeed.

The remainder of the kernel after extraction of the oil is converted into meal, which is the second most valuable by-product of the seed. Cottonseed extraction, with about 40% protein, is one of the standard components of animal feed. It can be fed directly to cattle and, after treating to reduce its gossypol content, to pigs and poultry. Currently, cottonseed cake is the second highest used high protein supplement for animal feed – after soyabean meal, forming about 10% of world output, some 4.5 million tonnes. At present, the solid part – after the extraction of the oil – is used only for ruminant feed; it is not possible for humans to eat because of the gossypol, which the human digestive system treats as poison and is unable to digest. A certain amount of cottonseed meal is also used as fertiliser for lawns, flowerbeds and other gardens.

Glandless cotton such as GL7 (see the section on 'Cotton agronomy' in Chapter 3) produces protein-rich seeds which can be processed into animal feed, especially for poultry and pigs, or can be used for human consumption in the form of flour and biscuits. Milling of glandless seeds must conform to strict international standards in order to give high-quality products, and specialised equipment is also required to extract the protein. This of necessity demands further investment from the cotton companies, whose interest in GL7 and the possibilities of glandless cotton is beginning to accelerate. This would produce cottonseed without gossypol, and could be achieved either through genetic manipulation or through mechanical methods, notably the liquid cyclone process.

Bread enriched with glandless cottonseed kernels is already being sold as far apart as Texas and Bremen. In the early 1990s, Togo had just one factory where cotton proteins could be extruded and expanded. This factory also processes seeds from Benin. Côte d'Ivoire has begun to process cotton flour with 50% protein, which is being sold to Europe to supply the large and price-sensitive European cattle feed market; such flour is also used as a partial substitute for wheat flour in bread-making.

Funding for studies into the nutritional value of glandless cotton is coming from the EU. The product is especially useful as a weaning food for babies and for malnourished children. When dehusked, grilled and pounded, the cottonseed can be eaten directly: farmers in west Benin, one of the places where the cultivation of glandless cotton is most advanced, eat cotton flour fritters already. Seed from older cotton varieties is eaten when food is short in Burkina Faso and gossypol-free seed can be expected to spread there. The benefits of edible cotton will be considerable: Morris commented that 'Clearly, if gossypol can be economically removed by those means, the potential in human nutrition is enormous' (1988, p. 15).

Trends in consumption and stocks

Consumption estimates make an extremely important contribution to sentiment in the international trade of cotton. There are three main sources of estimates: USDA, ICAC and Cotton Outlook. Even statistical cotton consumption figures, however, must be treated with some caution. They are usually based on returns from the spinning mills, and these returns do not always cover all the spindles known to exist in each country. Assumptions are made that the spindles not returning consumption figures consumed cotton at the same rate as those that did, but there is no evidence on which to decide the correctness or otherwise of this hypothesis. Moreover, statistics for China – one of the world's largest cotton-producing countries – are missing or incomplete for long periods. Tables 5.3–5.5 and Fig. 5.2 give figures for utilisation, and consumption.

In August 1992 the ICAC predicted that mill consumption worldwide would rise 4% over the previous season, to reach 19.4 million tonnes. The USDA said that global use would be 88 million bales, a record high, and in December 1992 the estimate was 86.7 million bales, the decrease coming from estimates revised downwards

Table 5.3 Cotton utilisation (million tonnes)

	1966–70	1971	1972	1973	1974	1975	1976	1977	1978	1979	1980	1981
Developed												
USA	1.876	1.762	1.749	1.653	1.500	1.392	1.540	1.438	1.379	1.396	1.377	1.228
Western Europe	1.496	1.380	1.401	1.391	1.302							
Others	0.907	0.952	0.977	0.998	0.900	2.040	2.252	2.063	2.047	2.284	2.203	2.092
Developing												
Latin America	0.731	0.790	0.855	0.913	0.895	0.931	0.967	0.969	1.015	1.103	1.088	1.014
Near East						0.674	0.716	0.728	0.745	0.725	0.738	0.800
Far East	1.919	2.044	2.227	2.387	2.384	2.333	5.402	5.122	5.358	5.764	6.162	6.463
Africa						0.246	0.258	0.247	0.254	0.240	0.250	0.235
USSR/Central Asia	2.250[1]	2.470[1]	2.518[1]	2.544[1]	2.570[1]	1.886	2.526[1]	2.510[1]	2.530[1]	2.544[1]	2.581[1]	2.597[1]
China	1.700[2]	2.208[2]	2.276[2]	2.357[2]	2.430[2]	2.676	2.860	2.750	2.852	3.114	3.377	3.636
India												
Pakistan												
Other developing	0.587	0.683	0.724	0.795	0.835							
Total	11.466	12.289	12.727	13.038	12.816	13.039	13.661	13.077	13.328	14.056	14.399	14.429

Table 5.3 Continued

	1982	1983	1984	1985	1986	1987	1988	1989	1990	1991	1992	1993	1994
Developed													
USA	1.129	1.278	1.986	1.276	1.477	1.706	1.59	1.91	1.89	2.09	2.23	2.24	2.39
Western Europe								1.28[3]	1.26[3]	1.14[3]	1.10[3]	1.09[3]	1.10[3]
Others	2.124	2.176	2.254	2.257	2.382	2.498	2.43						
Developing													
Latin America	1.019	1.015	1.014	1.186	1.304	1.390	1.46						
Near East	0.792	0.822	0.865	0.954	1.013	1.073	1.10						
Far East	6.555	6.882	6.810	7.576	8.308[6]	8.714[6]	8.83	1.76[4]	1.67[4]	1.60[4]	1.37[4]	1.26[4]	1.26[4]
Africa	0.250	0.274	0.290	0.318	0.309	0.326	0.33						
USSR/Central Asia	2.548[1]	2.624[1]	2.634[1]	2.691[1]	2.668	2.605	1.99[7]	2.65[5]	2.34[5]	1.91[5]	1.65[5]	1.38[5]	1.63[5]
China	3.805	3.632	3.638	4.370	4.717	4.918	4.79	4.15	4.23	4.25	4.35	4.60	4.70
India								1.88	1.96	1.90	2.03	2.16	2.26
Pakistan								1.10	1.34	1.44	1.48	1.53	1.61
Other developing													
Total	14.417	15.011	15.073	16.258	17.461	18.312	18.31	18.80	18.70	18.60	18.80	19.10	19.70

Source: USDA and ICAC
1 Including Eastern Europe 2 Including N Korea and N Vietnam 3 EC only 4 Japan and NICs 5 E Europe and CIS 6 Asia and Pacific 7 USSR only

Table 5.4 World cotton consumption (thousand 480 lb) bales

	1978/88	1988/89	1989/90	1990/91	1991/92 Estimate	1992/93 Estimate
World total	84 108	85 273	86 675	85 467	84 989	85 005
China	20 500	20 500	20 000	20 000	19 500	21 000
USA	7 617	7 782	8 759	8 657	9 608	9 800
India	7 990	8 115	8 667	9 018	8 674	8 825
Pakistan	3 351	3 721	4 801	5 649	6 482	6 800
FSU - 12	8 667	8 570	8 856	8 379	7 225	5 750
EC - 12	6 565	6 265	6 324	5 747	5 359	5 045
S.E. Asia[a]	2 834	3 282	3 485	3 734	4 237	4 105

[a] Includes Indonesia, Malaysia, Philippines, Singapore, Thailand and Vietnam.

for the CIS countries, Argentina and Brazil. The gap between production and consumption, on the ICAC estimate, would fall from 2.1 million tonnes in 1991/92 to 0.7 million tonnes in 1992/93. But this would still, even on what by December 1992 could clearly be seen to be an optimistic forecast, result in an increase in end-stocks, some of it occurring in importing countries.

The ICAC believed that the textile industry would be taking advantage of the 'below-average' world prices to increase reserves, but throughout 1992 the mills in fact bought just for their immediate needs, certainly throughout Europe, with the result that a number of trading companies and also companies that had acquired cotton (especially from CIS states) as barter exchange for chemicals, food or other goods, ended up storing the cotton in European warehouses. As a result of the international recession, cotton consumption growth has

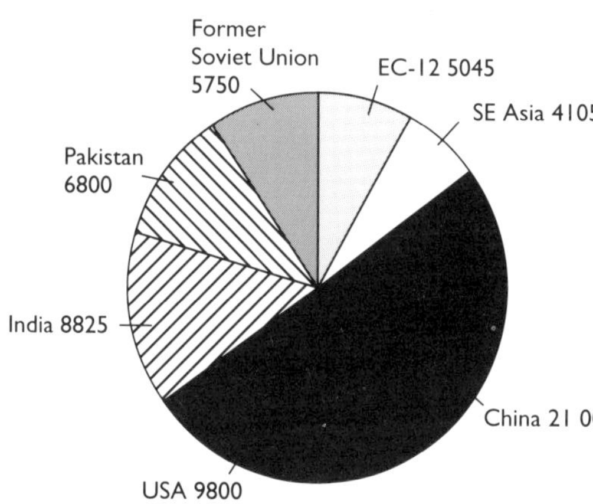

5.2 World cotton consumption: 1992/93 estimate (source: *USDA/FAS).*

Table 5.5 Textile fibre consumption (end use, thousand tonnes)

	World	Eastern Europe and USSR	Industrial countries	Developing countries: All	Africa	Asia	Europe	Middle East	Latin America and Caribbean
1960	15 152.8	2 645.6	7 440.6	5 066.6	451.9	3 124.1	332.9	278.9	878.3
1961	15 102.1	2 669.8	7 601.2	4 831.1	491.8	2 737.9	354.9	331.9	914.6
1962	15 339.0	2 763.5	7 785.6	4 789.9	481.7	2 683.8	362.5	345.5	915.3
1963	16 002.5	2 876.4	8 185.9	4 940.2	488.5	2 795.6	393.6	343.1	919.4
1964	17 255.8	3 280.2	8 725.0	5 250.7	597.5	2 921.5	398.0	369.6	964.1
1965	18 182.4	3 587.2	9 051.0	5 544.1	619.6	3 089.9	440.1	401.0	993.5
1966	18 795.5	3 715.5	9 498.4	5 581.6	596.8	3 117.2	465.6	410.6	990.3
1967	19 212.3	3 990.2	9 481.5	5 740.6	603.9	3 222.2	456.9	416.9	1 040.7
1968	20 434.4	3 954.0	10 306.7	6 173.7	663.4	3 467.0	493.6	470.6	1 079.1
1969	21 248.3	4 182.1	10 784.9	6 281.3	723.7	3 481.5	506.5	471.5	1 099.1
1970	21 739.6	4 216.1	10 496.2	7 027.3	735.7	4 173.7	510.2	477.1	1 131.6
1971	23 028.8	4 570.5	11 131.6	7 326.7	759.2	4 305.9	544.0	496.8	1 219.8
1972	24 412.4	4 659.4	11 936.4	7 816.7	755.5	4 619.9	572.6	544.6	1 324.1
1973	26 031.5	4 770.4	13 005.8	8 255.4	788.2	4 888.4	631.2	563.2	1 384.4
1974	25 270.6	4 583.5	12 602.4	8 084.7	773.8	4 784.4	612.3	555.8	1 358.3
1975	24 719.6	4 154.2	12 572.2	7 993.3	732.8	4 734.9	474.1	682.3	1 369.2
1976	26 538.0	4 468.7	12 216.5	9 852.9	914.5	5 786.5	591.7	851.5	1 708.7
1977	27 029.1	4 523.0	12 275.1	10 231.0	822.4	6 317.4	583.7	835.9	1 671.6
1978	28 260.4	4 609.9	12 648.6	11 001.9	856.4	6 941.7	602.4	855.6	1 745.9
1979	29 459.7	5 049.3	13 791.6	10 618.8	895.1	6 409.8	550.5	891.5	1 871.8
1980	29 577.6	5 292.3	12 769.9	11 515.5	955.2	6 979.4	592.8	1 039.5	1 948.5
1981	29 731.3	5 273.9	12 375.2	12 082.2	970.9	7 587.1	532.6	1 166.3	1 825.3
1982	28 880.6	5 266.5	11 915.9	11 698.2	940.2	7 336.6	536.5	1 116.5	1 768.4
1983	30 165.8	5 208.9	13 059.6	11 897.3	885.8	7 413.9	563.2	1 307.1	1 727.3
1984	31 252.7	5 263.7	13 766.3	12 222.7	890.9	7 722.9	586.1	1 168.4	1 854.3

Table 5.5 Continued

	World	Eastern Europe and USSR	Industrial countries	Developing countries					
				All	Africa	Asia	Europe	Middle East	Latin America and Caribbean
1985	32 813.6	5 525.3	13 743.7	13 544.6	811.4	8 915.3	669.0	1 139.1	2 009.7
1986	35 033.4	5 679.8	15 118.5	14 235.1	772.3	9 408.5	732.8	1 152.5	2 169.0
1987	36 616.1	5 607.8	16 086.2	14 922.0	770.6	10 284.0	601.6	1 098.1	2.167.7
1988	37 413.8	5 588.1	16 444.2	15 381.5	734.5	10 807.2	575.4	1 078.5	2 186.0
1989	38 360.4	5 641.8	16 570.5	15 968.0	702.2	11 369.2	571.2	1 067.7	2 257.8
1990	38 479.8	5 641.4	16 852.1	15 986.3	685.3	11 409.1	560.2	1 166.4	2 165.2
1991	38 932.0	5 249.3	17 341.3	16 341.3	656.5	11 662.6	567.9	1 245.7	2 208.7
1992	39 753.3	4 930.3	17 892.9	16 930.1	641.6	12 078.5	600.7	1 305.4	2 304.0
1993	40 612.0	4 853.8	18 648.2	17 110.0	648.4	12 206.8	607.0	1 319.3	2 328.5

Sources: ICAC, Commonwealth Secretariat, Fiber Economics Bureau and FAO.

slowed. A small ($5 million in the mid-1980s and virtually nothing in the 1990s) budget for international cotton promotion through the International Institute for Cotton (IIC) may be partly responsible for this. In some developing countries it is still rising but this no more than just compensates for slightly reduced consumption in Europe. US consumption has recently resumed an upward path aided by the advocacy of the fibre from Cotton Incorporated and Cotton Council International (CCI). Treatments have been developed which incorporate easy-care qualities into cotton textiles, and blends of cotton and man-made fibre have been created.

The FAO has published for a number of years a world apparel fibre consumption survey. The survey works by adjusting mill consumption by exports and imports of yarn, fabrics and manufactured goods, converted into fibre equivalents for 137 countries. The FAO estimated that the average worldwide consumption of cotton was about 7 lb/yr per capita in 1962–64 (in apparel end-uses). This concealed a vast difference in consumption between countries: in the USA consumption was about 22 lb/person, in the rest of the industrialised world about 10–15 lb/head and in Africa only 2.5 lb/head - in Indonesia even less. The surveys revealed that per capita consumption increased by 9% between 1975 and 1980, and after a reversal during the recession of the early 1980s, by another 4% by 1986 (Table 5.6). Cotton continued to be the most important fibre consumed - 30% more in 1986 than in 1975, although its share fell from 51% to 48% during the same period. World per capita consumption of cotton did not increase substantially from 1974, because of world population growth (Table 5.7).

In 1989 world consumption resumed its upward growth, reaching 18.6 million tonnes (FAO). This increase came primarily from China, which, despite closing some older mills, was using ever more. In the USA, increased attention began to be paid to consumers' preferences for 'all cotton' and 'cotton rich' textiles compared to synthetics and ordinary blends, which combined with a rising demand for denim and apparent linear elasticity of demand with respect to rising incomes. Had the recession of the 1990s not intervened, it seems certain that this 'environmental consumer preference structure' (ECPS) would have developed in specificity and size. But in 1989 ominous signs of downturn were already visible from the USSR and Europe in general, where almost all countries reduced their consumption. The decline in imports in some countries was compensated by much greater purchases of cotton textile products. Mill use and fibre imports in developing countries which imported cotton, principally the NICs, increased, as these countries responded

Table 5.6 Apparel fibre: apparent world consumption

	1974	1975	1979	1980	1981	1982	1984	1985	1986
Total availability (thousand tonnes)									
Cotton	13 363	13 032	13 968	14 284	14 239	14 389	14 678	15 627	16 960
Wool	1 459	1 551	1 768	1 770	1 792	1 826	1 880	1 904	1 926
Flax	816	745	767	709	696	719	779	758	777
Cellulosic fibres	3 538	3 033	3 497	3 332	3 273	3 069	3 111	3 016	2 978
Synthetic (non-cellulosic) fibres	7 459	7 400	10 576	10 498	10 606	9 933	11 503	12 176	12 764
Total, all fibres	26 635	25 761	30 576	30 593	30 606	29 936	31 951	33 481	35 405
of which:									
natural fibres	15 638	15 328	16 503	16 763	16 727	16 934	17 337	18 289	19 663
man-made fibres	10 997	10 433	14 073	13 830	13 879	13 002	14 614	15 192	15 742
Availability per caput (kg)									
Cotton	3.4	3.2	3.2	3.2	3.2	3.2	3.1	3.3	3.5
Wool	0.4	0.4	0.4	0.4	0.4	0.4	0.4	0.4	0.4
Flax	0.2	0.2	0.2	0.2	0.2	0.2	0.2	0.2	0.2
Cellulosic fibres	0.9	0.8	0.8	0.8	0.7	0.7	0.7	0.6	0.6
Synthetic (non-cellulosic) fibres	1.9	1.8	2.4	2.4	2.4	2.2	2.4	2.5	2.6
Total, all fibres	6.7	6.4	7.1	7.0	6.8	6.6	6.8	7.0	7.3
of which:									
natural fibres	3.9	3.8	3.8	3.8	3.7	3.7	3.7	3.9	4.1
man-made fibres	2.8	2.6	3.3	3.1	3.1	2.9	3.1	3.1	3.2

Source: World Apparel Fibre Consumption Survey, 1989.

Table 5.7 Population (millions)

	World	Eastern Europe and USSR	Industrial countries	Developing countries: All	Africa	Asia	Europe	Middle East	Latin America and Caribbean
1960	3010.99	331.75	613.30	2065.94	227.41	1411.92	92.47	64.55	205.66
1961	3052.05	337.45	620.34	2094.26	233.02	1424.08	93.47	66.50	211.25
1962	3108.59	342.81	627.87	2137.91	238.83	1450.14	94.85	68.55	217.36
1963	3182.52	348.02	635.10	2199.40	244.81	1493.31	96.04	70.63	224.13
1964	3250.97	353.16	641.92	2255.90	250.88	1532.24	97.20	72.79	230.06
1965	3320.64	358.77	648.74	2313.14	257.16	1570.43	98.35	75.00	236.93
1966	3388.66	362.26	655.02	2371.38	263.86	1610.25	99.54	77.22	242.98
1967	3457.25	366.59	660.64	2430.03	270.82	1649.33	100.84	79.55	249.52
1968	3528.05	371.34	665.42	2491.29	277.98	1690.70	102.38	81.90	256.19
1969	3603.81	375.53	672.33	2555.94	285.33	1734.86	103.77	84.34	262.94
1970	3677.05	379.37	678.80	2618.88	288.24	1780.17	105.06	86.91	270.99
1971	3754.44	383.79	686.25	2684.41	296.03	1824.77	106.38	89.52	277.57
1972	3833.12	388.30	692.61	2752.22	304.07	1870.55	107.75	92.22	284.69
1973	3910.09	392.81	697.88	2819.41	312.48	1915.07	109.13	95.14	291.83
1974	3988.62	397.51	704.58	2886.53	321.02	1958.87	110.61	98.16	299.40
1975	4063.89	402.24	709.31	2952.34	329.81	2000.31	112.63	101.26	307.00
1976	4142.11	406.71	713.36	3022.04	344.25	2040.58	114.30	104.45	314.30
1977	4215.43	411.25	718.47	3085.71	353.63	2079.57	115.83	107.76	321.85
1978	4289.24	415.70	722.70	3150.83	363.47	2119.28	117.32	111.18	329.55
1979	4364.93	419.71	727.54	3217.67	373.57	2159.68	118.82	114.74	337.42
1980	4441.75	423.71	732.75	3285.29	384.80	2199.33	120.20	118.43	345.68
1981	4519.33	427.92	737.54	3353.87	396.65	2239.79	121.77	122.08	353.54
1982	4601.61	432.29	742.03	3427.30	408.89	2284.27	123.30	125.84	361.61

1983	4679.62	436.93	746.53	3496.16	421.55	2323.34	124.84	129.69	369.85
1984	4758.50	441.57	750.53	3566.40	434.62	2363.03	126.36	133.68	378.26
1985	4837.62	446.13	754.57	3636.92	446.98	2403.94	127.95	137.72	386.57
1986	4915.32	450.52	757.99	3706.81	461.23	2446.59	129.35	141.81	395.12
1987	5002.64	455.06	761.48	3786.09	476.62	2491.65	130.81	146.22	403.15
1988	5089.96	459.61	764.98	3865.37	492.08	2536.91	132.28	150.62	411.93
1989	5177.28	464.16	768.47	3944.66	507.53	2582.16	133.75	155.03	420.70
1990	5264.61	468.70	771.96	4023.94	522.98	2627.41	135.21	159.44	429.40
1991	5347.59	472.84	774.53	4100.22	538.92	2669.73	136.58	163.48	437.29
1992	5432.06	477.02	777.10	4177.94	555.35	2712.73	137.96	167.61	445.33
1993	5518.05	481.23	779.68	4257.14	572.28	2756.42	139.35	171.85	453.52

Source: United Nations.

Table 5.8 World production and consumption (million bales)

Year	Opening stocks	Production	Consumption
1989	32.4	84.2	83.5
1990	33.1	80.1	84.7
1991	27.1	87.2	88.0
1992	29.5	95.0	85.0
1993	40.0	90.0	87.0

Source: ICAC (1991).

to increased overseas demand for their finished products.

World end-user textile fibre consumption was estimated by the ICAC to be 38.5 million tonnes in 1990, an increase of 0.3% from 1989. On a per capita basis, world fibre consumption decreased from 7.4 kg in 1989 to 7.3 kg in 1990, mainly due to the decline in the countries of the former Soviet Union. However, cotton's share declined only from 49% to 48.9%, as cotton consumption increased only slightly in 1990 – prices had increased in 1989 and there were low stocks which reduced the availability of cotton in world markets. But by 1992, the ICAC acknowledged that 'world economic growth has declined to the level of a recession from the standpoint of the textile industry', which was echoed by every trader and producer in 1992–3. Three factors were the dominant reasons for the slowdown. Firstly, and most importantly, the link between Central Asian cotton and Russian textile production was broken with the collapse of the Soviet Union. Russian and East European cotton consumption began its long-term slump. Secondly, textile production in the West, particularly in Europe, was hit by a long-lasting and damaging recession which shows only faint signs of abating in 1993/94. Thirdly, the Gulf crisis enforced a pause in Middle Eastern markets and encouraged other consumers to postpone their investment decisions.

World economic growth in the early 1990s was negligible, reflecting the deepening of the crisis in the former Soviet Union – living standards have collapsed and the textile industry is producing about half of its Soviet output levels – as well as economic recession in the West and a collapse in producer prices.

Still, the recession did not take world consumption down to the levels of the 1970s. In 1980 world consumption was 66 million bales and in 1991 it was 86 million bales (source: NYCE). ICAC figures are similar.

Geographically, there has been a very clear trend in the past ten years: towards Asia (Table 5.9). Economic growth in developing

Table 5.9 Textile fibre consumption per capita (end use; kg per capita)

				Developing countries					
	World	Eastern Europe and USSR	Industrial countries	All	Africa	Asia	Europe	Middle East	Latin America and Caribbean
1960	5.03	7.97	12.13	2.45	1.99	2.21	3.60	4.32	4.27
1961	4.95	7.91	12.25	2.31	2.11	1.92	3.80	4.99	4.33
1962	4.93	8.06	12.40	2.24	2.02	1.85	3.82	5.04	4.21
1963	5.03	8.27	12.89	2.25	2.00	1.87	4.10	4.86	4.10
1964	5.31	9.29	13.59	2.33	2.38	1.91	4.09	5.08	4.19
1965	5.48	10.00	13.95	2.40	2.41	1.97	4.48	5.35	4.19
1966	5.55	10.26	14.50	2.35	2.26	1.94	4.68	5.32	4.08
1967	5.56	10.88	14.35	2.36	2.23	1.95	4.53	5.24	4.17
1968	5.79	10.65	15.49	2.48	2.39	2.05	4.82	5.75	4.21
1969	5.90	11.14	16.04	2.46	2.54	2.01	4.88	5.59	4.18
1970	5.91	11.11	15.46	2.68	2.55	2.34	4.86	5.49	4.18
1971	6.13	11.91	16.22	2.73	2.56	2.36	5.11	5.55	4.39
1972	6.37	12.00	17.23	2.84	2.48	2.47	5.31	5.91	4.65
1973	6.66	12.14	18.64	2.93	2.52	2.55	5.78	5.92	4.74
1974	6.34	11.53	17.89	2.80	2.41	2.44	5.54	5.66	4.54
1975	6.08	10.33	17.72	2.71	2.22	2.37	4.21	6.74	4.46
1976	6.41	10.99	17.13	3.26	2.66	2.84	5.18	8.15	5.44
1977	6.41	11.00	17.08	3.32	2.33	3.04	5.04	7.76	5.19
1978	6.59	11.09	17.50	3.49	2.36	3.28	5.13	7.70	5.30
1979	6.75	12.03	18.96	3.30	2.40	2.97	4.63	7.77	5.55
1980	6.66	12.49	17.43	3.51	2.48	3.17	4.93	8.78	5.64
1981	6.58	12.32	16.78	3.60	2.45	3.39	4.37	9.55	5.16
1982	6.28	12.18	16.06	3.41	2.30	3.21	4.35	8.87	4.89
1983	6.45	11.92	17.49	3.40	2.10	3.19	4.51	10.08	4.67

Table 5.9 Continued

				Developing countries					
	World	Eastern Europe and USSR	Industrial countries	All	Africa	Asia	Europe	Middle East	Latin America and Caribbean
1984	6.57	11.92	18.34	3.43	20.5	3.27	4.64	8.74	4.90
1985	6.78	12.38	18.21	3.72	1.82	3.71	5.23	8.27	5.20
1986	7.13	12.61	19.95	3.84	1.67	3.85	5.67	8.13	5.49
1987	7.32	12.32	21.12	3.94	1.70	3.96	5.87	8.26	5.64
1988	7.35	12.47	21.49	3.94	1.68	3.97	5.93	8.20	5.64
1989	7.41	12.66	21.93	3.96	1.38	4.40	4.27	6.89	5.37
1990	7.31	12.57	21.90	3.90	1.31	4.34	4.14	7.32	5.04
1991	7.28	11.10	22.39	3.99	1.22	4.37	4.16	7.62	5.05
1992	7.32	10.34	23.03	4.05	1.16	4.45	4.35	7.79	5.17
1993	7.36	10.09	23.92	4.02	1.13	4.43	4.36	7.68	5.13

countries in 1982–92 has virtually all been in Asia. Asian developing countries – although the name is difficult to apply in the same terms – include Thailand, Myanamar, Indonesia, Malaysia and the Philippines. Their rate of growth averaged almost 4% in 1988–92, which, although a considerable slowing down from the previous period of 1983–88 (8%), still made the area enormously attractive for investment. Despite a worldwide slowdown of economic growth in the early 1990s, Asian GDP continued to grow at about 5% annually – more than the previous period. This increase came from the NICS, and strong domestic demand in those countries, as well as in Indonesia, Malaysia and Thailand.

Unfortunately the level of per capita income in the Middle East, Africa and Latin America has on average in real terms actually declined (Table 5.10). In the Middle East, GDP declined between 1.5–3% annually in the early 1990s. Recovery in 1992 has been slow: the Iraqi crisis has had substantial economic ramifications. In 1993 signs of economic recovery in the developed countries were faint. In late 1993 the ICAC expected consumption to rise 2% to more than 19 million tonnes, and cotton has, perhaps surprisingly, lost market share in some markets.

The effect of the interrelationship between consumption and production is on stock levels and prices (Tables 5.3 and 5.11). In the early 1990s, with cotton relatively abundant, and production exceeding consumption throughout the 1990/91 and 1991/92 seasons, world cotton stocks were rising. They were 6.8 million tonnes (31.2 million bales) at the beginning of the 1991/92 season and 9.5 million tonnes (43.6 million bales) on 1 August 1992, according to the ICAC – the USDA put them at 40 million bales in December 1992, falling to 38.8 million bales by May 1993, because of lowered stock estimates for Brazil, China, Colombia, Italy, Pakistan and Peru, primarily due to lowered production forecasts. Stocks have been falling throughout the 1992/93 season, except in the USA where they have stabilised at 4.2 million bales (Table 5.12). In November 1993 the ICAC expected stocks to fall by 1.3 million tonnes during the season.

The ICAC had originally forecast that stocks would be down to 7.6 million tonnes (34.9 million bales) by August 1992, but raw cotton stocks in China have proved to be substantially greater than previously thought: the unexpectedly large crop of 1991 led to stockpiling and the existing stock levels were already much higher than previously thought. Even the 1992/93 end stocks : use ratio for China was 0.6 (ICAC) and 46% (USDA), lower than in the mid-1980s. In Pakistan, end stocks are expected to decline during 1992/93, although the amount is not certain, owing to lower than expected production and

Table 5.10 Real GDP per capita index (index numbers, 1980 = 100)

				Developing countries					
	World	Eastern Europe and USSR	Industrial countries	All	Africa	Asia	Europe	Middle East	Latin America and Caribbean
1960	61.8	39.5	54.1	45.9	72.1	59.7	41.6	47.1	54.5
1961	63.6	41.6	56.1	50.2	72.8	61.3	43.5	48.1	56.1
1962	65.9	43.6	58.3	50.8	73.1	62.2	44.7	50.4	57.3
1963	67.5	45.7	60.5	55.4	76.9	64.1	48.9	53.0	56.8
1964	70.0	47.6	63.7	58.2	78.7	66.3	52.3	55.2	58.0
1965	72.5	49.5	66.2	60.4	80.1	64.0	53.7	58.3	62.6
1966	75.2	53.1	68.8	62.5	79.8	64.5	57.6	61.0	64.4
1967	76.7	56.6	70.7	63.0	78.3	66.7	59.4	63.4	65.4
1968	79.3	59.9	74.0	65.1	81.4	67.8	61.6	67.8	68.0
1969	81.9	63.2	77.2	69.7	86.2	71.1	65.3	73.0	70.5
1970	83.2	66.6	79.0	74.6	92.4	73.1	69.0	75.4	71.8
1971	84.8	70.4	80.9	76.9	95.4	74.0	73.7	79.2	74.7
1972	87.6	74.3	84.2	79.3	96.9	73.8	77.6	85.8	77.8
1973	90.8	78.1	88.3	83.2	96.8	76.9	81.7	87.8	81.7
1974	91.2	81.6	88.1	86.0	101.3	78.4	84.1	93.3	85.2
1975	90.6	85.2	87.1	87.6	99.5	82.0	86.3	93.3	86.1
1976	93.5	88.9	90.6	90.4	100.5	84.8	90.2	102.0	89.3
1977	95.7	92.4	93.2	94.3	101.2	89.4	94.8	106.2	91.2
1978	97.9	95.2	96.3	96.3	96.9	96.2	96.4	101.8	93.0
1979	99.5	97.2	98.7	98.7	97.4	96.6	99.4	102.5	96.9
1980	100.0	100.0	100.0	100.0	100.0	100.0	100.0	100.0	100.0
1981	100.1	101.5	100.9	99.7	98.8	104.0	98.6	95.7	97.6
1982	98.8	103.4	100.1	100.3	97.7	107.7	98.6	94.1	94.4
1983	99.9	106.0	102.0	100.1	93.7	114.3	99.7	92.3	89.7
1984	102.3	107.9	106.2	102.7	91.6	121.7	103.4	89.6	90.8

Table 5.10 Continued

	World	Eastern Europe and USSR	Industrial countries	Developing countries					
				All	Africa	Asia	Europe	Middle East	Latin America and Caribbean
1985	104.0	110.3	109.2	104.9	92.6	127.8	104.9	88.4	91.9
1986	105.6	113.6	111.7	111.1	91.3	134.1	108.3	85.3	94.1
1987	107.1	115.1	114.8	108.1	89.5	142.2	109.7	82.8	94.5
1988	109.6	118.4	119.3	110.2	89.2	152.3	109.9	84.1	92.6
1989	111.1	119.4	122.5	111.2	89.4	157.2	109.0	84.3	92.1
1990	111.5	115.6	125.1	111.8	88.4	162.9	107.9	80.8	89.3
1991	111.5	107.3	126.9	113.7	87.5	168.3	110.0	76.2	88.6
1992	112.8	106.7	130.1	115.6	87.0	174.2	112.2	80.6	89.8
1993	114.4	107.5	134.0	119.8	89.6	183.8	114.4	84.7	93.0

Table 5.11 Consumption and beginning and end stocks (thousand 480 lb bales)

Marketing year	Beginning stocks	Consumption	End stocks
1960/61	19561	46224	18875
1961/62	18875	45195	18667
1962/63	18667	43984	22508
1963/64	22508	47810	25499
1964/65	25499	51244	28557
1965/66	28557	53790	32069
1966/67	32069	56034	28076
1967/68	28076	56220	23245
1968/69	23245	56371	23620
1969/70	23620	55982	22448
1970/71	22448	57082	21470
1971/72	21470	58412	21999
1972/73	21999	59504	24018
1973/74	24018	60306	26819
1974/75	26819	56996	33004
1975/76	33004	61564	25608
1976/77	25608	60177	22106
1977/78	22106	61065	25356
1978/79	25356	63348	21708
1979/80	21708	65985	21282
1980/81	21282	64979	20780
1981/82	20780	63234	25775
1982/83	25775	66806	25771
1983/84	25771	68503	24466
1984/85	24466	68987	43995
1985/86	43995	76892	48297
1986/87	48297	82776	35735
1987/88	35735	84108	32838
1988/89	32838	85273	31908
1989/90	31908	86675	26332
1990/91	26332	85467	28706
1991/92	28706	84989	40660
1992/93	40660	85005	38763

Source: USDA/FAS.

increased consumption. Central Asian stocks have not grown as fast as expected due to reduced production and strong offtake, especially into European markets.

Table 5.12 US beginning and ending stocks (thousand 480 lb bales)

Marketing year	Beginning stocks	End stocks
1960/61	7501	7056
1961/62	7056	7699
1962/63	7699	11136
1963/64	11136	12351
1964/65	12351	14249
1965/66	14249	17028
1966/67	17028	12344
1967/68	12344	6584
1968/69	6584	6544
1969/70	6544	5843
1970/71	5843	4203
1971/72	4203	3258
1972/73	3258	4221
1973/74	4221	3808
1974/75	3808	5708
1975/76	5708	3681
1976/77	3681	2928
1977/78	2928	5347
1978/79	5347	3958
1979/80	3958	3000
1980/81	3000	2668
1981/82	2668	6632
1982/83	6632	7937
1983/84	7937	2775
1984/85	2775	4102
1985/86	4102	9348
1986/87	9348	5026
1987/88	5026	5771
1988/89	5771	7092
1989/90	7092	3000
1990/91	3000	2344
1991/92	2344	3704
1992/93	3704	4200

Source: USDA/FAS.

Regional cotton consumption

USA

The mid-1970s saw a reduction in utilisation in mills; in 1974 inventories rose but production of cotton goods declined as manufacturers lost out to imports. Cotton again demonstrated its ability to hold up relatively well in demand even when overall fibre consumption was shrinking, as demand for domestically produced cotton denims and corduroys was maintained. Demand did remain relatively weak, declining in the late 1970s partly in response to intense competition from synthetic fibres. Demand for heavyweight fabrics such as corduroy, which represent about one-quarter of mill usage, declined in 1980. Imports were sustained at high levels. Only the general economic recovery which began in about 1983 brought an end to this depressing situation, and consumption of cotton rose through the 1980s. What makes the USA different from Japan and Europe is that the worldwide recession has not halted the relentless growth in cotton consumption (see Fig. 5.3).

The USDA reported that in 1987/88 consumption was 7.617 million bales, 9% of world consumption. By 1992/93 they estimated that this had risen to 9.8 million bales (considerably less than the ICAC estimate), 11.5% of world consumption and a high per capita figure compared with international competitors as confirmed by year-end figures. By mid-1993 it was apparent that US cotton consumption was actually increasing, showing greater buoyancy than analysts had expected. The ICAC reported that US cotton consumption seasonally adjusted and annualised was 10.25 million bales in January 1993, compared with 9.73 million bales in December 1992. When combined

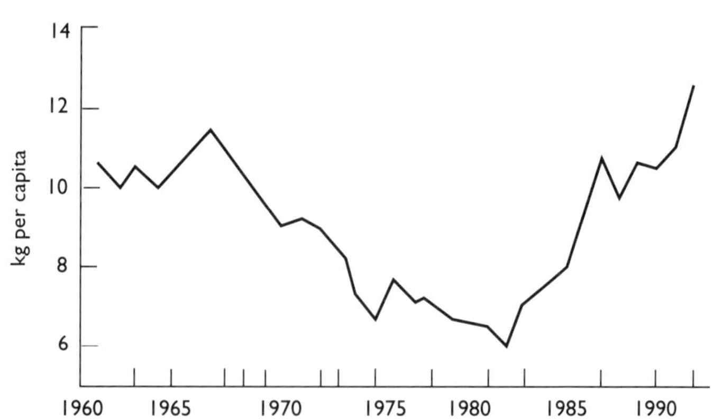

5.3 Net US cotton consumption.

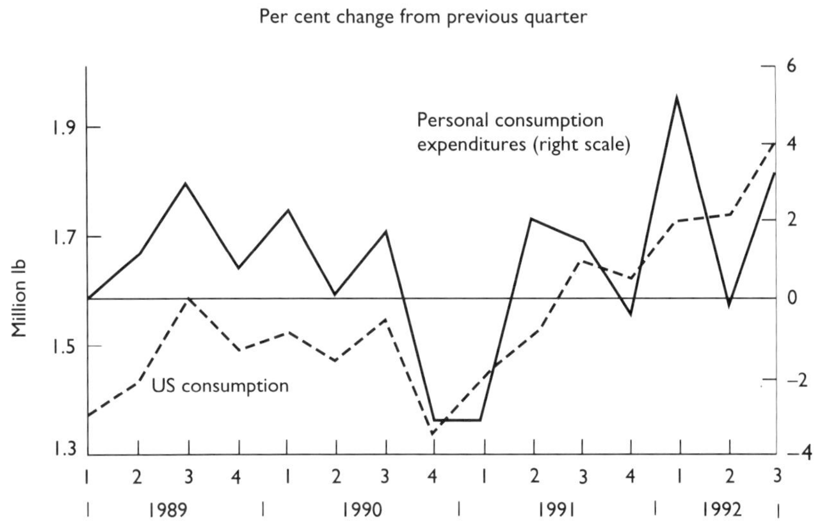

5.4 US cotton consumption follows personal consumption expenditure growth.

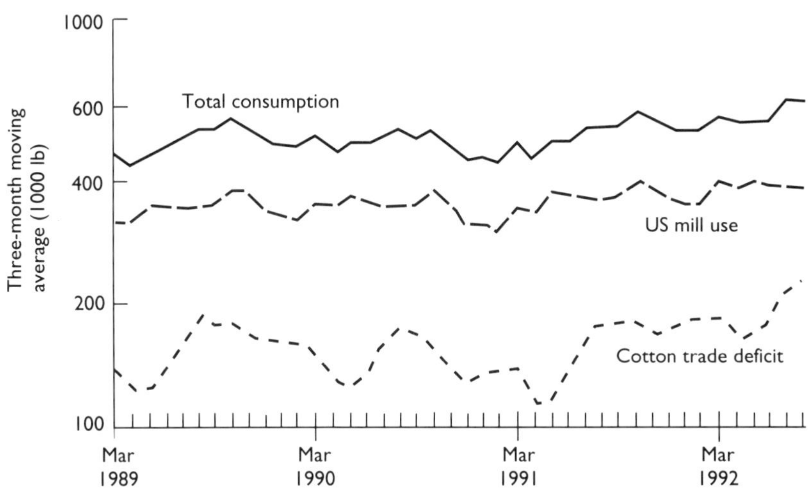

5.5 Cotton consumption rises steadily.

with exports this produced cotton stocks in consuming establishments of over 600 000 bales; stocks held totalled 10.2 million bales although this was expected to rise somewhat during 1993. During the 1990s, therefore, US cotton consumption continued to rise despite sluggish economic growth (Fig. 5.4 and 5.5). Data on textile trade suggest that US imports of textiles and apparel increased by 34% in 1992.

It appears that the cause is effective marketing amongst US consumers, paid for by the US cotton industry and the US government.

> The funds are spent on research and promotion through a private company owned by US cotton farmers, Cotton Incorporated, and through Cotton Council International, the international arm of the National Cotton Council. In total, more than $50m is being spent annually, mostly in the USA, on research. (ICAC, Nov.–Dec. 1992 *Review of the World Situation*, p. 7)

The EC

Demand was subdued during the mid-1970s - credit was relatively tight and inflation rates were high. Textiles from developing countries were able to enter under the GATT Multi-Fibre Agreement (MFA), with the result that from the late 1970s onwards the European textile industry had substantial overcapacity, which was shaken out only gradually during the 1980s during the continuation of subsidies by many national governments and the refusals by many companies to admit the necessity of moving production out of the high wage EC economies. The EC's overall consumption depended on the general health of the EC economies, with the result that consumption reached 6.565 million bales in 1987/88, but thereafter fell progressively, reaching only 5.045 million bales in 1992/93.

Japan

The effect of the oil crisis was pronounced and prolonged in Japan, with demand for all apparel fibres declining in the late 1970s. The textile industry had surplus capacity, and the revival of 1979 was brief - in 1980 a textile industry curtailment plan was launched, along with a programme to reduce excess capacity for synthetic fibres. This programme was timely, as Japan was set, even in the recovery that started in 1983, to suffer from high production costs. Japan's consumption then rose, reaching 2.3 million bales by 1992/93, all of it imported. This was, however, an enormous fall from the 2.783 million bales of 1991/92, signalling the impact of the recession on the Japanese spinning industry.

The former Soviet Union

As with production statistics, and perhaps more so, consumption statistics from the former Soviet Union and its satellites in Eastern Europe must be regarded with suspicion. The statistics provided to the FAO say that the centrally planned economies processed slightly more cotton in 1974/75 than the year before. Negligible increases were reported in the period to 1981 although the Soviet spinning industry was slightly expanded at the beginning of the 1980s. The serious collapse of consumption began in 1990 and has continued through the early 1990s. For example, USDA figures for SE Asia as a whole show that consumption was at 617 000 bales in 1987/88, reached 923 000 bales in 1991/92 but fell to 894 000 bales the next year.

Newly industrialised countries

Steady growth in cotton processing was already underway in these countries, especially Hong Kong and South Korea, by the 1960s, making the textile industry one of the trail-blazers of Far Eastern industrialisation. New records were reached in 1975 amongst the export-oriented textile industries of both these countries, and then promptly surpassed as the 1970s continued. In the early 1980s there were fewer export opportunities in developed countries and output in both Hong Kong and Korea was accordingly reduced.

Developing countries

Lower domestic milling costs resulted in less competition from imports during the recession of the mid-1970s in most countries other than the industrialised West. In 1974/75 declines in milling were slight in South Africa and the Mediterranean countries, and more severe in India, Pakistan and elsewhere in the Far East, as well as in Brazil, which were barely offset by increases in South Korea, Indonesia and Hong Kong and small increases in Africa generally. Utilisation was limited in both India and Pakistan in 1975/76 by supply constraints; elsewhere in the developing world the long-term upward trend in domestic cotton milling and usage continued, reviving especially in 1978/79, and in countries such as Egypt where there was substantial investment in spinning machinery. By 1981 Hong Kong had begun itself to feel the heat of competition from lower production countries, notably Thailand, Korea still, India and

Pakistan. Indian milling declined in 1982 because of strikes and power shortages but recovered the following year when the year-long strike ended. In the early 1980s generally textile manufacturing capacity and hence mill cotton use increased in Africa and the Middle East, but declined in Latin America.

By the late 1980s it was the worldwide recession which was restricting consumption growth. For example, Brazil's consumption was forecast by the USDA at 3.35 million bales, Zimbabwe's at 250 000 and South Africa down to 275 000. India had increased its consumption from the 1987/88 level of 1.74 million bales to 1.921 million bales in 1992/93, but the increase in Pakistan was very noticeable, up from 730 000 tonnes in 1987/88 to 1.481 million tonnes in 1992/93 - an increase of over 100%. Consumption in the former Soviet Union successor states fell dramatically, from 1.887 million bales in 1987/88 (for the old Soviet Union) to 1.252 million bales in 1992/93.

China

Statistics on Chinese mill use in the mid-1970s are not reliable. It is known that mill consumption began to rise in the late 1970s, probably more as a result of the end of the period of uncertainty following Mao's death than anything else. More than 750 000 new cotton spindles were installed between 1979–81, the ICAC claims, and consumption rose to 3.4 million tonnes, 23% of the world total. By 1981 the international textile and cotton-trading communities were fully alert to the new importance of the Chinese textile industry, which consumed 3.6 million tonnes the next year as a result of the Chinese government policy of increasing output of textiles for the domestic market as well as for export. A record output of yarns, fabrics and textiles for the domestic market was seen in 1983.

Cotton consumption in China rose throughout the 1980s, and despite difficulties in obtaining accurate statistics, it appears that by the 1987/88 season, consumption had reached 4.463 million bales. This compares with 4.8 million tonnes (ICAC) or 4.572 million tonnes (USDA) for 1992/93 and considerably less - 4.246 million tonnes - in 1991/92. The ICAC expects Chinese consumption to be 4.6 million tonnes in 1993 and 4.7 million tonnes in 1994.

CHAPTER

6

The international textile industry

Cotton spinning and manufacture

The saw gin of Eli Whitney (1793) resulted, within a decade, in cotton becoming the main fibre for clothing, and changed textile production from a domestic to a manufacturing industry. Sewing machines, invented by Hunt and Howe, were replaced rapidly by Singer's better machines. The textile industry had a need for efficient and durable machinery and a symbiotic relationship quickly developed with power suppliers and the machine tool industry, both in the USA and on the continent. Other industries, such as the automobile, took an increasing share of cotton production: the textile industry developed the cord to reinforce pneumatic tyres, for example. By 1960 cotton accounted for 78% of all textile fibres sold.

Cotton-spinning companies have themselves become well known. One of the best known in the UK, although relatively unimportant in size by comparison to Courtaulds, the industry leader in the UK, is still Shiloh Spinners, formed in 1874. The industry began to decline after World War I, enjoying a brief revival after World War II which was actually constrained by labour shortages. Cheap imports gathered momentum during the 1950s, however, and in 1959 the UK government introduced a fund to finance the closure of mills and the scrapping of surplus machinery. During the 1960s mills closed at a rate of one a week and 12 000 jobs were lost annually. By 1972 imports were capturing over 50% of the UK home market. Shiloh Spinners is one of the few UK spinning companies to remain in business: in Oldham, Lancashire and district alone there were once 470 mills but now there are only 4.

Intense lobbying continues with respect to alleged dumping. In their report for the year to March 1992, for instance, Shiloh Spinners reported that in March the EC imposed anti-dumping tariffs on cotton yarn and polyester from Brazil and Turkey. However, in the four years that the judgement took to be processed, 32 UK mills closed, forcing a 50% reduction in industry capacity. Only very cautious optimism could be expressed by Shiloh about prospects for the spinning industry in the 1990s.

After the cotton has been delivered to the spinning mill, it is first weighed and sampled again for quality. The mixing together of several varieties and staples of cotton is needed in order to obtain the right blend.

Spinners have been aware throughout the twentieth century that foreign matter found inside cotton bales could often be eliminated by hand, so long as spinning was still mainly a labour-intensive processing technology. Now it has become largely mechanised and is increasingly so. Within-bale contamination is a highly contentious issue. Some analysis suggests that cleanliness is the largest problem for automated cotton production in the 1990s, as existing automated equipment in widespread use is not able to detect contamination or foreign matter, let alone reduce or eliminate it. Plastic is one of the worst forms of contamination, and the evidence is only visible when the fabric leaves the final finishing process, by which time it is too late to do anything. Such contamination not only adversely affects the quality and appearance of the final textile product, but may also damage the processing machinery itself. The worst contamination is found in cotton from India, Turkey, Sudan, Brazil and Egypt (20% or more significantly contaminated); Israel on the other hand scores (ITMF figures) less than 2% contamination. Most cotton averages somewhat less than 10%. Efforts to improve cleaning techniques for cotton are continuing. The coupled lint cleaner began testing in 1991 – this technology is thought likely to enhance length and reduce short-fibre contamination. 1993 was estimated to be the date of introduction for the first cage gin.

The spinning method, called 'open end', where the lap or sliver (see below) is processed into yarn by feeding it through a disc which spins at 70 000 rpm, can currently be used only for short, fine, tough fibres that are completely clean. Cotton Incorporated reported in 1991 that

> open end yarns are of great interest to the textile industry because they can be spun ten times faster than with ring-spinning. Spinning research is now being focused on the development of soft-spun rotor yarns to improve fabric hand and evaluations of 100% cotton air jet spun yarn, both of which lines of research have proved promising. Improvements have been made in improving the opening and cleaning of bleached cotton for nonwoven uses. An air-doffed cleaner has been developed which fully opens kier-bleached cotton without a tremendous increase in neps.

Cotton non-woven fabrics are being improved also: a method of air-doffing cotton at 450 lb/h has been used since the beginning of the 1990s to improve its appearance.

Retailers throughout the USA and internationally are demanding lighter-weight cotton fabrics that are of high quality, yet can be produced on existing machinery. The industry is also demanding faster yarn production at the same time as improved quality. Efforts are being made to improve rotor and air jet spinning to spin faster, perhaps up to 120 000 rpm. Twin-spun yarns can also produce fabrics that approach the quality of ring-spun fabrics at lower cost, and this system is now being marketed commercially. Many spinning mill operations have become completely automated – for example, the opening of cotton bales with the EFS (engineered fibre selection) process of Cotton Incorporated in the USA. This enables blending to be carried out in order to produce the most desirable blend of fibre properties. The blended lint is blown by air from the feeder through chutes to carding machines which can comb, and which straighten and shape the fibres into a thin web. The web of fibres is then drawn through a funnel-shaped device and moulded into soft rope-like strands called slivers. Roving frames draw the slivers out more thinly and then twist them as the first step towards yarn itself. The spinning machines twist the roving and draw it out continuously, making it tighter and thinner until it reaches the thickness – called 'count' – required for weaving or knitting fabric. Strands are twisted as many as 10–30 times/inch. In the USA the majority of spinning frames continue to be ring-spinning, but open-end spinning is becoming more widespread. After spinning, the yarns are tightly wound around bobbins or tubes and are ready for fabric forming. Ply yarns are two or more single yarns twisted together on twisting machines similar in design to ring-spinning frames.

Improvements in carding machines along with other machinery

Table 6.1 Machine capacities

Machine	1950	1970	Expected by 1980 in 1977	1993
Card	4 kg/h	30 kg/h	60 kg/h	80–90 kg/h
Drawframe	6.6 kg/h	84 kg/h	120 kg/h	160–170 kg/h
Comber	8.1 kg/h	29.5 kg/h	40 kg/h	60 kg/h
Speed frame	295 g/sp/h	1105 g/sp/h	1500 g/sp/h	2000 g/sp/h
Ring spinning	21.8 g/sp/h	30.3 g/sp/h	35 g/sp/h	40 g/sp/h

has been dramatic. The industry has responded to higher wage costs – not now just in the USA and UK, but in Hong Kong also. Table 6.1 shows some examples of performance increases in conventional spinning machinery. However, to achieve acceptable strength levels, twist traditionally had to be added to open-end yarns, which resulted in a lower quality fabric. Cotton Incorporated offers US mills and participants overseas a software program called Engineered Fibre Selection (EFS), which improves the cost-efficiency of mills and enhances fibre and yarn quality under conditions of open-end spinning. EFS contains four interrelated software programs: MILL-Net, QRNet, GINNet and CALNet, each aimed at improving quality selection at particular points in the cycle. EFS has enabled open-end yarns with extremely fine counts to be produced. Nearly 80% of cotton consumed by US mills is processed through EFS at present, and its use is expanding in mills in the Far East and Europe at an annual rate in excess of 60%. There is an annual conference at which attendance is becoming general amongst US milling companies. These trends have been similarly observed throughout the world, although without the direct assistance of such programs as EFS, as fibre and yarn quality improves in response to consumer demand.

Cotton textile manufacturing

The process starts with preparation of the yarn for weaving or knitting. Large spools of yarn are wound into a beam which is used for weaving, a process called warping.

Weaving

Weaving is the oldest way to produce fabric from yarn. The basic principle of interlacing yarns to produce fabric has not changed for centuries. On the loom, the lengthwise yarns are called the warp. These normally demand a higher degree of twist than the filling of weft yarns, which are interlaced at right angles to them. Historically, cloth has been woven by a wooden shuttle moving horizontally back and forth across the loom, interlacing the filling yarns with the vertical warp yarn. A sizeable percentage of mills now use high-speed 'shuttleless' looms which can operate at weft insertion rates of up to 1200 m/min and produce a wide variety of fabrics. There are several types of so-called shuttleless looms. Some have metal arms which pick up the filling thread, carrying it halfway across the loom to where another metal arm picks it up and carries it the rest of the way; other types use small projectiles; and yet others pick up the weft from large cones and then shoot it across the warp and back from the other side by means of compressed air. The majority of new looms are also quieter than their predecessors.

There are three types of weave, although there are many variations. Cotton can be used in all three:

1 The plain weave: the filling is alternately passed over one warp yarn and under the next. This is used for cotton ginghams, percales, chambrays, batistes and other such products.
2 The twill weave: the yarns are interlaced to form a diagonal ridge series across the fabric, which is used for sturdy cottons such as denim, gabardine, khaki and ticking.
3 The satin (or sateen): these are the least common. They make a smooth fabric with a high sheen; they are used for cotton satin and faille, they are made with fewer yarn interlacings and with either the warp or the filling yarns dominating the cloth face.

In more advanced cotton mills, optical scanners monitor the production process continuously, searching for flaws in the cloth emerging from the loom. Computers identify the source of the flaw and that part of the cloth is automatically removed before posting the order. Diagnostic systems then operate to rectify the fault.

Knitting

This is a way of constructing fabrics by using a series of needles to interlock loops of yarn. Lengthwise rows of these loops, comparable to the warp yarn in woven goods, are called wales and crosswise rows comparable to filling yarns are known as courses. There are many similarities between knitting done by hand and machine-knitting, but there are also key differences. A typical hand knitter using two needles and forming one stitch at a time may produce 60 stitches/minute, but a modern knitting machine may use up to 1600 needles and produce over 3 million stitches a minute.

In place of a single yarn cone, a knitting machine may have 8–148 cones, each feeding approximately 700 feet of yarn/minute into the fabric. Most cotton is knitted on circular machines which have needles fixed to the rim of a rotating cylinder. As the cylinder turns, so the needles work their way from stitch to stitch, producing a tubular fabric. The width is regulated by the size of the cylinder, which is usually between 1 and 2 feet in diameter. Another basic type is the flat knitting machine. This is designed with a flat bed, and has dozens of needles arranged in a straight line and producing a knit fabric that is flat like woven goods. Such machines produce over 1 million stitches/minute, can be set to drop or add stitches automatically, and can narrow or widen the fabric at certain points in order to make particular shapes. As with weaving machines, knitting machines can be programmed to produce a wide variety of fabrics and shapes.

Dyeing, printing and finishing

When they emerge from the loom in their rough state, cotton fabrics are known as 'grey goods'. Most cloth goes through a further series of chemical treatments and finishing processes to meet specific end-use requirements. Some mills undertake the job of dyeing and printing the fabric in addition to spinning and weaving. Some mills sell gray goods to converters, who finish cloth in separate plants. Finishing includes cleaning and preparing the cloth, dyeing or printing it, and then treating it with chemicals, or mechanically, to get the necessary texture and performance characteristics: there are numerous finishing processes reflecting the vast range of colours, textiles and qualities demanded by the competitive Asian, Western and also African retailing sectors. In order to produce a smooth surface for dyeing and finishing, the grey goods are passed quickly over gas jets or heated copper plates to singe off lint and loose threads. At faster than 200 m/min, the cloth is then bleached and scoured in a continuous hydrogen peroxide process. The cloth is then stacked in large, J-shaped boxes to allow the chemical to work before the goods are withdrawn from the bottom. For a more lustrous cloth, the gray goods are immersed under tension in a caustic soda solution and then neutralised later. This process is called mercerising, which makes the fibre swell permanently. It gives the fabric a silken sheen, increases its strength and increases its dye take-up.

Cotton fibre can be mercerised, which allows cotton to be dyed in deeper colours while using less dye. There is also liquid ammonia processing which allows for tone-on-tone colours to be dyed on 100% cotton. This was only possible before by using expensive yarn-dye processes or by cross-dyeing cotton with synthetics. Pad-batch dyeing, which needs no heavy metals, salts or other chemicals to work, increases the cost effectiveness of short runs. Cross-linking resins allow durable press finishing without the use of formaldehyde. Soft, combed, fine-rotor-spun yarns are increasing the production of mid/lightweight cotton fabrics to meet rising demand. Finally, cotton production of bleached and scoured fibre is being improved to increase market share in personal hygiene products.

To colour cotton it is either yarn-dyed or piece-dyed, and computers are increasingly being used to make and match colours more quickly and accurately. Yarn-dyeing is done before the cloth is woven or knitted and is used to produce gingham checks, plaids, woven stripes and other special effects. To make denim, for instance, indigo-dyed warp yarns are combined with white filling yarns. In 'package-dyeing', which is one of the most frequently used methods of yarn-dyeing, the yarn is wound around perforated cylinders or packages, and placed on vertical spindles in a circular dyeing machine. Dye solution is forced alternately from the outside of the packages inward and vice versa; the dye reaches the core of the fibre and is less liable to run than when using the alternative method, piece-dyeing.

Piece-dyeing, conversely, is usually used for fabrics that are to be all one colour. A continuous length of cloth is passed through a trough of hot dye solution. The cloth is then squeezed through rollers, spreading the colour evenly across the cloth and removing the excess liquid. In a variant of this method, the fabric is made into a rope-like coil which is processed on a reel passing through the dye beck or vat.

The printing of coloured designs on to cotton cloth is a technique similar to that used for printing on paper. Long runs of the same fabric design are usually made on a roller print machine, making 1–200 m/min, printing up to 16 different colours in one go. Such a machine has a large padded drum or cylinder, surrounded by a series of copper rollers, each with its own source of dye and a 'doctor' blade to remove excess dye. The number of rollers depends on the fabric design: each colour in the fabric is applied by a separate roller. The cloth moves between the rotating drum and the rollers under great pressure. It acquires colour from the engraved area of each roller one after the other. The printed cloth is dried immediately and moved to an oven where the dye is set.

The other main way of putting colour designs onto cotton is automatic screen printing. This is slower than roller printing, but is able to make larger and more intricate designs, elaborate shadings and other special effects. In flat-bed screen printing, the fabric design is reproduced on fine mesh screens, one per colour. On each screen, the area not to be covered by the dye is covered with lacquer or some other substance resistant to the dye. The screens are covered with dye on the back and placed in the appropriate sequence above a flat bed. As a belt takes the fabric from screen to screen, a roller or other device presses the dye through the open parts of the screen onto the fabric. The fastest flat-bed machines can operate at approximately 1200 m/h for a fabric with a 1 m design repeat. Rotary screen printing machines can operate up to three times faster. This system combines roller and screen printing, using perforated cylinders in lieu of flat screens. The colour dye is fed inside the cylinders and a small metal roller forces the colour through the holes in the roller onto the fabric moving continuously under the cylinders. Up to 12 colours at present can be printed on one fabric using this method - it can be expected that this number will soon be exceeded, as use of this method is increasing. The reason is cost - the screens or cylinders cost less to produce than the engraved copper rollers employed in roller printing.

Finally the fabric is finished. There are hundreds of different mechanical and chemical finishes that can be applied to textiles, and many are applied to the same fabric; there are hundreds of different methods of application. It has been suggested that cotton can be finished in more ways than any other fabric; for example, the 'stonewashed' appearance of much denim. The aims are to improve the appearance of the fabric in many cases, and also to improve such characteristics as durability, water or fire resistance, control of shrinkage, or colourfastness.

The scope of the industry worldwide

The cotton industry began as a domestic, 'cottage industry' in Asia and America well before 3000 BC. It stayed in areas of cotton production, with the exception of some westward spread from India, until after the Crusades - between 1096 and 1270. As a result of the Crusades, Indian cotton products were introduced to Western Europe. A large demand immediately followed, but European wool growers and manufacturers forced the introduction by West European states in the period after 1300 of severe restrictions on these imports. This stimulated the passage of the cottage system of cotton manufacturing from India to Western Europe, and was the genesis of the international cotton trade. High demand within Europe for cotton goods, and the relative expense of making yarn from cotton were the stimulants which caused research into the mechanisation of cotton manufacturing (see the section on 'From first cultivation to the end of World War I' in Chapter 1).

Mechanisation brought the centre of the industry to Europe and put trade in cotton on a world basis. The industry stayed predominantly in Europe until the close of the nineteenth century, when the abundance of cheap labour and the ample cotton supplies available persuaded entrepreneurs to establish cotton mills in the USA itself, and in India, the first mill there being built in 1850, in Japan (1866) and in China (1890). In the USA, the geographic shift of the cotton-manufacturing industry after 1850 was pronounced: from 74% in New England in 1850 to 53% in 1915 and 14.7% in 1951, compared with 6.6%, 40% and 81.3% for the cotton-growing states. Spinning and weaving were integrated to a large extent in the same mills and companies: this was dictated by population and geography; the trend towards vertical integration accelerated during and immediately after World War II as mills began to convert their cloth to raise profits whilst, similarly, converters bought mills. By the 1960s the majority of US mills converted their own cloth, usually in their own finishing plant.

By mid-century many other African and Asian countries, for example Egypt and Pakistan, were also beginning to develop their own cotton industries. Hong Kong installed its first spindles in 1948 with refugee Chinese capital, seeing the birth of a national industry with phenomenal growth. Such countries doubled their spindle numbers between 1929 and 1952. The Indian industry developed, with power loom weaving largely integrated with spinning, and with a separate hand loom industry which has enjoyed government support. In Pakistan the structure has been similar with self-sufficiency being attained as early as 1957. Cotton manufacturing developed rapidly in Communist China also.

The spread of the cotton industry from 1875 onwards benefited many countries, but it also caused great hardship in the traditional cotton manufacturing countries, including the UK, the USA, India,

China and Brazil. Textile manufacturing associations, including internationally, began to concern itself with this problem when Japan started to manufacture in the 1930s, but its conferences in the 1960s paid increasing attention to the problem. Machine utilisation differentials were in part responsible. Until 1959, single-shift working was the usual practice in Lancashire, whilst three-shift working was almost universal in the USA and Hong Kong. The average annual hours worked by spindles and looms in 1966–67 were: UK, 3000; Western Europe, 3–3500; Japan, 5000; USA, 7000; and Hong Kong, 8400.

Complaints from the Europeans and in the USA about the rising volume of textile exports from developing countries due to lower wage costs - despite the very high percentage of lower-paid, mainly female, workers at European mills (52% of the UK workforce in spinning and weaving in 1968 were women, 20% in the Netherlands, 66% in Italy, 75% in Finland and 80% in Japan) - had begun in earnest by the 1961 Vienna Conference of the IFCATI (International Federation of Cotton and Allied Textile Industries) organisation. In 1954 India began to export goods for consumption in the UK, not for re-export, and soon Pakistan and Hong Kong did the same. In 1955 the Japanese cotton industry was persuaded to put a voluntary limit on its exports to the USA, but from 1959 imports began to flood in from other countries - Hong Kong, India, Pakistan, South Korea, Spain, Portugal, Brazil, Mexico and Taiwan. IFCATI was instrumental in the Kennedy Seven-Point Programme for the textile industry of May 1961. Such concerns were voiced alongside the concerns as to the impact of the division of the European trading area into the competing blocs of the European Free Trade Association (EFTA) and the European Community (EC). Already by 1961 Hong Kong was making substantial exports. International negotiations on the subject led to a short-term agreement (1961–62) and then the General Agreement on Tariffs and Trade (GATT) arrangements from 1962–73 held, as piecemeal quotas such as those introduced by the UK were seen to fail; these were followed by the Multi-Fibre Agreement (MFA). But the trend continued: by 1968 under the GATT 11% of US imports of cotton products constituted 11% by weight of all US consumption.

These arguments paralleled a decline in the quantity of machinery installed. IFCATI statistics showed that the number of spinning spindles worldwide peaked at 165 million in 1928, declining thereafter to 148 million in 1939, 126 million in 1960 and rising to 134 million in 1967. Most of this was accounted for by replacing mule spindles by ring spindles such that there were fewer than 500 000 mule spindles in operation by 1969. Power loom numbers fell from 3.159 million in 1930 through 3.07 million in 1936 and 2.685 million in 1967, with the proportion of automatic looms increasing from 23% to 58% over the whole period. This was explained by the higher number of hours worked by each machine and the increased speed of the machinery as a secondary factor.

During the 1960s it was proposed that there should be an international commodity control agreement for cotton, but this was formally rejected at the Lima meeting of the ICAC in 1966. The 1960s saw the establishment instead of the International Institute for Cotton, which was set up to finance and research cotton projects internationally, partly in response to pressure from synthetics. Then came the oil crises.

The FAO annual world fibre consumption survey showed that by 1986 almost 50% of all textile fibres were spun in developing countries, compared with 37% in 1975. The share of cotton processed by developing countries was 63% in 1986 (FAO) compared with 54% in 1975. Having milled the cotton, developing countries sought increasingly to export the finished products: from 0.73 million tonnes fibre equivalent in 1975 to 2.5 million tonnes in 1986. The Asian textile industries grew in response to the needs of their fast-growing populations. Further international liberalisation and cost differentials provided the impetus, fuelled by rising demand in the industrialised countries, for a rapid growth in textile demand.

In the 1990s the industry worldwide has gone into a major recession. Attention has been focused on when a recovery can be expected, but the news from analyses of 1990 and 1991 was pessimistic throughout the developed world. As the ITMF Director-General, Herwig Strolz, pointed out at the 1992 ITMF Annual Conference

> the warning signs were already up in the latter part of the '80s
>
> * fast rising investments in new machinery
> * insufficient scrapping of obsolete production capacity against the background of rapidly increasing productivity
> * mounting yarn and fabric stocks
> * falling growth rates in textile consumption.

The world knows of the difficulties that the textile industry is facing.

The key figures are that world production declined by 1.7% in 1990 and 3.8% in 1991 (Tables 6.2–6.4, Fig. 6.1). Yarn stocks have continued, despite a fall in yarn inventories during 1991, to be on an upward curve, now reaching 85% higher than was the case in 1987 by the first quarter of 1992. The result has been that shipments of textile machinery fell by 25–30% in 1991 compared with 1990 and the low

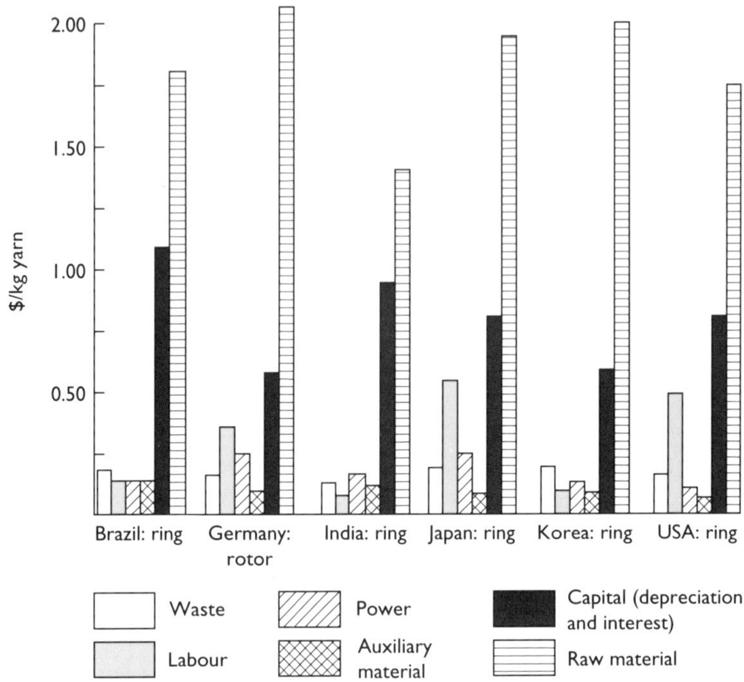

6.1 Total yarn costs 1991.

level of deliveries internationally continued into 1992. Almost all regions were affected by the downturn except for South America which recorded an increase of 26% in shipments in 1991 for spinning machinery. Weaving machinery shipment to Asian countries continued at about the 1990 level, but the rest of the world fell by almost 50%. Comitextil observed that this decline was only exceeded in savagery by the situation in 1981, and that a series of factors - the end of the Gulf War, the UK General Election (for the UK), the US Presidential election - would all signal the beginning of the recovery in demand. Their hopes have yet to be fulfilled.

In 1990 the number of ring spindles in ITMF countries declined by approximately 2%, and in 1991 this number was virtually unchanged. About 10% and 5% respectively more open-end rotors were, however, installed, so this represents more of a technical change than evidence of demand decline. The trend towards shuttleless looms (up 6% compared with shuttle looms (down 11%) also reflects this technical change. Shuttle looms are indeed disappearing in Western Europe and especially in the USA, with overall capacity falling by 4% in 1991. In Italy, which has a relatively efficient textile industry, the

Table 6.2 1991 cost factors: weaving

Size of mill:	140 Sulzer type P7100 B 360 N1-1EP Looms (tappet treading motion) & weaving preparation & cloth inspection	**Fabric data:**	Cloth type	-	sheeting
			Construction	-	60/60 treads per inch
			Yarn type	-	100% cotton
			Yarn count	-	Ne 20 warp & weft
			Grey width	-	63 inch
Rate of output:	17.5 yd of fabric per hour per machine				

	Brazil	Germany	India	Japan	Korea	USA[a]
Hourly wage for skilled personnel	707.60 Czs	32.60 DM	23.58 Rs	2 893 Y	2 938 W	$12.97 [b, c]
Hourly wage for machine tenders	375.83 Czs	30.45 DM	22.85 Rs	1 466 Y	1 420 W	$11.83 [c]
Hourly wage for unskilled personnel	302.41 Czs	28.83 DM	16.82 Rs	1 303 Y	1 105 W	$8.65 [c]
Operating hours (per year)	7180	5500	8420	5 960	8 300	7 000
Cost of electric power (kW h)	13.60 Czs	0.16 DM	1.26 Rs	13.57 Y	39.38 W	$0.045 [d]
Cost of water (m^3)	180.29 Czs	2.20 DM	12.00 Rs	35.00 Y	195.00 W	$0.702 [d, e]
Cost of steam (lb yarn)	2.38 Czs	0.108 DM	0.25 Rs	9.10 Y	25.60 W	$0.040
Sizing cost (lb yarn)	41.55 Czs	0.38 DM	2.92 Rs	19.60 Y	156.00 W	$0.120
Cost of heating oil (kg) or gas (cu. ft)	-	0.48 DM (oil)	-	45.60 Y (oil)	135.00 W (oil)	$0.0027 [d] (gas)
Cost of buildings (m^2)	165 616 Czs	1 100 DM	3 300 Rs	182 640 Y	303 000 W	$656
Building space (m^2)	4 700	4 700	5 500	4 700	5 000	4 700
Annual building maintenance (%)	1.0	1.0	1.0	1.0	1.0	1.0
Depreciation period for machinery (years)	10	10	9	10	9	11

Table 6.2 Continued

	Brazil	Germany	India	Japan	Korea	USA[a]
Depreciation period for accessories (years)	8	5	5	7	9	5
Depreciation period for buildings (years)	25	30	30	35	40	32
Customs, sales tax, etc. (% of machine price)	40	0	25	0	13	5
Capital interests rate (%)	15.2	10.5	23.0	8.1	10.5	9.0
Raw material costs	0.390	0.446	0.305	0.422	0.429	0.375

[a] Based on conditions in southeastern United States.
[b] Rate for loom technician.
[c] Includes 25% fringe benefits.
[d] Rate for industrial user in Greenville County, South Carolina.
[e] Includes sewage at 70% of cost.

Table 6.3 1991 cost factors: spinning

Type:	Rieter ring spinning (incl. winding) Rieter open-end rotor spinning	**Yarn data:**	Raw material Staple length Yarn count	- 100% cotton - 11/16 inch - Ne 20 carded
Rate of output:	450–500 kg per hour			

	Brazil	Germany	India	Japan	Korea	USA[a]
Type of mill	Ring	Rotor	Ring	Ring	Ring	Ring
Hourly wage for skilled personnel	719.36 Czs	32.60 DM	18.58 Rs	2 893 Y	2 938 W	$11.95 [b,c]
Hourly wage for machine tenders	396.34 Czs	30.45 DM	15.67 Rs	1 466 Y	1 420 W	$9.87 [c]
Hourly wage for unskilled personnel	302.24 Czs	28.83 DM	15.04 Rs	1 303 Y	1 105 W	$8.23 [c]
Operating hours (per year)	7 180	6 900	7 900	5 960	8 300	7 000
Cost of electric power (kW h)	14.04 Czs	0.16 DM	1.26 Rs	13.57 Y	39.38 W	$0.045 [d]
Cost of buildings (m^2)	165 616 Czs	1 100 DM	3 300 Rs	182 640 Y	303 000 W	$656
Annual building maintenance (%)	1.0	1.0 DM	1.0 Rs	1.0 Y	1.0 W	1.0
Depreciation period for machinery (years)	10	10	9	10	8	7
Depreciation period for accessories (years)	9	5	5	10	8	5
Depreciation period for buildings (years)	25	30	30	35	40	32
Customs, sales tax, etc. (% of machine price)	50	0	0	0	0	5
Capital interests rate (%)	15.2	10.5	23.0	8.1	10.5	9.0
Raw material costs (US $ per kg cotton)	1.79	2.05	1.40	1.94	1.97	1.72

[a] Based on conditions in southeastern United States. [b] Rate for spinning technician. [c] Includes 25% fringe benefits. [d] Rate for industrial user in Greenville County, South Carolina.

Table 6.4 1991 manufacturing costs: spinning (units of national currency per kg of yarn)

Cost element	Brazil	Germany	India	Japan	Korea	USA
	Ring	Rotor	Ring	Ring	Ring	Ring
Waste[a]	44	0.27	2.66	25	141	0.16
	11%	11%	9%	10%	17%	10%
Labour[b]	33	0.60	1.69	72	76	0.48
	8%	25%	6%	29%	9%	29%
Power[c]	34	0.40	3.04	32	94	0.11
	8%	16%	10%	13%	11%	7%
Auxiliary material[d]	34	0.16	2.22	12	70	0.08
	8%	7%	8%	5%	9%	5%
Depreciation[e]	127	0.55	7.80	62	265	0.54
	30%	23%	27%	25%	33%	33%
Interest[f]	148	0.43	11.59	46	168	0.27
	35%	18%	40%	18%	21%	16%
Tool manufacturing costs[g]	420	2.41	29.00	249	814	1.64
	100%	100%	100%	100%	100%	100%
Tool manufacturing costs (US$ per kg of yarn)	1.65	1.40	1.40	1.86	1.10	1.64
(Index: Germany = 100)	(118)	(100)	(100)	(133)	(79)	(117)

[a] Waste: in spinning, revenue from the sale of waste (waste from slivers, filters, flats and grid droppings, etc.) is considered when calculating the waste costs.

[b] Labour: the wage costs are calculated on the basis of the wages paid to operatives and to skilled and unskilled labour for maintenance work. All social charges and shift-work premiums are included. For reserve personnel a percentage figure is added.

[c] Power: the energy costs include the costs relating to the actual power consumption of the machines, the illumination, the air conditioning, steam and heating. It is assumed that the mill is lit for the entire production time.

[d] Auxiliary material: the costs for spare parts, lubricants, cleaning materials, and maintenance work on the buildings represent the costs for auxiliary material. In weaving this cost element also includes yarn preparation costs.

[e] Depreciation: this element includes depreciation of machines, accessories and buildings. The machinery costs are inclusive of free delivery to the mill, erection and, where applicable, of customs duty and taxes.

[f] Interest: costs for capital interests.

[g] Total manufacturing costs: the sum of the above group of costs gives the total manufacturing costs.

average number of spindles is 19 800 and 65 looms per mill (Fig. 6.2 and 6.3).

Fewer people are being employed in the textile industries of ITMF countries – the number fell by 1.3% in 1990, and was 3.8% when India and Pakistan were excluded. The international textile industry is no longer a net supplier of new jobs although some countries such as Pakistan have noted increases. There is, as expected, a wide range of hours worked and wages paid in ITMF countries. Examining these gives the reasons why there is such a wide geographical disparity in

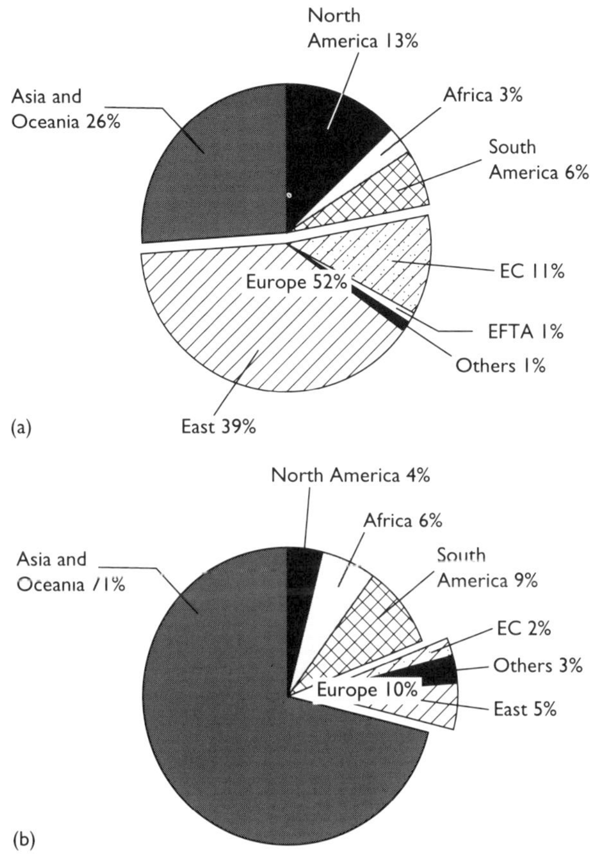

6.2 Weaving machinery: installed capacities 1991. (a) Shuttleless looms, world total 642 450; (b) shuttle looms, world total 2 128 380.

the performance of the textile industry which shows a close correlation with overall economic performance. In Belgium in 1990 the average annual hours worked was 1470 and in most European countries it is decreasing; in Malaysia it was 3120. The hourly wage paid in the early 1990s varied from US$0.78/h in India (with similar levels in Colombia, Egypt and the Philippines) compared with US$18.8 in the Netherlands. Some countries such as Turkey experienced a rapid rise in labour costs - 75% in 1991, for example. This is bound to affect their international competitiveness for textile exports, and in 1992 had the effect of increasing Turkish raw cotton exports.

Comitextil suggested that there were three groups even within Europe: countries that were slightly increasing their textile manufacturing activity during the early 1990s, such as the Netherlands, Italy

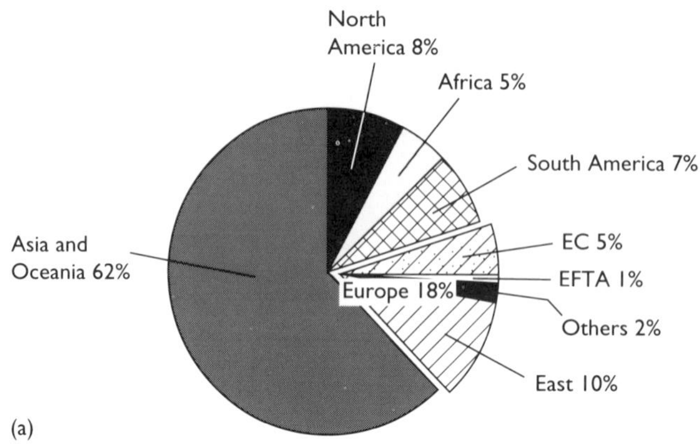

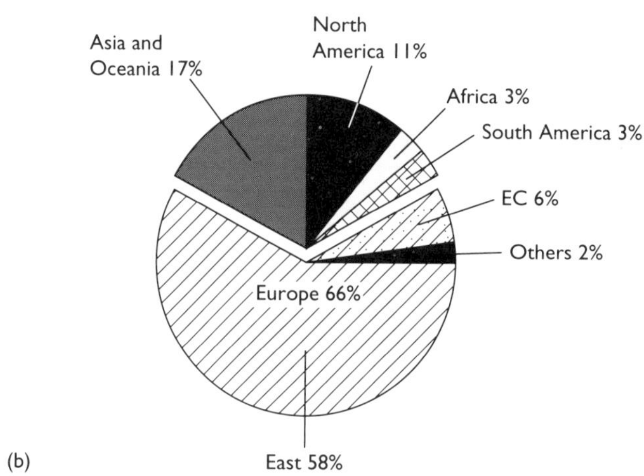

6.3 Spinning machinery: installed capacities 1991. (a) Ring spindles, world total 169 058 000; (b) open-end rotors, world total 8 064 000.

and Denmark; the second group began to enter 'textile recession' in 1991 - Germany, Portugal and Belgium - and the final group - France, the UK and Spain - where there have been two continuous years of industrial contraction, only just showing slight signs of reversal in 1993. Information on Greece was not available, but based on its trade figures, production declined over both years. The most significant fall, and perhaps comparable to declines in other East European countries, was in East Germany where production fell dramatically, by 46.7% in 1991 and 34% in 1990, leaving the East German textile industry operating at a fraction of its former level. The German statement to the ITMF was abrupt:

> Since the middle of 1990 the yarn market has been characterised by a still growing discrepancy between supply and demand. The result has been a murderous price war, which has led to considerable financial losses in particular for those spinners who produce large quantities of standard qualities.

This continued throughout the early 1990s, allowing for no price increases in either raw cotton or yarn prices. European weaving was also in economic difficulties by the end of 1990; the market for dyed fabrics continued to deteriorate, with that for printed fabrics in an improved position.

Clothing production during 1991 fell by over 10% throughout Europe as a whole compared with 1990, according to Comitextil. The ITMF reports that textile output fell by 3.5%, having already fallen by 1.4% in 1990, but it believes that apparel production actually fell by less, by 1.8%. The key facts are that production is now well below its level in the mid-1980s and was still falling well into 1992/3. Some economists define a recession as when production falls below the average level of the past five years, and a depression when it falls below the average level of the past ten. Europe, by these standards, is perilously close to a textile depression. Production declined from the end of 1990 onwards, and industry experts are currently reluctant to say when a revival can be expected. Again, there were wide geographic variations: in Portugal and Belgium there were increases in production, Denmark and Germany succeeded in maintaining their production levels, and production fell in the UK, Ireland, France and Spain. Amongst the different textile and apparel products there have been wide variations in performance. During 1991, synthetic yarns, carded and worsted wool yarns, fabrics from synthetic and artificial filaments, covers and carpets did relatively well: as for clothing, knitwear has an above average performance, especially for trousers, suits and baby clothes. Cotton's dominance as a material for clothing has, in fact, been continually reinforced throughout the 1990s, which may well mean that all the scepticism about depressed future demand even when there is an economic recovery is false. The main reason for this has been the positive 'natural' image of cotton and wool as clothing materials compared with synthetics. The demand curve for cotton may therefore be sharply kinked at certain income levels.

For the present, however, the message from the EC and other industrialised countries is one of almost unrelieved gloom. Yarn stocks in Europe generally reached their high point in mid-1991, but they are still well above 1987 levels (Fig. 6.4).

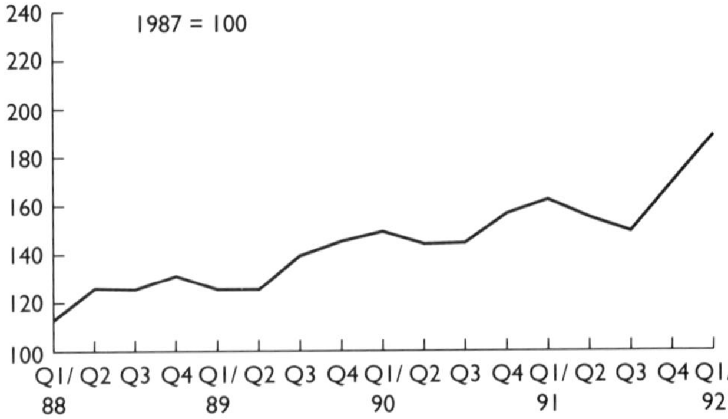
1987 = 100
240
220
200
180
160
140
120
100
Q1/ Q2 Q3 Q4 Q1/ Q2 Q3 Q4 Q1/ Q2 Q3 Q4 Q1/ Q2 Q3 Q4 Q1/
88 89 90 91 92

(a)

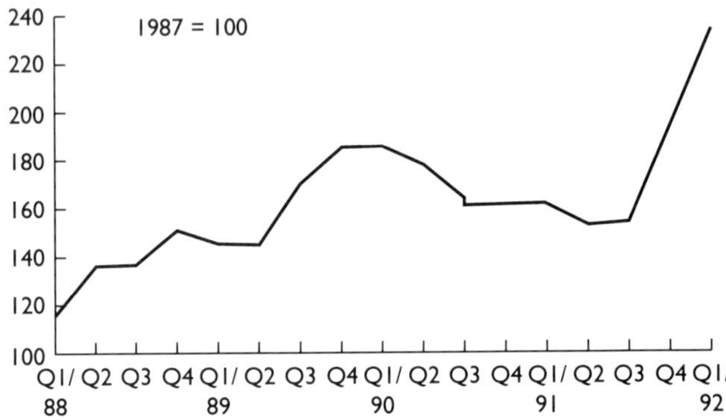
1987 = 100
240
220
200
180
160
140
120
100
Q1/ Q2 Q3 Q4 Q1/ Q2 Q3 Q4 Q1/ Q2 Q3 Q4 Q1/ Q2 Q3 Q4 Q1/
88 89 90 91 92

(b)

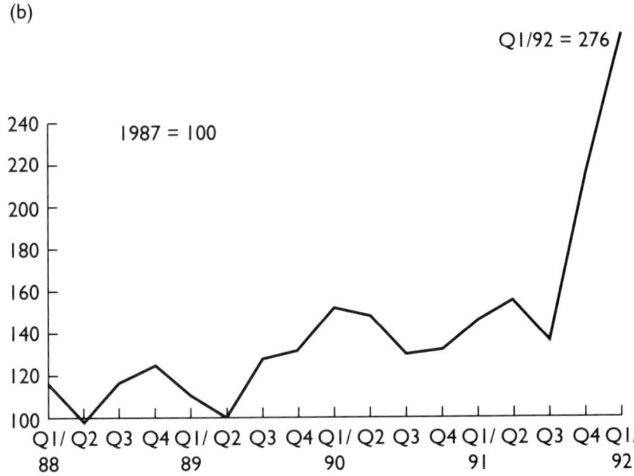
Q1/92 = 276
1987 = 100
240
220
200
180
160
140
120
100
Q1/ Q2 Q3 Q4 Q1/ Q2 Q3 Q4 Q1/ Q2 Q3 Q4 Q1/ Q2 Q3 Q4 Q1/
88 89 90 91 92

(c)

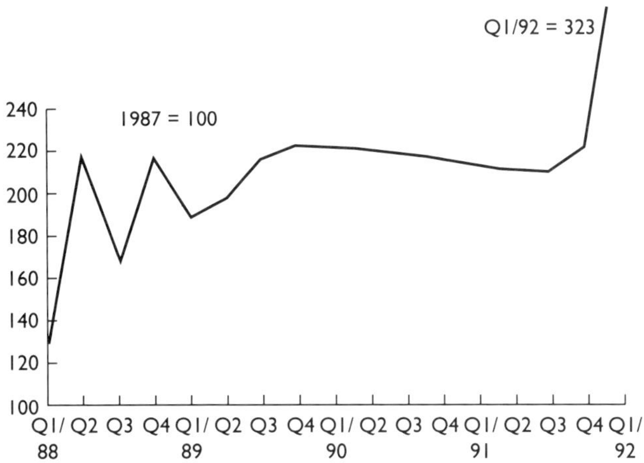
Q1/92 = 323
1987 = 100
240
220
200
180
160
140
120
100
Q1/ Q2 Q3 Q4 Q1/ Q2 Q3 Q4 Q1/ Q2 Q3 Q4 Q1/ Q2 Q3 Q4 Q1/
88 89 90 91 92

(d)

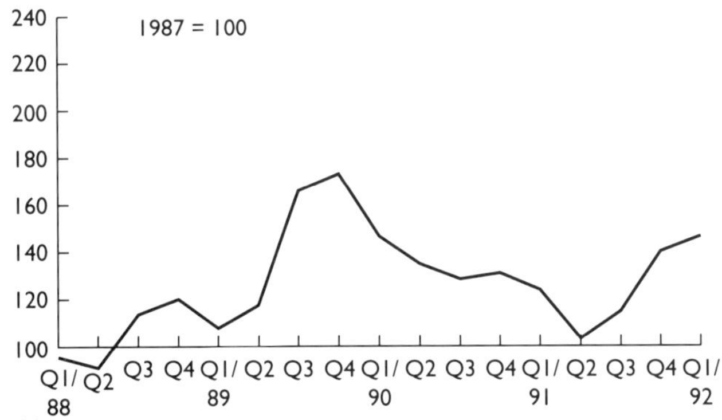
1987 = 100
240
220
200
180
160
140
120
100
Q1/ Q2 Q3 Q4 Q1/ Q2 Q3 Q4 Q1/ Q2 Q3 Q4 Q1/ Q2 Q3 Q4 Q1/
88 89 90 91 92

(e)

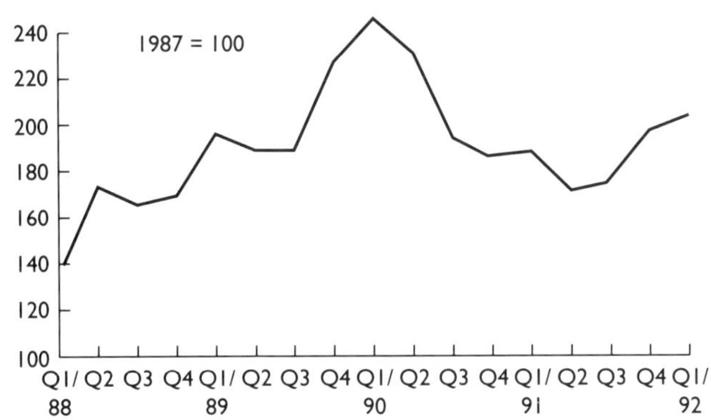
1987 = 100
240
220
200
180
160
140
120
100
Q1/ Q2 Q3 Q4 Q1/ Q2 Q3 Q4 Q1/ Q2 Q3 Q4 Q1/ Q2 Q3 Q4 Q1/
88 89 90 91 92

(f)

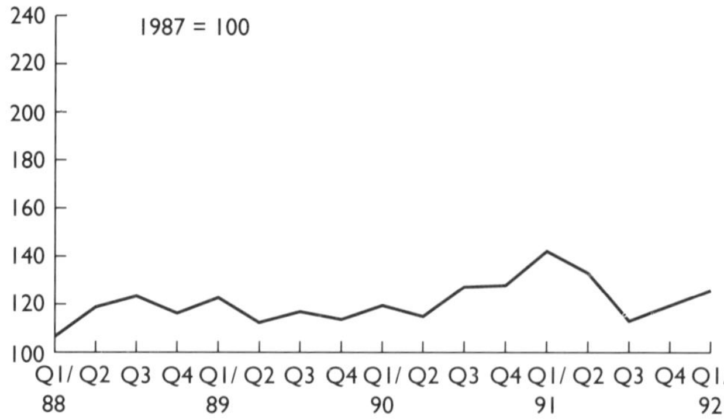

(g)

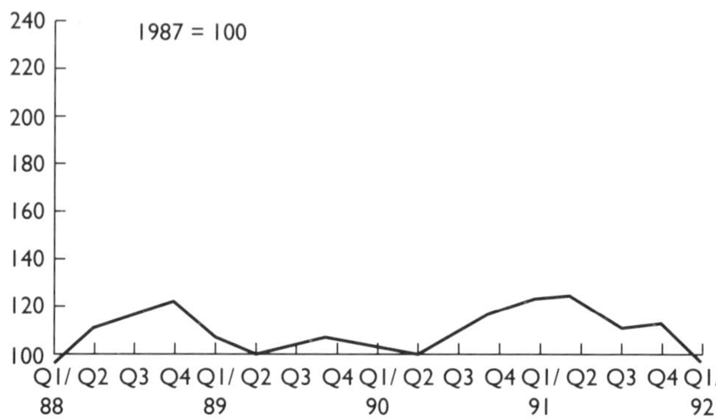

(h)

6.4 Yarn stocks 1988 - 92: (a) World; (b) Asia; (c) Pakistan; (d) Korea; (e) Taiwan; (f) Japan; (g) Western Europe; (h) Germany (Federal Republic).

Eurostat figures suggest that the textile industry of the EC had a sales reduction of over 1% in nominal value during 1991, and this decrease undoubtedly continued during 1992. It was at a level of ECU 102.4 billion in 1991 and may even have been below 100 billion in 1992. Germany, Italy and the Netherlands experienced less severe conditions than other member states in 1991, but the Italian crisis of 1992 reduced consumer demand generally, including that for textile products. The effect of the recession throughout consumer industries has been to reduce the price levels achievable. Overall in 1991, prices increased by 1.2%, considerably less than the average rate of inflation, but in France and Belgium there was actually a decline in prices. Exceptional factors contributed to an increase of 8% in Greek production prices: this was not matched by any increased demand. The consequence of falling prices and slack demand was a reduction in employment. 1991 was an especially bad year for job losses, which throughout the EC amounted to some 60 500 workers. RER estimated that in 1991 worldwide over 100 000 workers in the textile industry lost their jobs, including a sizeable percentage in the Far East export-oriented mills and factories. The trend has been towards accelerated job losses since 1988 and there was no let-up in 1992. To some extent even in the 1980s automation has been responsible for the job losses, but these do not amount to more than 10% of the total. A similar picture pervaded investment in textile manufacturing, which fell by 9.7% in the EC in 1991 compared with 1990. Further reductions of a more substantial size occurred during 1992. These were the first years of investment reduction since 1981 (Fig. 6.2, 6.3 and 6.5).

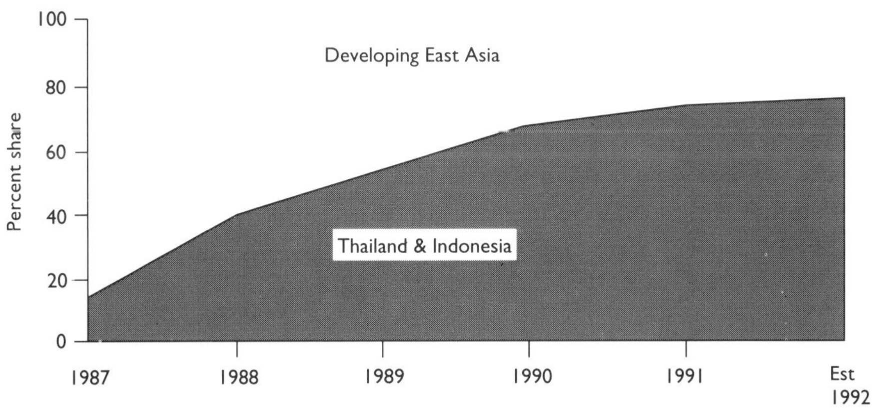

6.5 Investments, 1987 - 91: ring spindles.

In 1992 European yarn output remained stable, although orders showed a slight decline, with the consequence that stocks rose, in the first quarter, by 6.1%. Fabric output was down 7.5%, but inventories declined to an equivalent level to 1987. The European yarn order position is now over 20–25% lower than it was in 1987, a measure of the decline in the industry. Confidence is low, continues to be low, and shows no sign of recovery (Fig. 6.6). As a result, textile manufacturers are seeking to implement changes, for example in Germany to work on Sundays by running machines continuously. This would decrease the number of yarn faults that occur when machines are turned off and thus the yarn would be more likely to meet quality guarantees demanded by customers. Werner International calculated that machines currently operate annual hours as follows: Germany, 6334; Switzerland, 6869; Italy, 6841; France, 7000; and Indonesia/ Singapore, 8496.

To these declines and problems must be added the collapse of the

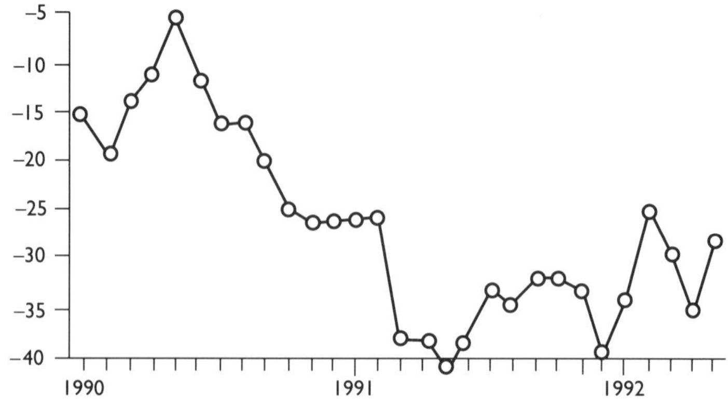

6.6 EC confidence indicator for the cotton industry, 1990 - 92 (*source:* OETH).

textile industry in the former Soviet Union. As recently as the 1991 Annual Conference of the ITMF, the Soviet representative, from GOSKOMLEGPROM (the State Committee for Light Industry) stated

> there are 545 enterprises with more than 1m (1.12m) employees ... According to our data, the USSR holds one of the first places in the world in the production of textile fibres and the commodities made from them. Thus, our country is ... in second place in the production of cotton fabrics ... During the last ten years the total production of fabrics has increased from 8.9bn m^2 in 1970 to 12bn m^2 or by nearly one fourth.

By 1992 this already looked outdated and inaccurate: Russian textile production was under half that claimed for the old Soviet Union, and was still falling. If Europe has a recession, then Russia has an economic disaster. This has already had extremely important consequences for the international cotton trade, as the states of the former Soviet Union have entered the world market to sell their cotton to obtain hard currency and basic commodities that cannot be provided through barter trade with Russia. Similar extreme difficulties beset the other countries of the former Soviet bloc, with Romania operating at only 50% capacity owing to lack of foreign currency to import cotton for milling. At present, 32% of Romanian cotton imports come from the US GSM-102 and PL-480 programmes.

Traditional payment arrangements with the former Soviet republics have broken down because the republics want to be paid in either US dollars or meat – both of which are difficult to supply from Eastern Europe. In 1993 Romania imported no cotton, and prospects for US sales continue to be tied to export credit assistance. In the former Soviet Baltic republics, Latvia, Lithuania and Estonia, prospects are a little brighter. February 1993 saw the first purchases of US cotton by Estonia, where the USDA's Export Credit Guarantee of $5 million has opened a new market for US cotton. But the US exports are clearly destined to form the kernel for the rejuvenation of the Baltic spinning mills.

Asia

Cotton traders themselves agree that no part of the European textile industry is likely to recover swiftly. They point out that in general, textiles can be produced in developing countries more cost effectively - especially in respect of labour costs - than in developed countries. And given modern machinery and sensible purchases of raw cotton, the end product in quality terms is not inferior to that produced in developed countries. Moreover, the governments of developing countries normally give substantial financial support to their textile industries as a route to foreign currency earnings. Far East developed countries, such as South Korea and Taiwan, are making efforts to support the textile industry in other countries, by, for example, establishing joint ventures in new textile-producing countries such as Vietnam, Indonesia, Thailand and Bangladesh. The combined cotton yarn production of Thailand and Indonesia alone equalled that of the entire EC in 1992, and is almost certain to surpass it in 1993/4. In these countries, Asian expertise and capital from Japan and Korea is being used to produce textiles on a viable basis and textile machinery is still being imported at a considerable rate: almost all the shuttled looms delivered in 1991 went to Asian countries; South Korea alone absorbed one-fifth of all the shuttleless looms produced in 1991, and was the largest textile investor worldwide in 1991. Thailand and Indonesia were also major importers, including of highly productive air-jet looms, although these two countries have yet to assume a similarly dominant position in weaving as they do in spinning: they are still steadily importing more fabrics. To some extent, therefore, the rise in yarn stocks in South-East Asia reflects the development of the industry, especially in Thailand and Indonesia, rather than its decline.

Japanese spinners and weavers, however, have started to reduce production from 1990 onwards, faced by declining profitability, with an estimated fall of 5% in 1992/93: the margin between Japanese futures prices for 40-count cotton and the 'A' Index fell from nearly 200 yen/lb in December 1991 to less than 100 yen/lb in November

1992. During the previous 12 years, such a small spinning margin was only observed between 1990 and 1991. In 1991 itself, Japanese spinning margins were increased as cotton prices fell whilst the members of the Japan Spinners Association maintained yarn prices by reducing production. Import growth and falling domestic demand stopped this strategy. The number of spinning mills in operation in Japan fell from 169 in 1980 to 120 in 1991 and is still declining. The USDA said in March 1992 that this could be explained by

> 1) steep production costs and labor shortages resulting in increased competition from imports of cheaper cotton textile goods. In particular, the flood of cheap foreign cotton textiles and knit imports, primarily from China, are causing serious damage to Japanese textile makers
> 2) a general economic slowdown lowering consumer purchases and
> 3) an increasing share for polyester filament woven fabrics chipping away at cotton's share of the Japanese textile market.

This is expected to worsen as technology, machinery and marketing improve in countries exporting to Japan: Pakistani yarn amounts to three-quarters of imports and is now competing in better-quality yarns. Japan plans to retaliate through technical expertise, niche marketing and yet more automation, for example by automating sewing. A measure of overseas relocation can also be expected. Japanese overseas investment in textile industries was estimated at $3.2 billion in 1989, almost twice the 1980 level (Fig. 6.7).

In south Asia, Pakistan has become a dominant supplier of yarn to

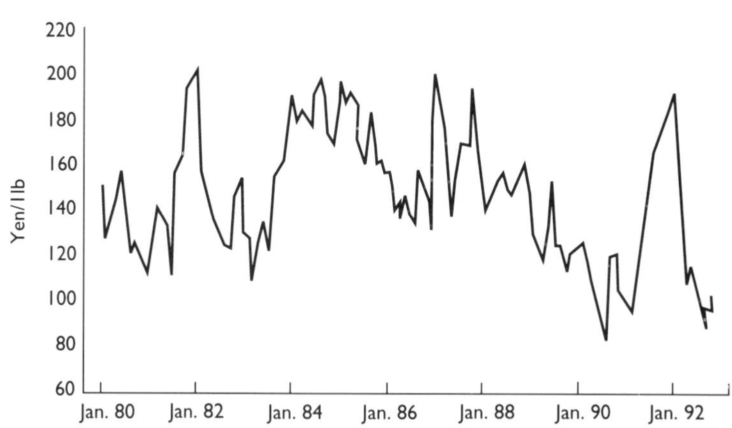

6.7 Yarn-cotton price spread: Japan futures, '40s minus the Cotlook 'A' Index.

Asia. The Asian Development Bank recently concluded that Pakistan supplied 80% of the Asian market of coarse and medium count yarn. Cotton has been available to Pakistani mills at prices significantly below world market levels. The price differential has been maintained through an export tax fixed by the government. This differential led to a price advantage of over 40% in recent years when cotton prices increased, although it has since fallen from that level to about 10%.

1992 saw the first signs that the international recession was seriously affecting the Asian textile industry. Although, the ITMF reported, world yarn production in the first quarter of 1992 remained stable, for the sixth successive quarter - which was a level 4-5% higher than the average of 1987 - this bald figure disguises an 18% fall in Korea and a general fall in Asia of 8.2%. Moreover, production figures do not tell the whole story by far. Inventories leapt upwards in 1992, reflecting the weak demand for yarn, increasing by 50% in 1991/92 and probably rising throughout 1992. The largest increase occurred in Pakistan (which partly explains their aggressive marketing of raw cotton during 1992), followed by Korea, Taiwan, and Japan. Asian fabric production fell by 6.9% in the first quarter of 1992, making it 5.1% lower than the first quarter of 1991. The dominant theme of Asian textile production in the 1990s appears to be a repositioning away from the relatively high wage economies of Japan and Taiwan towards cheaper locations. The overall effect, however, is almost certain to increase the Asian share of world textile production and of cotton of necessity within this.

The USA

In the USA, mill consumption has continued to gain ground throughout the period 1989-92, buoying the price of US raw cotton and surprising analysts worldwide. In 1990 the US textile industry employed 634 000 people with a total payroll income of $11.1 billion. Active cotton system spindles in place totalled 8.6 million bales, compared with 9.4 million bales in 1991, of which 3.4 million and 3.6 million respectively were dedicated to 100% cotton.

Yarn production actually increased by 11% in the first quarter of 1992 compared with 1991, and total mill fibre consumption rose by 14%. Cotton improved its share of mill fibre consumption from 25% in 1985 to 32% in 1991, resulting in a nearly 6% rise in cotton consumption in 1991, and another 8% was recorded in 1992. The fact that yarn prices rose during 1991/92 despite the steady increase in supply shows how sustained is the rising demand for cotton yarn in

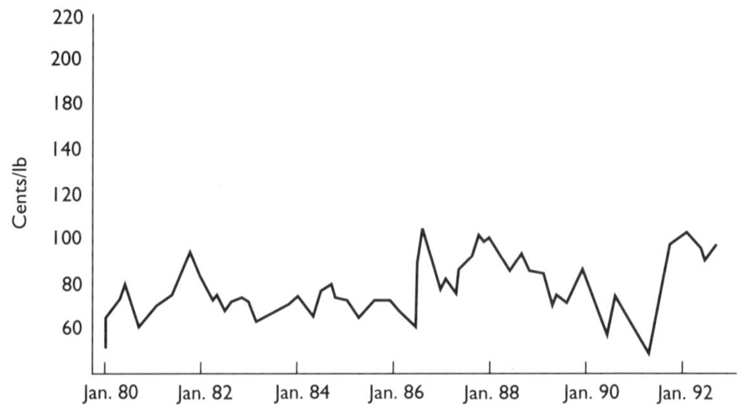

6.8 Yarn-cotton price spread: USA, carded '20s minus Memphis spot price.

the USA, being used for denims, cotton underwear, knits and fleece goods; and as capacity utilisation has increased, profit levels have increased with it (Fig. 6.8).

All this lent weight to the opinion that the USA was emerging from recession earlier than its competitors. Fabric production in the USA increased by 11.7%, making it 19% higher than in the first quarter of 1991, stocks fell by 6% and the Order Index for Fabrics, which is a key US indicator, gained 8% over the fourth quarter of 1991 and stood at 54% higher than the first quarter of 1991. For 1991/92, the US textile industry consumed 9.6 million bales of cotton, the highest level in 42 years. The US textile industry has performed extremely impressively over 1989–92 and is poised either to expand again when the recession properly ends, or at least to maintain its position throughout the 1990s.

The corporate net profits of textile companies as a percentage of sales have increased in 1990–92, and now stand generally above the market for the manufacturing sector as a whole. Indeed, textile manufacturing profits in the second quarter of 1992 were the highest in 20 years, and mills are using the profits to modernise and expand.

In the USA during 1991/92 efforts to exploit the continued satisfactory mill rates were intense. Textile mills in the USA produce approximately 19 billion sq. yards of woven cotton goods and more than 5 billion sq. yards of cotton knits annually. In 1991 over 50 companies used the Cotton Seal trademark for the first time, joining the many hundreds who already use it. Improvements in denim appearance, including fabric trials for enzyme treatments to create a stone-washed appearance at reduced cost, were widely important –

10% of the crop annually, and still increasing percentage of the cotton market - especially for slacks. The denim market worldwide was worth some $6 billion at retail in 1991, and cotton takes 95% of the denim market. Mid/lightweights are increasing their market share and improving appearance, with cotton's share currently at 52%. A series of retail campaigns, a good example being the 'Fabric of Our Lives' campaign in 1992, has maintained and increased consumer awareness of cotton. The USA is a market where clever, expert marketing really works, and some of the best has been carried out in favour of cotton.

Attention is now being focused on blending, for example with silks, to produce new appearances, and also the infusion of fashion colours and novelty yarns and discharge printing techniques. Cotton knits are a strong growth area for cotton fabrics. Improvements in shrinkage resistance and face-finishing techniques can be expected during the 1990s. Major mills have introduced all-cotton bedding programmes in the early 1990s, including the fashion of duvet covers imported from Europe and the 'bed in a bag' concept. Threadcounts in cotton sheets have increased in both the 180 and 200 count percales markets. Cotton towels continue to be the market leaders with a 95% market share; the trend is towards thicker towelling. Similar improvements have been noted in blankets, decorative fabrics, bath and accent rugs. Sophisticated market research carried out by Cotton Incorporated has given mills accurate appreciations of purchase preferences and plans at the retail level. The industry is researching the reapplication of cotton technology to carpet and rug production. Shedding has historically been the difficulty with cotton carpets; through blending of 90% cotton with synthetics, this has been reduced as a problem. Cotton is now able to comply with flame-resistance guidelines without the addition of topical finishes; such rugs are already in production. Research is being undertaken to improve drapability and softness of woven cotton fabrics. Shrinkage control remains a high priority in knitted cotton, dependent on the introduction of more sophisticated dyeing technology, which, coupled with compaction, is now holding shrinkage in finished garments to less than 5%.

Tables 6.5-6.8 and Fig. 6.9 give details of US production.

International textile trade

Overall in 1991, trade in textiles increased by 4%, almost half the growth registered in 1990 (7.5%) but still, most analysts believe, greater than any growth experienced in 1992/93. The international

Table 6.5 US textile employees: textile mill products (thousands)

Year	Jan.	Feb.	March	April	May	June	July	Aug.	Sept.	Oct.	Nov.	Dec.	Annual
1976	925	929	928	932	928	922	919	916	913	907	905	905	919
1977	905	908	912	916	915	914	914	908	907	909	909	908	910
1978	909	908	905	901	899	895	895	894	897	896	896	897	899
1979	899	897	893	889	888	887	886	883	881	885	887	889	889
1980	888	888	888	884	858	837	828	844	833	845	844	842	857
1981	841	839	838	842	843	846	856	849	827	814	804	794	833
1982	780	777	760	773	759	741	741	736	734	733	727	727	749
1983	726	726	730	733	736	743	745	751	753	758	760	762	744
1984	768	767	769	766	762	759	755	751	744	735	731	727	753
1985	720	712	706	708	703	700	698	698	697	699	701	702	704
1986	702	705	707	699	702	704	706	707	707	709	714	715	706
1987	718	722	725	719	721	724	730	728	730	731	733	733	726
1988	732	732	729	731	732	731	730	723	726	726	725	724	728
1989	728	728	729	728	728	729	730	727	723	724	721	719	721
1990	715	711	709	699	694	693	692	690	685	678	677	671	691
1991	667	661	660	660	665	665	671	670	670	672	673	674	667

Table 6.6 US textile employees: apparel and other textile products (thousands)

Year	Jan.	Feb.	March	April	May	June	July	Aug.	Sept.	Oct.	Nov.	Dec.	Annual
1976	1 329	1 332	1 341	1 341	1 333	1 330	1 321	1 311	1 302	1 294	1 296	1 294	1 319
1977	1 300	1 309	1 310	1 310	1 314	1 319	1 315	1 318	1 322	1 323	1 329	1 332	1 317
1978	1 328	1 330	1 334	1 338	1 336	1 335	1 322	1 333	1 336	1 331	1 332	1 336	1 333
1979	1 337	1 328	1 324	1 323	1 318	1 311	1 316	1 305	1 298	1 302	1 294	1 296	1 313
1980	1 305	1 313	1 316	1 302	1 264	1 261	1 254	1 263	1 261	1 256	1 253	1 250	1 275
1981	1 244	1 243	1 243	1 250	1 258	1 264	1 278	1 272	1 253	1 243	1 235	1 222	1 250
1982	1 201	1 201	1 186	1 165	1 165	1 161	1 141	1 151	1 149	1 148	1 141	1 140	1 162
1983	1 150	1 148	1 143	1 144	1 149	1 156	1 171	1 170	1 174	1 186	1 195	1 202	1 166
1984	1 207	1 213	1 218	1 226	1 217	1 209	1 206	1 200	1 181	1 178	1 178	1 186	1 202
1985	1 182	1 175	1 167	1 126	1 119	1 109	1 122	1 117	1 121	1 121	1 122	1 130	1 134
1986	1 133	1 122	1 117	1 107	1 105	1 101	1 103	1 102	1 102	1 104	1 101	1 110	1 109
1987	1 106	1 101	1 103	1 092	1 095	1 098	1 116	1 098	1 104	1 106	1 110	1 106	1 103
1988	1 105	1 104	1 106	1 096	105	1 093	1 091	1 085	1 085	1 093	1 088	1 090	1 094
1989	1 092	1 096	1 101	1 098	1 095	1 093	1 094	1 095	1 088	1 084	1 084	1 081	1 079
1990	1 053	1 045	1 037	103	1 048	1 045	1 041	1 039	1 039	1 032	1 023	1 017	1 043
1991	1 012	1 010	1 009	1 005	1 013	1 017	1 032	1 031	1 034	1 039	1 043	1 044	1 022

Source: US Dept of Commerce.

Table 6.7 Average weekly gross hours per US production worker: textile mill products (seasonally adjusted)

Year	Jan.	Feb.	March	April	May	June	July	Aug.	Sept.	Oct.	Nov.	Dec.
1976	41.6	41.1	40.8	39.3	40.4	40.1	40.3	39.5	39.0	39.3	39.6	40.0
1977	39.9	40.4	40.4	40.4	40.3	40.2	40.3	40.3	40.3	40.6	40.6	40.5
1978	40.0	40.4	40.5	40.9	40.5	40.4	40.4	40.5	40.5	40.4	40.3	40.5
1979	40.5	40.1	40.3	39.0	40.0	40.2	40.3	40.3	40.7	40.8	41.0	41.0
1980	41.5	41.1	40.8	40.3	39.7	39.1	38.8	39.2	39.7	39.9	40.1	40.6
1981	40.0	40.0	39.9	39.8	40.5	40.2	40.4	40.3	38.8	39.0	38.7	37.8
1982	32.3	38.2	37.6	37.7	37.9	37.8	37.7	38.1	38.1	38.3	38.8	38.9
1983	39.7	39.0	39.6	40.9	40.5	40.7	40.8	40.9	41.3	40.8	40.6	40.7
1984	40.6	40.8	40.6	41.2	40.0	40.0	39.8	39.4	39.2	38.7	39.0	39.1
1985	39.2	38.8	39.1	38.9	39.2	39.5	39.2	40.0	40.5	40.7	40.8	41.0
1986	40.8	40.6	40.7	41.1	41.0	40.8	41.0	41.2	41.4	41.4	41.4	41.6
1987	41.6	42.0	42.1	41.5	42.0	42.1	42.3	42.0	41.4	41.8	41.6	41.5
1988	41.5	41.6	41.2	41.5	40.9	40.8	41.0	41.0	41.0	41.0	41.0	40.5
1989	40.9	40.8	41.1	41.7	41.4	41.4	41.2	41.0	40.6	40.7	40.5	40.2
1990	40.3	40.2	40.0	40.0	40.2	40.4	40.2	40.0	40.0	39.9	39.3	39.4
1991	39.4	39.2	39.4	39.6	40.2	40.8	41.0	41.4	41.3	41.3	41.4	41.6

Source: US Dept. of Commerce.

Table 6.8 Average weekly gross hours per US production worker: apparel and related products (seasonally adjusted)

Year	Jan.	Feb.	March	April	May	June	July	Aug.	Sept.	Oct.	Nov.	Dec.
1976	36.9	36.6	36.4	35.5	36.2	35.9	35.5	35.3	35.3	35.3	35.4	35.4
1977	34.4	35.9	35.6	35.5	35.6	35.8	35.5	35.6	35.5	35.8	35.8	35.8
1978	33.9	35.4	35.8	36.3	35.8	35.7	35.7	35.7	35.8	35.3	35.7	35.5
1979	35.4	35.4	35.3	34.4	35.1	35.2	35.3	35.3	35.2	35.4	35.3	35.6
1980	36.0	35.9	35.3	35.8	35.3	35.2	35.1	35.1	35.1	35.4	35.2	35.5
1981	36.1	35.6	35.7	35.5	36.0	36.1	35.9	36.1	35.2	35.3	35.5	35.1
1982	31.4	35.5	35.0	34.7	34.8	35.1	35.1	35.0	35.2	35.1	35.0	35.1
1983	36.6	35.2	35.6	36.5	36.1	36.2	35.9	36.3	36.7	36.6	36.7	36.6
1984	36.6	36.9	36.7	37.4	36.5	36.4	35.8	36.0	35.9	35.9	36.0	36.0
1985	36.2	35.9	36.1	36.0	36.2	36.3	36.4	36.6	36.6	36.6	36.8	36.8
1986	36.7	36.3	36.5	36.8	36.6	36.6	36.6	36.6	36.8	36.8	36.9	37.0
1987	37.0	37.4	37.0	36.4	37.1	37.0	37.2	37.2	36.4	37.3	37.1	37.1
1988	36.8	37.0	37.0	37.3	36.9	37.0	37.0	36.9	37.1	36.9	37.0	36.8
1989	37.0	37.1	36.9	37.6	37.1	37.1	37.0	37.0	37.0	36.9	36.8	36.3
1990	36.6	36.6	36.3	36.4	36.5	36.7	36.6	36.6	36.6	36.4	36.3	36.6
1991	36.3	36.5	36.6	36.4	36.7	36.9	37.0	37.3	37.3	37.4	37.3	37.5

Source: US Dept. of Commerce.

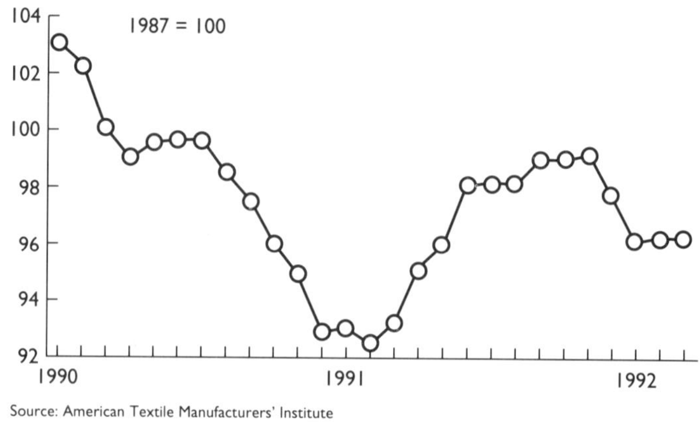

6.9 US apparel production 1990–92 (source: *ATMI*).

textile industry is adjusting to the idea that indefinitely continued growth in textiles has stopped, and is unlikely to resume with the same ferocity in the remainder of the century. 1992 saw an increased pressure on ITMF members from non-ITMF members. In Western Europe the dominant concerns are the steady contraction of the indigenous apparel industry and the increase in outward processing business. By contrast to the domestic production picture, the EU still represents a potent import market. Imports increased by 17% in value terms in 1991, up to ECU 39 billion, and this without including imports from the former East Germany. Import volume increased by less, 9%, accounted for by import inflation and possibly a shift to lighter and more expensive products. The net result however was a deficit of nearly ECU 12.6 billion by the end of 1991 (Fig. 6.10).

Textile imports increased less (4%) than clothing (26%), which was a matter of grave concern to the European industry, especially as this occurred throughout the EC. There was some geographical dispersion of the results: textile imports decreased in value in France, Italy, the UK and Portugal, by 1–5%; imports rose in Germany, which imports 43% of the EC's textiles and clothing, by 10.9% for textiles and 28.4% for clothing, probably to meet new demands from the former East Germany (producing a textile deficit of $15 billion, 50% up on 1990), and there was a 25% rise in Spain, which did not really begin to feel the impact of the recession until 1993. The Yugoslav civil war imposed its own problems, resulting in Yugoslavian imports falling by 10% in textiles and 15% overall in 1991, and undoubtedly far more in 1992, before virtually collapsing in 1993, except in Slovenia. This caused European companies to change their outward processing trade

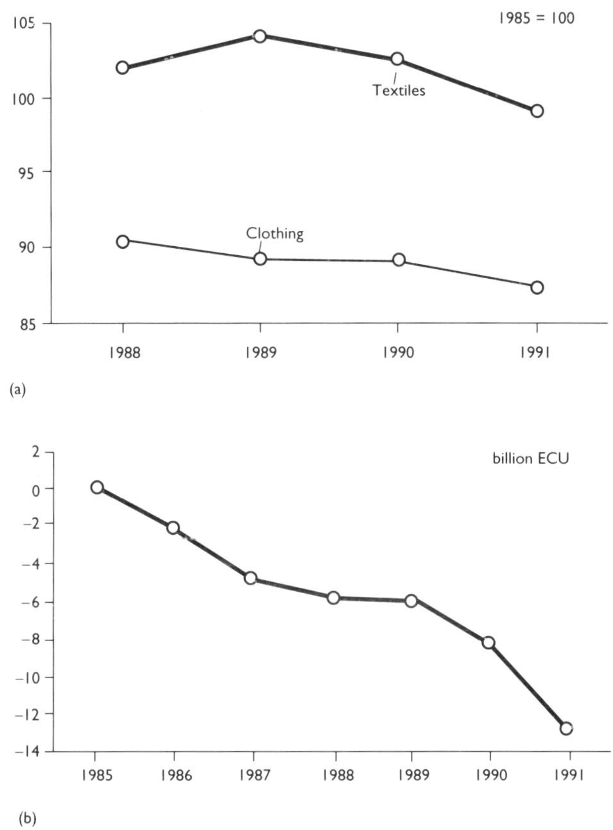

6.10 (a) EC textile and clothing production, 1988–91. (b) EC trade balance, textiles and apparel, 1985–91 (*source*: OETH).

to Central Europe. Outward processing imports have increased as a share of EC imports in recent years, especially for Germany, enabling the EC's textile and clothing industry to benefit from lower-cost manufacturing in Eastern Europe and North Africa.

Amongst the importers to Europe, Turkey's rate of export growth slowed, and US exports continued, aided in both 1991 and 1992 by a weak dollar. Austria and Switzerland amongst EFTA countries saw their textile exports decline, although they did achieve some slight growth in clothing exports. Growth was seen from China, Pakistan and Bangladesh in both textiles and clothing, and from India in clothing. The Asian countries of the Association of South East Asian Nations (ASEAN) continued their push into European markets, only Thailand amongst the major producers falling back slightly, with Hong Kong and Taiwan maintaining their expansion of clothing exports. The recession appears to have had the effect of consolidating the position of the major exporters, with the minor exporters such as Macao (level) and Singapore (strong increase) enjoying more mixed fortunes in their export campaigns.

EC exports at a time of intense price competitiveness and world recession could hardly be expected to be anything other than disappointing, and in fact on the face of the figures the EC did relatively well. What is suspected is that the figures themselves conceal a large amount of dumping to remove unwanted stocks, without which the decline would have been more serious. Exports to the USA fell by 11% in 1990 and again by 7% in 1991, which was especially serious given that there was no major textile recession in the USA. EFTA countries received 1% more clothing but 6% fewer textile products. Exports to Japan fell by 9%, confirming the effects of the Japanese recession. It appears that European textile products and clothing are subject to import elasticities of demand in respect of income well in excess of unity, and are 'luxury goods'. The EC continues to make efforts to export to other Asian countries, and succeeded in increasing exports to South Korea by 6% and to Hong Kong by 3%, but not to Taiwan, where the figure fell by 6%. Any unification of Korea can be expected to serve as a major boost to demand. The EC also brooked fierce price competition to increase exports to the Middle East in 1991, by 16%. This was probably accounted for by an increased acceptance of Western textile goods in the countries forming part of the coalition against Iraq.

As a result of these trends, the EC remained resolutely in debit for its international trade in 1991 and 1992. It was ECU 12.4 billion in 1991 and RER estimated that it reached at least ECU 14 billion in 1992. This was a growth of 68% in 1991 and a further 10% at least in 1992. Two years have seen the deficit grow by ECU 7.3 billion and 1992 will add a further ECU 1.6 billion at least. This recession has created a long-term trading deficit for the EC in clothing in particular, some ECU 4.9 billion in 1991 and more in 1992. It is likely that as economic recovery starts, the income elasticity of demand for imports will not be as great as for domestic products and therefore the deficit should stabilise at the ECU 6 billion mark for at least a few years. In the short term, the EC, as most industrialised countries, has a bleak outlook, with no signs of recovery during 1992/93. Consumption remains relatively weak. Import demand has been sustained.

The difficulties being faced by European countries in their textile industries and trade can be illustrated especially well by what has happened in the UK. Textile output continued to fall in the early 1990s, although the rate of decline may be slowing. Recession is hitting the UK very hard with textile demand reduced across a range of industries. The sharp swings in value of the US dollar, offset to some extent after September 1992 by the devaluation of the pound, led to variance in competitiveness against Asian competitors which was difficult to predict. The recession caused a fall in imports overall in value terms by 6%, but Asian cotton imports continued to be buoyant. Exports continued to fall throughout 1991 in every major market except Germany, and evidence suggests that 1992 showed a similar pattern.

Evidence from Eastern Europe and the former Soviet Union is mixed. It is known that the former Soviet textile industry, which was 85% based in Russia, has experienced very severe output decline. Accurate statistics are difficult to find. The USDA said in May 1993 that:

> Russian distribution channels continue to experience problems in delivering export goods to market. According to industry sources, an export bottleneck has developed at Russian seaports, with thousands of railcars waiting to be unloaded. In far eastern Russia alone, 11,300 railcars, three times the normal number carrying exports, have accumulated at the ports of Vladivostok, Nakhodka, Vaninn and Kholmsk. The railcars will continue to stand idle until a central distribution network is set up to handle the planning of the export of goods from Russian ports and to restore lost ties with foreign countries. This inability to properly control the flow of exports could (i) delay hard currency earnings (ii) create internal shortages where barter arrangements were made and (iii) damage relationships with Russia's foreign trading partners. (USDA *Monthly Report*, May 1993)

But in 1992 with those former Soviet countries that the EC practises outward processing (with the finished goods being reimported), there was significant textile trade expansion: Poland (up 50%), Hungary (up 14%) and the Czech and Slovakia Republics (up 76%). Some expansion was seen in Maroc countries, of 12%.

US imports increased during 1991, up 4.7% from 1990, reaching a new record of 12.2 billion sq. metres, and representing a very welcome home for numerous textile exports from Asia in particular. These imports continued to rise in 1992, probably by about 30% over the year. There was little change in the structure or price of the imports. Exports continued to surge, however, rising 17% in 1991 compared with an even more impressive 24% in 1990 and rising continuously through 1992/93, but the net result is an increase over time in the size of the US cotton textile trade deficit. The trend is towards more apparel exports and less fabrics. This reflects the increase in '807' trade in which US pieces are exported to Mexico and

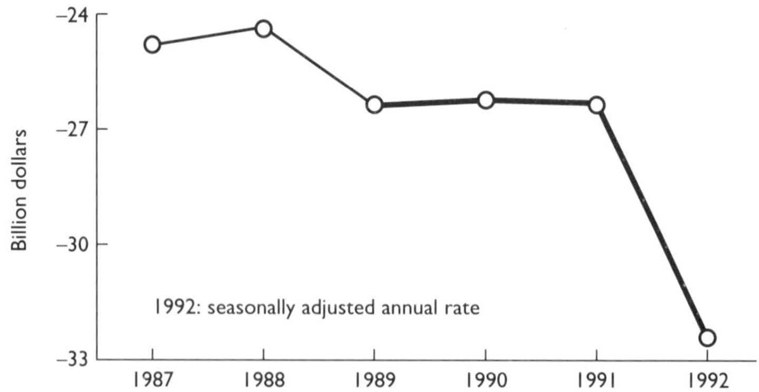

6.11 US trade balance, textiles and apparel, 1987-92 (*source: ATMI*).

the Caribbean, assembled, and then shipped back to the USA for sale as completed apparel. By the end of 1993, arrangements for the creation of the North American Free Trade Agreement were well advanced, and both textile and apparel agreements had been reached (Fig. 6.11).

The NAFTA Agreement was finally passed in late 1993 and this is almost certain to boost all textile trade in North America, probably, despite assurances to the contrary from the US Administration, at the expense of US domestic manufacture.

Competition for cotton

Cotton has been the main fibre in the textile market for over 200 years - accounting for about half of total world fibre production - and looks set to continue its dominance into the twenty-first century. Throughout the world the main competition for cotton is man-made fibre. The first artificial silk was manufactured in France in 1870 by Count Chardonnet; towards the end of the nineteenth century artificial fibres were produced from cellulose, and during the 1920s artificial staple fibres were produced. 'Viscose' now refers to the artificial silks, made from cellulose (wood pulp), which include acetate and triacetate. Then there are the synthetic fibres: acrylic, elastane, modacrylic, nylon, polyester and polyolefin. Nylon was the first true synthetic, invented in 1938. Shortly afterwards, ICI developed polyester. At first, nylon was used for military needs - parachutes - and nylon stockings.

Production of man-made fibres has increased dramatically since 1945, and by comparison wool, silk and linen production is statistically insignificant. Before World War II production was minimal, but by the early 1950s man-made fibres supplied one-sixth by weight of the world's textile needs, by the 1960s a third, and by the 1970s over half. Europe saw the most dramatic growth. In the UK for example, at that time a shortage economy where growth was production-led, people bought what could be made available in the immediate post-war years. The middle class, now bereft of the domestic help to which it had had access before World War II on a wide basis, valued the new easycare textiles and clothing, for their durability and their availability also. Cotton was not available in the quantities required. During the 1950s and 1960s, polyester, acrylic and nylon enjoyed extraordinary growth in consumption. Nylon was used for end uses to which cotton would have been much better suited; the best examples are bed linen, underwear and shirts. Crimplene, for example, was used for suits. Polyester, and the other synthetic fibres marketed under such names as Kodel, Teron and Vycron, appeared to match consumer needs for convenience. In 1968 'permanent press' clothing was being introduced, and in the optimistic atmosphere of the 1960s it seemed impossible that man-made fibres would not relatively quickly take over ever more substantially from cotton. In 1968 total worldwide production of nylon and polyester reached 3.2 billion and 2.5 billion lb respectively; cotton production fell to 22 billion lb. Prentice in 1972 suspected as much when he pointed out that production of man-made fibres was under the control of a few firms willing to spend on research and promotion, providing technical and marketing support throughout the production process, and outspending the cotton industry 15 : 1. This compared with the diversity of the cotton-producing industry and its dependence on mills, manufacturers and retailers for marketing. In the USA, development differed because there was no such cotton shortage. Synthetics therefore developed more logically and in response to specific needs. Polyester/cotton blends swiftly became popular, and nylon never caught on to the same degree as in Europe.

Attention in the USA was already being paid to the issue in the 1960s, with the passing in 1966 of the Research and Promotion Act. In 1971 the formation of Cotton Incorporated demonstrated the determination of the national industry to fight back with approved standards and integrated marketing. Two years later the 'Seal of Cotton' logo was introduced. Television advertising for cotton, aimed especially at women aged 18–49, was launched.

Events then conspired to assist cotton: the first major headaches for synthetics began with the oil crisis of 1973. Consumer reaction, moreover, was beginning to set in. By 1976 the share of nylon in the shirt market was down to 25% and falling. Polyester/cotton blend, already in widespread use in the USA, was a more satisfactory product for shirts in particular than nylon: it was durable, easy to iron, and yet much more comfortable to wear than nylon: 60% cotton/40% polyester blends were introduced in 1974 by Spring Mills in the USA which reached the shops the following year. Cotton Incorporated technology was built in to products such as sheets. Recognition of cotton products increased steadily during the late 1970s and on into the 1990s - by 1992 recognition of the Cotton Seal was almost 80% in the USA. Towels became a cotton market again, and cotton broke into new markets such as non-woven personal hygiene products, mattress pads, cotton window coverings, upholstery fabrics and wall coverings. The baby boom generation, US cotton advocates justifiably claim, just adore cotton products with which to pamper themselves in their few free moments; they do indeed as the 1987 advertisement campaign claimed, 'take comfort in cotton', just as they did in the 1960s - although this grossly simplifies matters - when they rebelled by wearing cotton jeans (see Reich, 1986). For the 1990s, cotton's position looks both promising and more importantly unchallenged, with a fresh generation of youngsters growing up dressed in cotton and ever more fashion-conscious. Cotton Incorporated reported that in 1991 cotton retail sales grew by 2.3%, compared with a fall in other fibre sales of 1.9%, and a total retail market growth of 0.4%. In 1992 cotton took 56% of the textile (excluding carpet) and apparel markets in the USA, compared with 55% the year before. In home fabrics cotton reached a 61% share, and in childrenswear the share reached 59%, up from 56%.

Meanwhile the position for synthetics deteriorated accordingly. Plans for a blended denim jean collapsed. The chemical companies worldwide found after 1973 that they had invested in too much capacity, and the dynamic growth of man-made fibres reversed. Products became debased as competition increased across all synthetic textiles, with a circle of consumer discontent resulting. Consumers cited pilling, snagging, static electricity and yellowing as reasons for rejecting synthetics. The synthetic fibre industry had to fight a series of unfavourable images about their products: as Newsweek commented in 1986 'by the mid-1970s, synthetics had become synonymous with double-knit leisure suits and a certain all-American sleaze' (Newsweek, 21 July 1986). Yet amidst this retreat some new sense was emerging, based on a rational assessment of which uses synthetics were best suited for and which they were not.

In addition, the synthetic manufacturers sought to gain 'fashion' status for their products, forming the Polyester Fashion Council in the USA – their aim was not fully realised until the widespread introduction of the lightweight fabric Lycra in the 1990s. But some uses became clearer. Carpets, for instance, could make good use of nylon's resistance and molecular strength. Nylon now has 10% of the volume and 20% of the value of the world carpet market. About 98% of women's fine hosiery is nylon. Kevlon, a derivative, is used for personal body armour. Anti-flam clothing is oil-based. For mountain wear, experts choose certain types of microporous nylon; similarly for certain types of linings. What now determines the choice of fibre is its suitability for the task, and there is now competition between cotton and synthetics only in those areas where there is no clear choice. There are therefore certain end uses where cotton has gained ground: nightwear, for instance – although this is a declining market.

Certainly, man-made fibres have the advantage of uniformity, freedom from waste, and relative stability in supply and price – which does mean that cotton's erratic price sometimes takes it below man-made fibres. Moreover, the raw materials of man-made fibres themselves – wood pulp, oil, coal and gas – are themselves subject to wide price fluctuations. Secondly, the price of the fibre is not a significant percentage of the cost of the finished article. The supply of man-made fibre cannot be adapted to changing circumstances so easily as cotton, because the factories have a substantially fixed output, at least to remain economic, and take several years to come into production; cotton field planting can simply be adjusted on an annual basis, usually with a wide range of available alternative crops. The shortage of available land, of water, competitive crops and government policy limit the production of cotton, whilst it is long-term investment that limits man-made fibre production.

Most of the regenerated fibres are made from natural cellulose, although some protein fibres are still produced. Pure cellulose from cotton linters was originally used; now the main source is wood pulp, which is 50% cellulose and is cheaper. The most common regenerated cellulose fibre blend is viscose rayon. This material is reasonably strong, between wool and cotton, but loses half its strength when wet, and it accepts dye and moisture. It is a low-priced competitor for cotton and wool, and many types of staple which have been developed are widely used in blends with other fibres. In addition there is cuprammonium rayon, with a finer filament but expensive to produce; it has been replaced by synthetic nylons in widespread use.

Some types of viscose rayon have additional properties, such as chemical crimp and additional strength. This makes the material suitable for industrial uses, such as reinforcing fabrics for tyres and conveyor belts. 'Polynosic' or 'Modal' rayons have a wet strength equivalent to cotton, and have properties more akin to good quality cotton, but they are more expensive to produce. Many of the man-made fibres are suitable for carpets, carpet backing, some textile uses, sacking and ropes. All these tend to be based on oils and oil derivatives, defined by their molecular structure.

In the 'acetate' process, the cellulose is converted into cellulose acetate, which remains chemically unchanged when extruded; acetate fibre is thus chemically different from cellulose. Cellulose acetate is not as strong as rayon nor as resistant to abrasion, but it is more resistant to general wear and handles like silk. It is naturally crease-resistant, but has a low melting point and needs to be handled at temperatures well below the boiling point of water. Its dye affinity is quite different from that of viscose or cotton, and special 'disperse' dyes are needed.

Regenerated protein fibres have been made from milk casein, groundnuts, soyabeans and maize. They have some of the properties of wool, but when wet their strength is negligible, and production of this type of fibre has almost ceased.

The last decade

The threatened obliteration of the cotton industry has not happened. Cotton has fought back with a vengeance. Why? Firstly, since 1980 the competitive position of cotton has dramatically improved. This has been due to favourable moves in customer choice for clothing in the developed world, and increasingly in all Western countries - although not in Asia - away from man-made fibres. It is extremely important to watch what happens in Eastern Europe and the former USSR in terms of relative consumption. It seems likely that after a prolonged period of low consumption, these countries will adopt very similar demand patterns to those of the other developed countries. In Africa and Latin America, the existing domestic cotton industries will have an ever greater demand for the product.

The use of cellulosic fibres, the FAO reported, declined from the late 1970s onwards; but total synthetic fibre consumption increased rapidly, increasing its share in total fibre use from 29% in 1975 to 36% by the mid-1980s. Per capita use of synthetics increased more rapidly than cotton, reaching 2.6 kg/person/year by the mid-1980s. Developing countries also expanded their exports of synthetics more rapidly than of cotton, from 80 000 tonnes in 1975 to 1.35 million tonnes by 1986.

Cotton's main synthetic competitor remains polyester, which is also subject to rapid price fluctuations. It took 22% of the US domestic market in 1991. In 1988/89, for instance, short supplies of ethylene glycol and paraxylene, the base chemicals for polyester, pushed costs up and made cotton highly competitive on price. This trend has continued: although polyester prices have been constant or falling since August 1990 (as reported by Cotton Outlook), the relatively much greater fall in cotton prices has resulted in a rise of polyester prices as a percentage of the Cotlook 'A' Index from 94% to 141% over two years (US polyester quotes) and similar figures (58–90%) and (106–131%) for Taiwan and Italy were observed. This was expected to continue to bolster cotton's market share, but in fact cotton's share of world fibre use fell in 1992. Despite a 20 c/lb decline in cotton prices in 1991 cotton consumption's 1% rise led to a fall in its share from 48.7% to 48.2%.

The trend towards lighter fabric weights across all sectors of the textile industry is very likely to continue. Immediately after World War II, the average textile weight was 14 oz/sq. yard; by the 1990s it was 7-9 oz/sq. yard. The reasons lie in the development of microclimates inside offices, homes and factories which have decreased the need for protection against a generally cold environment and increased the demand for comfortable clothing.

Another important aspect to the competition between cotton and synthetics is the environmental aspect. There are numerous vested interests on both sides of this debate. ICI has concluded that there is no 'green' fibre. Cotton is notorious for the amount of fertiliser and chemicals that it demands for economic production. Cotton uses more energy per square metre of finished textile than synthetics: the consumption of oil by the textile industry is less than 2% of the total worldwide. There is now intense regulation and competition to reduce emissions from production. Moreover man-made fibres are generally longer-lived in use and require less replacement. Even nylon is biodegradable if reduced into its separate components and remade. At present this process is expensive but it is highly likely to reduce in price over time. It seems certain that unless emotional considerations override factual ones completely, the environmental debate will not easily be settled in favour of cotton at a scientific level - nor in favour of synthetics.

Only about 50% of cotton is actually used for cotton fabrics and only 25% of the man-made fabrics. Man-made fibres have largely replaced cotton for tyre cords; plastic insulation has largely replaced cotton in electrical wiring, but information on industrial consumption is difficult to find. For example, statistics on production are frequently in area whereas consumption figures are given in weight, and the conversion figures vary substantially. Another example is the difficulty of ascertaining mill waste in cleaning, carding and combing.

The conclusion most analysts appear to reach is that since 1945 there have been three ages of synthetics: the first of novelty; the second, a rejection on aesthetic grounds; and the third, an era of consumer choice based on suitability for the task.

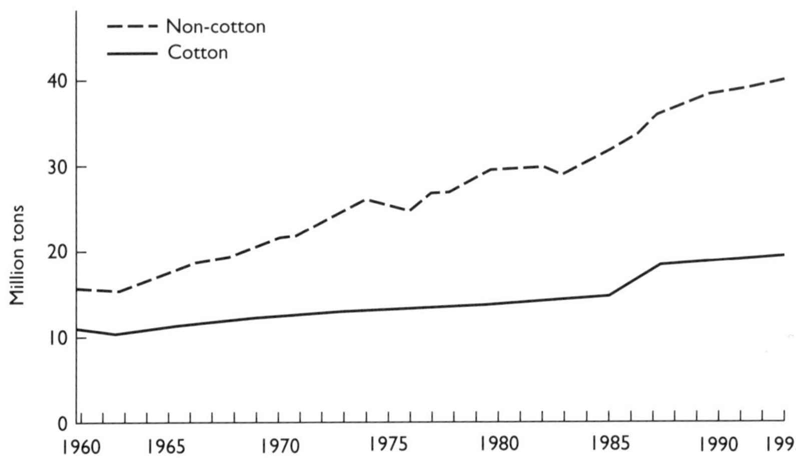

6.12 World textile fibre consumption (end use).

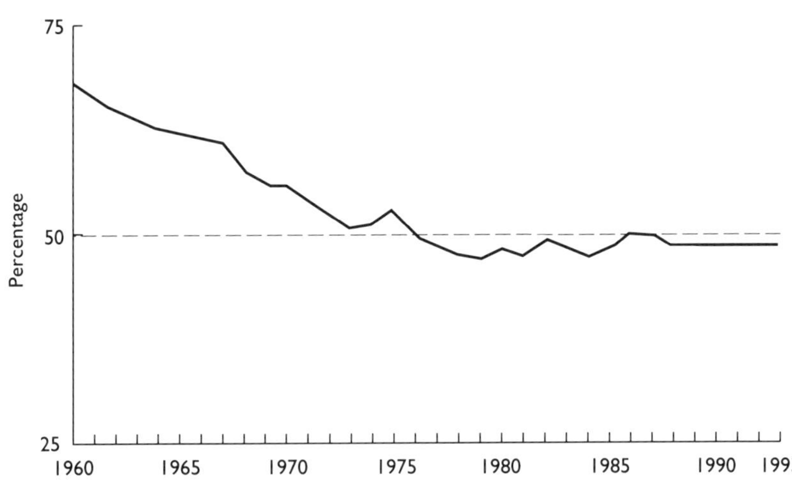

6.13 Cotton's share of the market: world end use consumption.

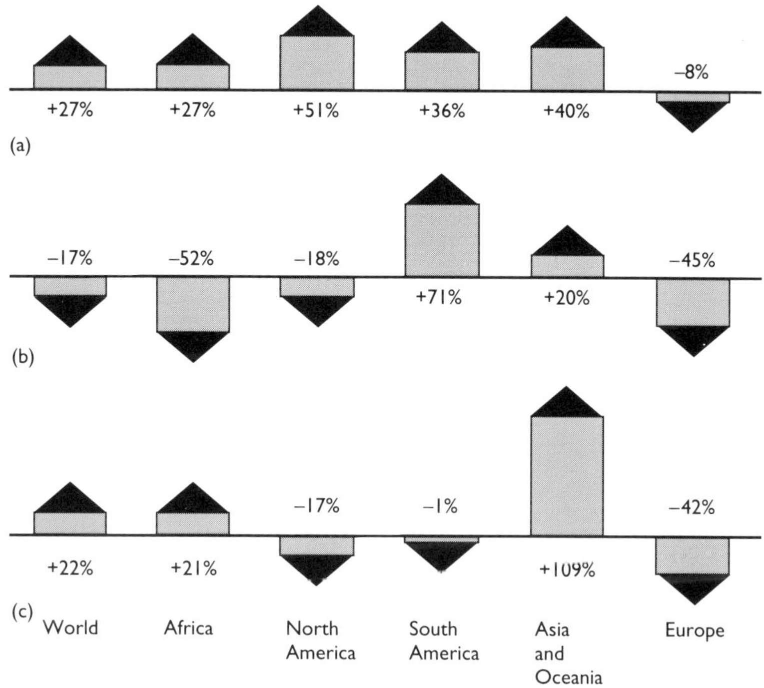

6.14 Changes in raw materials consumption, 1981-91: (a) raw cotton; (b) cellulosic fibres; (c) synthetic fibres.

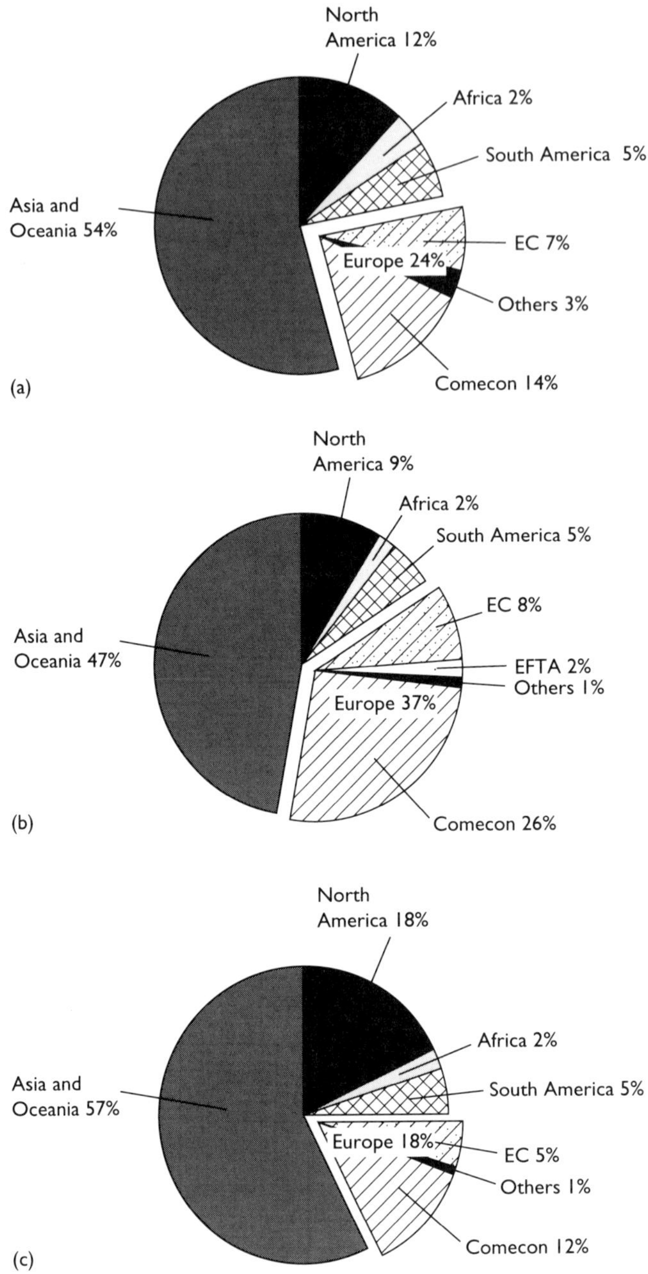

6.15 Raw material's consumption, 1990: (a) raw cotton, world total 17 612 000 tonnes; (b) cellulosic fibres, world total 1 528 900 tonnes; (c) synthetic fibres, world total 4 495 000 tonnes.

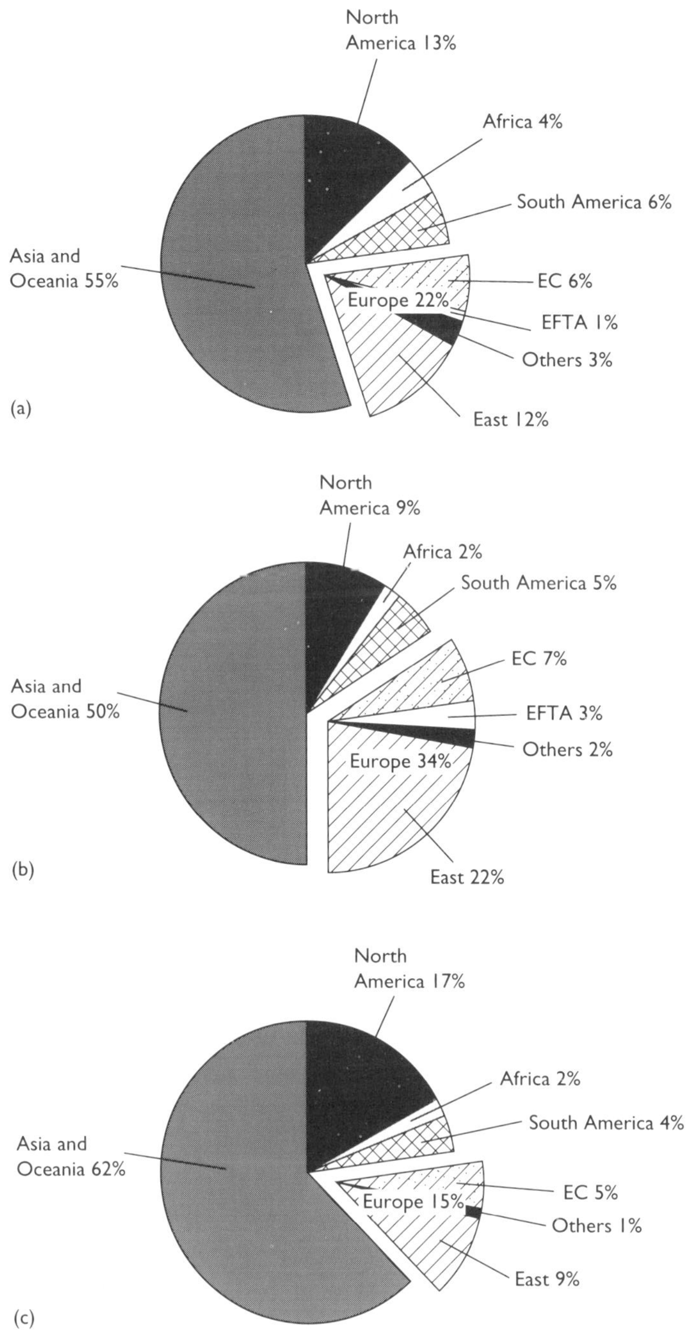

6.16 Raw materials' consumption, 1991: (a) raw cotton, world total 17 647 000 tonnes; (b) cellulosic fibres, world total 1 464 000 tonnes; (c) synthetic fibres, world total 4 546 200 tonnes.

Table 6.9 World consumption of major textile fibres

Year	Total (000 tonne)	Cotton (000 tonne)	Wool (000 tonne)	Chemical fibres[a] (000 tonne)			Cotton (%)	Wool (%)	Chemical fibres (%)		
				Subtotal	Non-cellulosics	Cellulosics			Subtotal	Non-cellulosics	Cellulosics
1960	15 153	10 356	1 495	3 302	702	2 600	68.3	9.9	21.8	4.6	17.2
1961	15 102	10 085	1 505	3 512	830	2 682	66.8	10.0	23.3	5.5	17.8
1962	15 339	9 902	1 501	3 936	1 080	2 856	64.6	9.8	25.7	7.0	18.6
1963	16 003	10 147	1 475	4 381	1 331	3 050	63.4	9.2	27.4	8.3	19.1
1964	17 256	10 830	1 460	4 966	1 687	3 279	62.8	8.5	28.8	9.8	19.0
1965	18 182	11 318	1 473	5 391	2 052	3 339	62.2	8.1	29.6	11.3	18.4
1966	18 796	11 539	1 545	5 712	2 371	3 341	61.4	8.2	30.4	12.6	17.8
1967	19 212	11 695	1 473	6 044	2 730	3 314	60.9	7.7	31.5	14.2	17.2
1968	20 434	11 763	1 565	7 106	3 578	3 528	57.6	7.7	34.8	17.5	17.3
1969	21 248	11 911	1 604	7 733	4 178	3 555	56.1	7.5	36.4	19.7	16.7
1970	21 740	12 104	1 500	8 136	4 700	3 436	55.7	6.9	37.4	21.6	15.8
1971	23 029	12 485	1 480	9 064	5 609	3 455	54.2	6.4	39.4	24.4	15.0
1972	24 412	12 898	1 578	9 936	6 377	3 559	52.8	6.5	40.7	26.1	14.6
1973	26 032	13 289	1 443	11 300	7 640	3 660	51.0	5.5	43.4	29.3	14.1
1974	25 271	12 990	1 262	11 019	7 487	3 532	51.4	5.0	43.6	29.6	14.0
1975	24 720	13 050	1 358	10 312	7 353	2 959	52.8	5.5	41.7	29.7	12.0
1976	26 538	13 212	1 515	11 811	8 601	3 210	49.8	5.7	44.5	32.4	12.1
1977	27 029	13 121	1 478	12 430	9 149	3 281	48.5	5.5	46.0	33.8	12.1
1978	28 260	13 429	1 481	13 350	10 032	3 318	47.5	5.2	47.2	35.5	11.7
1979	29 460	13 917	1 558	13 985	10 614	3 371	47.2	5.3	47.5	36.0	11.4

1980	29 578	14 293	1 567	13 718	10 476	3 242	48.3	5.3	46.4	35.4	11.0
1981	29 731	14 124	1 576	14 031	10 827	3 204	47.5	5.3	47.2	36.4	10.8
1982	28 881	14 234	1 556	13 091	10 145	2 946	49.3	5.4	45.3	35.1	10.2
1983	30 166	14 548	1 612	14 006	11 076	2 929	48.2	5.3	46.4	36.7	9.7
1984	31 253	14 832	1 621	14 800	11 804	2 996	47.5	5.2	47.4	37.8	9.6
1985	32 814	15 769	1 625	15 420	12 489	2 931	48.1	5.0	47.0	38.1	8.9
1986	35 033	17 502	1 745	15 786	12 927	2 859	50.0	5.0	45.1	36.9	8.2
1987	36 616	18 263	1 787	16 566	13 741	2 825	49.9	4.9	45.2	37.5	7.7
1988	37 414	18 306	1 830	17 278	14 394	2 884	48.9	4.9	46.2	38.5	7.7
1989	38 360	18 792	1 887	17 681	14 804	2 877	49.0	4.9	46.1	38.6	7.5
1990	38 480	18 801	1 964	17 715	14 869	2 846	48.9	5.1	46.0	38.6	7.4

[a] Production plus fibre trade balance is used as a proxy for chemical fibre consumption.
Sources: ICAC, Commonwealth Secretariat and Fibre Economics Bureau.

Table 6.10 World cotton and non-cotton textile fibre consumption (end use)

	Cotton (kg per capita)	Non-cotton (kg per capita)	Cotton (000 tonnes)	Non-cotton (000 tonnes)	Market share of cotton
1960	3.44	1.59	10 355.8	4 797.0	68.34
1961	3.30	1.64	10 085.1	5 017.0	66.78
1962	3.19	1.75	9 802.0	5 437.0	64.55
1963	3.19	1.84	10 146.5	5 856.0	63.41
1964	3.33	1.98	10 829.8	6 426.0	62.76
1965	3.41	2.07	11 318.4	6 864.0	62.25
1966	3.41	2.14	11 538.5	7 257.0	61.39
1967	3.38	2.17	11 695.3	7 517.0	60.87
1968	3.33	2.46	11 763.4	8 671.0	57.57
1969	3.31	2.59	11 911.3	9 337.0	56.06
1970	3.29	2.62	12 103.6	9 636.0	55.68
1971	3.33	2.81	12 484.8	10 544.0	54.21
1972	3.36	3.00	12 898.4	11 514.0	52.84
1973	3.40	3.26	13 288.5	12 743.0	51.05
1974	3.26	3.08	12 989.6	12 281.0	51.40
1975	3.21	2.87	13 049.6	11 670.0	52.79
1976	3.19	3.22	13 212.0	13 326.0	49.79
1977	3.11	3.30	13 121.1	13 908.0	48.54
1978	3.13	3.46	13 429.4	14 831.0	47.52
1979	3.19	3.56	13 916.7	15 543.0	47.24
1980	3.22	3.44	14 292.6	15 285.0	48.32
1981	3.13	3.45	14 124.0	15 607.3	47.51
1982	3.09	3.18	14 233.6	14 647.0	49.28
1983	3.11	3.34	14 548.2	15 617.6	48.23
1984	3.12	3.45	14 831.6	16 421.1	47.46
1985	3.26	3.52	15 768.8	17 044.8	48.06
1986	3.56	3.57	17 502.4	17 531.0	49.96
1987	3.65	3.67	18 263.1	18 353.0	49.88
1988	3.60	3.75	18 305.8	19 108.0	48.93
1989	3.63	3.78	18 792.4	19 568.0	48.99
1990	3.57	3.74	18 800.8	19 679.0	48.86
1991	3.53	3.75	18 864.5	20 067.4	48.46
1992	3.55	3.77	19 301.6	20 451.7	48.55
1993	3.58	3.78	19 731.4	20 880.7	48.59

Table 6.11 Net US consumption and market share of major fibres

Calendar year	Total bales	Cotton		Wool		Man-made fibres	
		Bales (million)	%	Bales (million)	%	Bales (million)	%
1972	23.978	8.670	36.2	0.584	2.4	14.724	61.4
1973	25.175	8.086	32.1	0.432	1.7	16.657	66.2
1974	21.876	7.117	32.5	0.296	1.4	14.463	66.1
1975	21.340	6.700	31.4	0.327	1.5	14.313	67.1
1976	23.482	7.675	32.7	0.427	1.8	15.380	65.5
1977	24.609	7.229	29.4	0.441	1.8	16.939	68.8
1978	25.262	7.355	29.1	0.250	1.0	17.657	69.9
1979	24.796	6.946	28.0	0.142	0.6	17.708	71.4
1980	23.559	6.987	29.7	0.422	1.8	16.150	68.6
1981	23.317	6.873	29.5	0.266	1.1	16.178	69.4
1982	21.612	6.522	30.2	0.450	2.1	14.640	67.7
1983	26.215	7.729	29.5	0.557	2.1	17.929	68.4
1984	26.957	8.237	30.6	0.709	2.6	18.011	66.8
1985	28.437	8.782	30.9	0.758	2.7	18.897	66.5
1986	31.321	10.286	32.8	0.827	2.6	20.208	64.5
1987	34.058	12.041	35.4	0.823	2.4	21.194	62.2
1988	33.040	10.955	33.2	0.717	2.2	21.367	64.7
1989	33.643	12.249	36.4	0.588	1.7	20.807	61.8
1990	32.475	12.178	37.5	0.579	1.8	19.718	60.7
1991	33.286	12.934	38.9	0.632	1.9	19.720	59.2

* Seasonally adjusted annual rate.
Source: National Cotton Council and USDA.

Table 6.12 US mill fibre consumption, seasonally adjusted rate (thousand 480 lb bales)

	1988	**1989**	**1990**	**1991**
Total, all fibres	26 764	27 882	27 487	28 107
Cotton	7 309	8 429	8 459	9 057
Wool	276	265	277	329
Man-made	19 178	19 188	18 662	18 721
Staple	9 674	9 570	9 058	9 060
Cellulosic	847	817	848	724
Rayon-acetate	804	705	605	532
Imports	43	112	244	192
Non-cellulosic	8 827	8 753	8 209	8 336
Nylon	1 969	1 990	1 980	1 740
Acrylic	878	865	734	665
Polyester	4 817	4 711	4 200	4 433
Olefin	758	786	857	950
Imports	406	401	439	548
Filament (yarn)	9 504	9 618	9 604	9 661
Cellulosic	417	445	411	439
Rayon-acetate	380	390	360	375
Imports	37	55	51	64
Non-cellulosic	9 087	9 173	9 193	9 222
Nylon	3 296	3 282	3 203	3 112
Polyester	2 410	3 314	2 176	2 321
Olefin	2 541	2 601	2 959	2 927
Glass	598	596	524	498
Imports	242	380	331	364
% cotton, all fibres	27.3	30.2	31.1	32.2
% cotton, staple fibres	42.3	46.2	47.8	49.1

Unadjusted data sources: US Dept of Commerce and Fiber Economics Bureau; Natural fibres, US Dept of Commerce, man-made fibres, Fiber Economics Bureau.

Table 6.13 Prices of textile fibres (cents/kg)

	Cotlook A-index	Australia greasy wool export unit value	US rayon staple	US polyester staple	Textile fibre index: World	Textile fibre index: Industrial countries	Textile fibre index: Developing countries	Non-cotton textile fibre price index	Relative textile fibre Price index[a]: World	Relative textile fibre Price index[a]: Industrial countries	Relative textile fibre Price index[a]: Developing countries
1960	65.0	134.4	61.7	277.8	81.1	121.5	92.2	116.0	184.3	206.1	803.8
1961	67.1	139.3	57.3	260.1	83.2	119.5	93.9	115.4	184.1	199.0	766.6
1962	65.7	122.3	57.3	251.3	82.7	119.4	85.0	113.8	175.5	193.9	633.5
1963	64.0	142.7	59.5	251.3	86.0	122.1	83.6	124.1	176.3	193.7	558.5
1964	64.2	150.6	61.7	216.1	85.9	113.4	76.1	122.4	168.8	175.7	439.8
1965	62.6	126.0	59.5	185.2	81.0	100.1	70.4	111.3	151.9	150.4	355.0
1966	60.6	136.3	57.3	178.6	81.1	97.5	68.0	113.7	144.9	141.7	301.6
1967	65.0	122.0	52.9	127.9	76.2	83.4	69.0	93.7	130.8	118.0	277.1
1968	67.2	116.6	55.1	121.3	78.4	82.5	70.5	93.5	129.0	112.3	264.7
1969	60.8	131.9	57.3	99.2	73.1	76.0	64.6	88.9	114.3	98.6	226.0
1970	63.2	98.1	55.1	90.4	70.2	69.8	65.7	79.0	103.6	85.8	212.6
1971	74.1	79.7	59.5	83.8	74.6	71.8	73.2	75.3	104.0	83.9	216.6
1972	793.	117.8	68.3	79.4	80.2	78.8	79.1	81.2	105.9	88.0	209.8
1973	135.5	305.3	72.8	83.8	120.9	114.0	127.5	105.7	145.8	118.3	282.8
1974	141.5	252.0	112.4	101.4	131.1	129.4	137.8	120.0	137.6	118.6	248.7
1975	116.1	182.4	112.4	110.2	117.5	118.7	116.9	119.2	109.0	97.9	172.3
1976	169.1	198.0	119.0	119.0	148.5	142.5	154.6	128.0	124.0	108.6	185.4
1977	155.4	227.0	132.3	123.0	145.5	145.5	149.6	136.2	109.2	102.2	144.7
1978	157.2	234.9	127.9	119.0	144.3	143.6	149.0	132.6	98.8	94.0	120.0
1979	168.9	254.5	143.3	130.1	156.5	157.3	162.6	145.4	95.3	94.5	106.6
1980	204.7	302.3	163.1	160.9	189.8	186.5	194.4	175.9	100.0	100.0	100.0

Table 6.13 Continued

	Cotlook A-index	Australia greasy wool export unit value	US rayon staple	US polyester staple	Textile fibre index			Non-cotton textile fibre price index	Relative textile fibre Price index[a]		
					World	Industrial countries	Developing countries		World	Industrial countries	Developing countries
1981	184.5	328.2	189.6	174.2	188.9	195.1	190.7	192.9	87.1	95.0	76.9
1982	159.9	306.1	185.2	169.2	173.6	182.4	172.8	187.0	71.1	82.6	53.8
1983	185.4	269.5	176.4	160.9	180.0	183.9	184.5	175.0	65.5	79.3	41.3
1984	178.5	281.3	185.2	173.8	182.7	187.5	184.2	186.5	58.4	77.1	28.2
1985	131.8	258.5	173.8	146.3	147.3	160.5	150.3	161.7	41.5	63.3	15.7
1986	105.6	237.8	167.0	137.4	128.9	144.7	129.9	152.2	33.5	55.7	10.4
1987	164.8	343.6	178.6	146.6	167.8	179.8	174.4	170.7	39.3	67.3	10.0
1988	140.0	567.1	199.9	162.8	174.3	194.5	175.3	207.9	35.6	70.5	5.9
1989	167.4	515.1	242.5	190.7	199.2	224.5	204.4	230.3	36.7	77.9	3.4
1990	182.2	458.8	264.6	182.1	201.4	237.0	217.8	221.7	35.0	78.4	1.6
1991	174.2	367.0	273.4	175.0	191.2	231.7	211.5	208.8	31.7	73.2	1.1
1992	172.0	345.7	237.1	175.8	186.6	212.9	198.5	201.8	29.4	64.7	.9

[a] Textile fibre index divided by Consumer Price Index for each country group, 1980 = 100.
Sources: Cotton Outlook, International Monetary Fund and USDA.

CHAPTER

7

Cotton prices

The measurement of cotton prices

The Liverpool Cotton Association has published cotton prices since the mid-1800s. Before World War I, the *Weekly Circular* of the Liverpool Cotton Association quoted the prices of 46 different varieties of cotton - 19 of them were Indian. In July 1914 their prices varied from *4d.*/lb to 400*d.*/lb - a greater variation between varieties than has existed in later decades, although it continues to be true that there is no one single price for cotton, and the relationships between growths vary. Probably the most well-known measurements of cotton prices are those published by Cotton Outlook in Liverpool. The present Cotton Outlook indices are published as part of the *Cotton Outlook Journal*'s activity. This journal (earlier called *The Cotton and General Economic Review*) published cif prices for the main growths of cotton since just after the end of World War II.

The first time that an average price, called the 'Liverpool CIF Index' price was published was 1966, and the first published index price was 31.05 cents/lb, based on cif Liverpool terms. The index was composed of a series of different growths: in 1966 annual imports through Liverpool still amounted to just under 1million bales and it was relatively straightforward for Cotton Outlook to collect the representative prices on a cif Liverpool basis. The producers of the index were surprised and delighted that international interest in the index increased very quickly and substantially. By 1972 Cotton Outlook had developed a daily information service, and from January 1973 onwards the index was calculated on a daily basis. It is a measure of the sophistication of the international cotton market that this date was so early: other commodity markets (rice, for instance) still lack anything resembling such an index in the 1990s. In January 1973 also, Cotton Outlook introduced the 'B' Index which measures prices for 'Coarse count' cotton – that commonly in use for the production of coarse count yarn.

In January 1974 Cotton Outlook recognised that cif Liverpool was no longer appropriate, owing to the reduced imports entering the UK, and changed the geographical basis of the quotations to cif North Europe. Prior to August 1981, quotations were for Strict Middling 1 1/16 in.; at the beginning of the 1981/82 season, changing agricultural practices, and their influence on the typical traded qualities, led to the adoption of Middling 1 3/32 in. as the quality used for the 'A' Index. At present, the Cotlook 'A' Index is derived from a group of medium staple cottons, and is intended to provide an average price for middling grade 1 3/32 in. cotton, cif Northern Europe. It is calculated as the simple average of the cheapest 5 of the Cotton Outlook quotations, of which there are at present 13:

1 Memphis Territory Middling 1 3/32 in.
2 Central Asian Middling 1 3/32 in.
3 Mexican Middling 1 3/32 in.
4 Turkish Izmir/Antalya St I Wht 1 White R.G. 1 3/32 in.
5 Tanzanian 'AR' Type 3.
6 California/Arizona Middling 1 3/32 in.
7 Pakistan Punjab SG 1503 1 3/32 in.
8 Australian Middling 1 3/32 in.
9 Central American Middling 1 3/32 in.
10 Indian H-4 1 3/32 in.
11 Paraguayan Middling 1 3/32 in.
12 Chinese Type 329.
13 African Franc Zone Middling 1 3/32 in.

The Cotlook 'B' Index is calculated as a simple average of the day's cheapest three of Cotton Outlook's quotations for these eight growths in the selection of coarse count, lower-grade cotton:

1 Orleans/Texas SLM 1 1/32 in.
2 Central Asian SLM 1 1/16 in.
3 Argentine Grade C-1/2.

4 Brazilian Type 5/6 1 1/16 in.
5 Turkish Adana Std 1 RG 1 1/16 in.
6 Chinese Type 527.
7 Pakistan Afzal 1 1/32 in.
8 Indian J-34 SG.

Growths are withdrawn or added to the selections, following a warning period beforehand, as the availability and quality of cotton internationally changes. Selections have been changed several times; for example, the Indian growths were added to the respective selections on the introduction of the 1991/92 indices.

The Cotlook Ltd procedure is that on every UK business day, they publish at about 1430h UK time, an extensive list of asking prices for the principal growths of raw cotton. Cotlook Ltd staff themselves are analysts who have spent the majority of their working lives connected with the cotton industry, and have no trading involvement.

A fairly recent change in the Cotton Outlook indices was the introduction of the Dual Index System. Cotton Outlook announced in March 1988 that the original practice of replacing old crop by new crop values gradually in a single series of prices would be replaced at the end of the 1987/88 season by a Dual Index system which had been explained for some months previously by the Cotton Outlook team. With the new system, there are two sets of indices, one reflecting quotations for nearby shipment - usually within the next month or two - and one for forward shipment. These run concurrently until the conclusion of the marketing season on the last business day of July. Then, the existing 'old' cotton indices cease to be quoted, and the forward indices become the current indices. Only one index number is then published for 'A' and 'B' values until early in the following year. As soon after January as information permits, a forward value is published for each growth, which refers to shipment no earlier than October/November of the coming cotton season. Cotton Outlook introduces the forward index as soon as they have sufficient confidence in its validity. When sufficient growths' forward values have been introduced, they are consolidated into 'A' and 'B' Indices. There is no pre-ordained date for this move, which depends on the international market: it has been as early as February, and in the case of the 'B' Index, as late as July. October/November rather than August/September was chosen for the initial shipment period as it is during these months that the northern hemisphere crop begins to move in volume. Southern hemisphere values are not included for the purpose of the Index until 1 January, whether or not they are separately published by Cotton Outlook (Fig. 7.1). Tables for the price of each

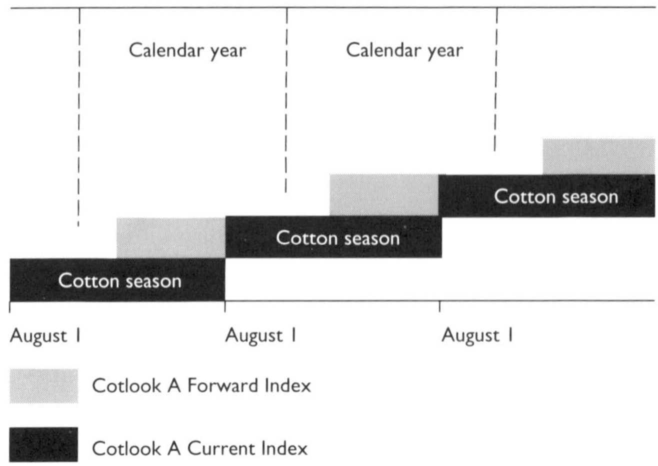

7.1 Cotlook 'A' Index and Cotlook 'A' Forward Index, dual index sequence (*source*: Cotlook Ltd).

class of cotton are published regularly by the New York Cotton Exchange.

The international base grade of US cotton delivered to northern Europe is usually 2–3 cents/lb above the 'A' Index. However, during the 1985/86 season, for example, US prices were supported above prices charged by competing exporters, with serious ramifications for US exports. In the past 20 years US cotton has only once sold cheaper than the 'A' Index, reflecting drought in several countries and the availability of payments in kind entitlements. Obviously, in general transport costs from the CIS states, Turkey and even West Africa are less than from the USA to Europe. During years when the demand for US cotton from Asia is strong, the Cotlook Index is frequently more than 5 cents less than the cheapest US component, which is usually Memphis. The spot prices in Memphis are frequently 10–17 cents below the 'A' Index US component, reflecting transport and marketing costs.

The Cotlook 'A' Index is referred to by the US government in calculating the US Adjusted World Price (AWP), which is a key value affecting the competitiveness of US cotton on the world market. The AWP and the CCA (coarse count adjustment), despite suggestions to alter it during 1992, are announced weekly each Thursday (Fig. 7.2).

US farmers placing their cotton in the US government farm loan programme are permitted repayment of the loan at the lower of the loan rate or the adjusted world price in effect for that week. This part of the US Food Security Act is called the marketing loan. It is intended

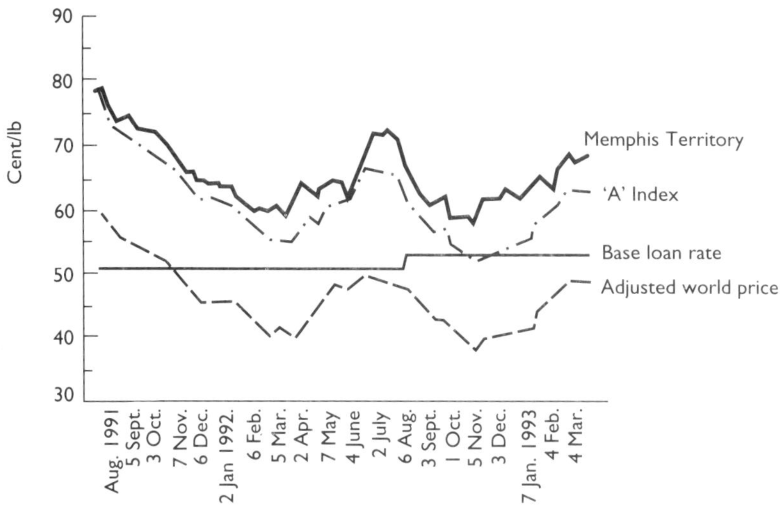

7.2 (a) Northern Europe cotton quotes and US programme prices (*source*: Cotlook Ltd and ASCS/USDA); (b) cotton production by major producers; (c) cotton area by major producers (*source*: PECAD/FAS/USDA).

to keep US prices competitive in world markets and keep cotton supplies as a continuous flow (Tables 7.1 and 7.2). The sales payable in different years under the USDA grants in operation are noted.

Some criticisms can be made of the indices. Cotton Outlook emphasises that the prices, from which the individual quotations are derived, are offers, not actual transactions. No attempt is made to weight the indices (virtually an impossible task), but the belief is held that the averaging at any one time of the cheapest elements serves to offset, at least partially, any distortions that might otherwise occur. So the Cotlook indices are not weighted by the quantity traded and shipment dates, which can vary by several months for different growths. The Cotlook indices share this with a number of other similar indices such as the Baltic Freight Index published by BIFFEX in London, also traded for a futures contract.

An analogy with UK house prices may make the point more clearly: Cotton Outlook does the equivalent of telephoning a large sample of estate agents and asks them what their vendors are offering their houses at. The Nationwide Building Society in the UK - and to be fair, this is the most sophisticated house price index yet operational in the world - employs actual recorded transaction prices over a variety of different types of houses, and adjusts them with complex formulae to derive pure price changes from a mass of different sample data each month. In an ideal world, cotton indices would be calculated in a similar way. However, the Nationwide has a tame data sample through

Table 7.1 CCC loan premiums and discounts for grade and staple length of 1992 crop american upland cotton, basis grade 41 staple 34 (SLM 1 1/16 in), net weight

		Staple length, inches								
Grade	Code	13/16 (26) to 29/32 (29)	15/6 (30)	31/32 (32)	1 (32)	1 1/32 (33)	1 1/16 (34)	1 3/32 (35)	1 1/8 (36)	1 5/32 (37) & longer
White:										
SM & better	(11 & 22)	−685	−525	−310	−265	−105	130	175	180	195
MID PLUS	(31)	−710	−540	−325	−275	−110	125	160	170	190
MID	(31)	−745	−550	−340	−285	−125	115	155	165	185
SLM PLUS	(40)	−770	−590	−380	−350	−190	50	95	105	120
SLM	(41)	−805	−630	−420	−410	−240	BASE	45	55	60
LM PLUS	(50)	−920	−795	−640	−640	−490	−260	−235	−230	−225
LM	(51)	−985	−865	−750	−750	−600	−375	−350	−345	−335
SGO PLUS	(60)	−1285	−1265	−1265	−1265	−1115	−1030	−1010	−1005	−1000
SGO	(61)	−1415	−1415	−1415	−1415	−1245	−1185	−1170	−1170	−1165
GO PLUS	(70)	−1595	−1575	−1495	−1495	−1345	−1345	−1345	−1345	−1345
GO	(71)	−1625	−1605	−1530	−1530	−1380	−1375	−1375	−1375	−1375
Light spotted:										
SM & better	(12 & 22)	−785	−595	−370	−325	−170	50	70	80	90
MID	(32)	−850	−670	−435	−405	−235	−5	40	45	90
SLM	(42)	−920	−735	−630	−630	−480	−280	−255	−245	−240
LM	(52)	−1145	−1100	−1100	−1100	−950	−935	−925	−925	−925
SGO	(62)	−1530	−1530	−1530	−1530	−1380	−1360	−1355	−1355	−1355

Spotted:										
SM & better	(13 & 23)	−1025	−860	−705	−655	−505	−480	−465	−460	−450
MID	(33)	−1075	−965	−825	−825	−675	−675	−675	−675	−675
SLM	(43)	−1320	−1320	−1320	−1320	−1170	−1160	−1150	−1150	−1150
LM	(53)	−1590	−1590	−1590	−1590	−1435	−1400	−1400	−1400	−1400
SGO	(63)	−1690	−1690	−1690	−1690	−1535	−1500	−1500	−1500	−1500
Tinged:[a]										
SM	(24)	−1685	−1565	−1540	−1515	−1290	−1245	−1245	−1245	−1245
MID	(34)	−1735	−1615	−1590	−1565	−1340	−1295	−1295	−1295	−1295
SLM	(44)	−1825	−1755	−1745	−1735	−1555	−1555	−1555	−1555	−1555
LM	(54)	−1980	−1950	−1895	−1885	−1735	−1735	−1735	−1735	−1735
Light gray:										
SM & better	(16 & 26)	−820	−720	−590	−585	−380	0	15	15	20
MID	(36)	−1090	−1090	−1090	−1090	−890	−425	−370	−365	−365
SLM	(46)	−2165	−2165	−2165	−2165	−1960	−1470	−1415	−1405	−1400
Gray:										
SM & better	(17 & 27)	−1100	−1100	−1100	−1100	−900	−435	−380	−375	−375
MID	(37)	−2175	−2175	−2175	−2175	−1970	−1480	−1425	−1415	−1410
SLM	(47)	−2275	−2275	−2275	−2275	−2070	−1580	−1525	−1515	−1510

[a] Cotton classed as 'Yellow Stained' (middling and better grades) will be eligible for loan, if otherwise eligible, at a discount 200 points greater than the discount for the comparable quality in the colour group 'tinged'.

SM – Strict Middling, MID – Middling, SLM – Strict Lower, LM – Lower, SGO – Strict Good Ordinary, GO – Good Ordinary.

Source: USDA, Agricultural Stabilization and Conservation Service.

Table 7.2 Upland cotton estimates, December 1992

Options for final ARP	1991-92	1992-93
(ARP/PLD)	(5% ARP)	(10% ARP)
Part/ARP	84%	89%
Part/PLD	0%	0%
Acreage (thousand)		
Reported base	14 575	14 861
Total planted + idled	15 038	14 878
Alloted/NPA	0	0
CRP	1 300	1 400
50/92 w/unpaid acs	627	423
0/92 prev. planting	179	9
ARP	609	1 299
Reduced −PLD	0	0
Permitted/cotton base	11 600	11 864
Payment acres	9 416	9 300
Acres elig. for loan	10 631	11 187
Pltd part	10 214	10 739
Pltd nonpart	3 171	1 969
UPL pltd on other crops	417	448
Other pltd on UPL base	266	319
Planted	13 802	13 156
Harvested	12 716	10 947
Yield (lb/ha)	650	691
Program yld	590	600
Supply/use (t. 480 lb bales)		
Beginning stocks 8/1	2 262	3 583
Production	17 216	15 758
Imports	13	2
Total supply	19 491	19 343
Internal use	9 543	9 630
Exports	6 348	5 725
Total use	15 891	15 355
Unaccounted	(17)	145
Ending stocks 7/31	3 583	4 133
CCC inventory	0	0
CCC loans out	297	1 133
Free stocks	3 286	3 000
Stocks/use	22.5%	26.9%
Cents per lb		
Target price	72.90	72.90
Cal. yr avg. price	62.80	
Defic. pay rate - limit	10.10	20.55
Loan def. pay rate	8.07	11.35
Disas. pay rate	0	0
Diver. pay rate	0	0

Options for Final ARP	1991–92	1992–93
Loan rate	50.77	52.35
Avg. loan repay rate	47.23	43.00
Eff. loan repay rate	42.70	41.00
Season avg. price		
Mkt yr avg. price	56.00	
Formula AWP	47.71	44.00
115% loan rate	58.39	60.20
Adjusted AWP	47.23	43.00
Avg. Coarse count	0.00	0.00
World supply/use		
Beginning stocks	28.7	40.1
Production	96.0	85.9
Imports	23.1	22.3
Total supply	147.8	148.3
Consumption	85.0	85.9
Exports	22.4	22.4
Total use	107.4	108.3
Ending stocks	40.1	40.0
Stocks/use	47.2	46.6
US% of trade	29.7	26.8
A-index	63.1	58.0
Thousand dollars		
Diversion payments	0	0
Deficiency payments	555 000	1 032 000
Loan def. payments	140 000	265 000
Disaster payments	0	0
Farm value	4 733 749	3 858 000
Seed value	481 825	514 000
Mkt loan gain/grop	232 000	414 000
Adj. gross income	6 142 574	6 083 000
Step 2 payments/FY certs		100 000
Step 2 payments/FY $	140 293	200 000
Net returns		
Costs/$:		
Var. costs/acre	282.01	287.28
Var. cost of cover	28.20	28.73
Mil $:		
Pltd * var.	3 892	3 779
CUA* var.	35	49
Tot. var. costs	3 927	3 828
Total net returns	2 216	2 255
Divided by (pltd + CUA) =		
Net return/acre($)	147	152

CCC = Commodity Credit Corporation AWP = Average World Price its mortgage application procedure: by contrast, over 120 countries trade raw cotton, and there are numerous different grading systems, measurement units and systems of transport, storage and packaging.

Nevertheless, some traders have suggested that the result of relying on offer prices only have been occasional distortions and inaccuracies in which the 'A' Index has not reflected trading prices. There are two reasons for this. One is that not all the traders spoken to by Cotton Outlook every day are necessarily offering significant quantities of the cotton at these prices - they may in fact be offering at higher prices, but the transactions may be taking place at much lower levels: 2–3 cents below the index is the usual figure cited. Secondly, a number of companies that have acquired CIS cotton through barter deals may be selling to European mills in particular at prices significantly lower than those quoted by the trading companies to Cotton Outlook. Much cotton has always been traded throughout the world in barter deals, which have always posed a problem for the correct determination of world prices. Many countries attempt to isolate their domestic cotton markets from international trading conditions (which would be a problem for a transactions-based index too, of course), and there are few cotton commodity markets, so there are in fact numerous opportunities to fix cotton prices on a basis different from that quoted by Cotton Outlook. US companies also point out that since most of their exports go to East Asia - Japan is usually the main importer of US cotton - using the price into Europe as a measure is not entirely satisfactory. Usually the price of US cotton Osaka cif is about 6 cents less than cif Northern Europe, reflecting lower transport costs, a structured market, and also, to a lesser extent, lower marketing costs because of higher traded volumes. Finally there is the point that the Cotlook indices are measured in dollars, which does tend to obscure currency fluctuations which can frequently be more important than the actual price changes themselves.

Notwithstanding all these criticisms, and noting the idea that a hedonic transaction-based analysis would be ideal - and no commodity market has one - the Cotlook indices are a splendid achievement for Cotton Outlook, and they play an important role in keeping Liverpool to the forefront of the cotton trade. They also mark out the cotton trade from other commodity markets which do not have the advantage of such regular and effective pricing.

A further point of reference for cotton prices are the US spot market prices, which have been quoted by the USDA's Agricultural Marketing Service (AMA) since 1915, from when staple lengths were quoted. Some information on cotton spot prices in the USA is available right back to 1790. The AMA price is the average quoted for the base quality in each of the 8 designated spot markets every day. These are: Greenville, SC; Montgomery, AL; Memphis, TN; Greenwood, MS; Dallas, TX; Lubbock, TX; Phoenix, AZ; and Fresno, CA. There were originally 12 such markets, but this was reduced to 8 between 1974 and 1983. Of them, by far the most famous and widely quoted is that in Memphis, which is where the US office of Cotton Outlook, sensibly, is situated. Even these spot prices are in fact gathered together by spot quotation committees in each market, whose assessments are made daily after the close of the New York futures market. Committee members use their judgement in assessing prices because there are not always futures market prices with which to make comparisons, and not all grades trade daily. The average is not weighted by the volume traded in any one market, any more than the Cotlook indices are weighted by the volume of cotton offered in any one staple or known to have been traded in any one period. Since 1973 the base grade and staple length has been strict low middling (SLM) 1 1/16 in., although by no means a large percentage of the US crop is this grade.

Farm prices, by comparison, tend to be averaged across all qualities and are representative surveys of prices paid to farmers for cotton lint at the point of first sale. They include forward sales, which range from 6% to 50% of the crop (see Chapter 8). The reported prices published by the US Agricultural Statistics Service are monthly average prices weighted by sales volume; hence, because about 75% of sales occur between October and January, this period largely determines the annual figure. Farm prices vary more across the USA because of inter-state quality differences and transport costs, although the shape of the distribution is much the same for both. The lowest farm prices in the USA are usually in Texas and Oklahoma, mainly for quality reasons; similarly, high quality and close proximity to the export terminals of Los Angeles and San Francisco explains why these growths are almost always the most expensive in the USA. Delta cotton and Carolina cotton are usually similar in quality, so price differentials reflect transportation costs to mills and export terminals.

There are also mill delivered prices, which in the USA represent the price of a specific quality of cotton delivered to mills in the western half of the Carolinas. This price includes all associated transportation and marketing costs. These prices are also influenced by forward contracting and therefore do not always exactly reflect one month's trading. The main use of these prices is to obtain a measure of to-mill transport costs, storage costs for cotton prior to delivery to mills, and marketing expenses. Over the 1980s there was a trend for mills to reduce their inventories, down by some 50% on average between 1980 and 1990, and the effect has been to increase transport costs as cotton is moved more quickly to react to mill demand.

Cotton price trends

The long-term chart, Fig. 7.3, shows the strong influence of worldwide economic and political conditions on cotton prices since 1790. The high prices during the US Civil War are especially noticeable - particularly given the fact that the diagram shows nominal, not real, prices. The downward path of prices during the 1920s and thereafter shows the weakness of the international economies at that time. Low prices before World War II were replaced by inflation during 1941–45.

Figure 7.3 does not easily demonstrate the course of prices since 1945. In general terms it is possible to identify three distinct phases. In phase I, prices were relatively stable for some years after World War II. This period can be said to have lasted roughly until 1973, when the price of both medium and long staple cotton doubled and phase II began. That year represents a watershed: after that time prices have fluctuated with considerable volatility. Why? World supply and demand now changes more rapidly, for one reason, and the whole

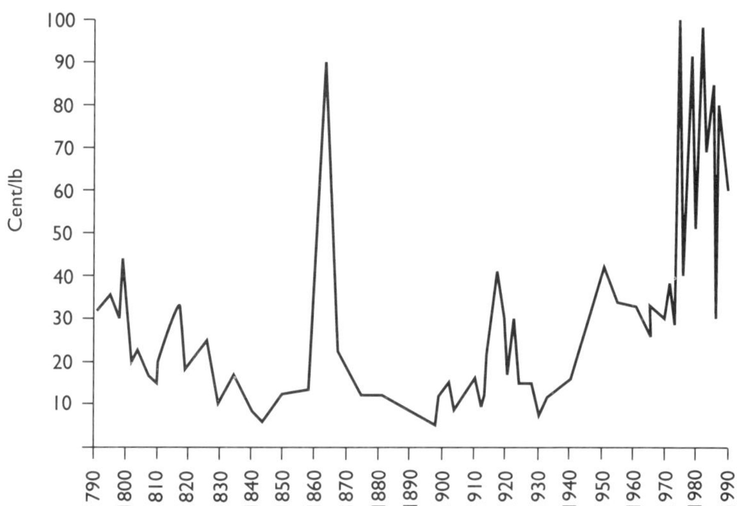

7.3 Development of cotton prices.

7.4 NYCE cotton weekly high, low, Friday close. Nearest Futures Contract June 1977–May 1992.

international trade responds more swiftly. Imports and exports fluctuate more quickly in various nations; for example, Pakistan is currently attempting to restrict its exports of raw cotton and concentrate on more value-added products such as clothing. Political, economic and social events such as the move of the former Soviet bloc countries to free market economies can have a major impact on demand and prices. Finally, there is the unpredictability of the weather, which is certainly not decreasing. Weather predictions are rarely accurate for more than a few days ahead. The progress of the Cotlook 'A' Index demonstrates this substantial volatility (Table 7.3). Fluctuations can be substantial even within a week, so weekly, monthly and especially annual averages have the effect of smoothing out the fluctuations, but they can still be seen in the annual figures given in Fig. 7.4.

World prices of all types of cotton increased throughout most of 1973, weakening during November/December but reaching a new peak in late January 1974. Throughout the rest of 1974 and the first two months of 1975, prices then proceeded to collapse. This was caused by cuts in production in textile industries throughout the world; mills became less interested in stockholding. This trend halted in March 1975, and prices strengthened during April/May 1975, as mills slowly began to rebuild their stocks, demand began to pick up in major industrial countries, and it became clear that plantings were drastically reduced for the next season. Synthetic fibre producers

Table 7.3 Northern European cotton price quotations and SDA's adjusted world price (cents/lb)

Date	Cotlook 'A' Index[a]	Memphis Territory	Diff. from 'A' Index	Lowest 'A' quote	Adjusted world price[b]
Yearly average		Middling 1 3/32 in.			SLM 1 1/16 in.
1986/87	61.77	61.71	−0.06	60.04	48.86
1987/88	73.37	76.39	3.03	71.57	60.39
1988/89	66.38	69.09	2.71	64.84	51.98
1989/90	80.44	80.93	0.48	78.49	64.97
1990/91	82.85	88.15	5.29	80.66	66.02
1991/92	63.05	66.39	3.33	58.20	47.59
1992/93 (Aug–Feb)	56.15	61.88	5.73	50.95	42.57
Monthly average					
Aug. 1992	59.55	63.40	3.85	53.91	45.84
Sept. 1992	56.23	60.26	4.03	52.23	42.74
Oct. 1992	52.92	58.06	5.14	49.50	39.29
Nov. 1992	52.77	60.63	7.86	48.55	39.11
Dec. 1992	54.30	61.93	7.63	49.48	40.76
Jan. 1993	55.80	62.77	6.98	50.79	43.51
Feb. 1993	57.43	63.81	6.38	52.16	46.75
Weekly average					
Sept. 3	57.11	60.50	3.39	52.94	43.20
Sept. 10	56.45	60.20	3.75	52.25	42.54
Sept. 17	56.43	60.45	4.02	52.20	42.52
Sept. 24	56.61	61.55	4.94	52.60	42.70

Oct. 1	54.55	58.30	3.75	51.15	40.64
Oct. 8	53.88	58.25	4.37	50.40	39.98
Oct. 15	53.29	58.20	4.91	50.70	39.35
Oct. 22	52.70	58.65	5.95	49.50	38.80
Oct. 29	51.85	57.30	5.45	48.40	37.70
Nov. 5	52.11	58.90	6.79	40.95	38.21
Nov. 12	52.78	60.95	8.17	48.50	39.10
Nov. 19	52.69	61.15	8.46	48.60	39.01
Nov. 27	53.15	61.00	7.85	48.80	39.42
Dec. 3	53.48	61.40	7.92	48.75	39.80
Dec. 10	54.10	62.45	8.35	48.85	40.42
Dec. 17	54.38	62.00	7.62	49.55	40.70
Dec. 24	54.39	61.10	6.71	49.90	40.73
Dec. 31	54.78	62.19	7.41	50.19	41.17
Jan. 7	55.18	62.50	7.33	50.50	41.50
Jan. 14	56.87	63.95	7.08	51.60	43.21
Jan. 21	58.01	64.45	6.44	52.45	44.33
Jan. 28	58.68	63.95	5.27	53.40	45.00
Feb. 5	58.73	62.35	3.62	53.65	45.05
Feb. 12	60.26	65.90	5.64	54.40	46.58
Feb. 19	60.94	66.80	5.86	55.20	47.26
Feb. 26	61.80	67.25	5.45	55.50	48.12

[a] The 'A' Index is the average of the five lowest quotes of the following descriptions (all 1 1/32"): Memphis Terr.; Calif. Ariz.' Mexico; Guatemala; Paraguayan; Turkish Central Asian; Pakistani; Punjab; Chinese Type 329; French African; and Australian.
[b] Reflects incorporation of forward shipment quotations, but does not include the Secretary's discretionary adjustments.

Source: Cotlook Ltd.

announced that they intended to raise prices when they could; cotton price support measures were introduced in several producing countries, and it was believed that both the loan rate and the cotton target price would be increased in the USA.

In the first half of 1976, prices rose rapidly, as the recovery in the international textile industry began and supplies became tighter. Prices stayed relatively high during the summer and autumn of 1976, but a reduction in textile output led to prices falling off towards the end of the year. The upward trend was resumed in January 1977 when supply tightened again and stocks declined quickly, at the same time as significant increases in the prices of synthetic fibres occurred in the USA and Western Europe. Responding to this demand, cotton production increased during the year and forced prices down during 1977, especially since demand for fibre was not increasing and therefore stocks were rising. Prices in 1977 were between 4 and 9% lower than in 1976, depending on the type of cotton - extra-long staple cotton excluded, which actually rose in price because of low output - an excellent example of market segmentation.

Prices generally began to recover at the beginning of 1978. The tight supply of 1978/79 caused prices to rise throughout 1978 for most types of cotton. They levelled off in early 1979 again in response to an increase in output, and oscillated around a mean in the 70 cent region for most of 1979. With an increase in textile demand in 1979/80, more spinning and no additional supplies, prices moved up quickly in early 1980. It was partly tightening of supply/demand and partly speculative purchases of cotton as an inflation hedge (the FAO claimed) which was pushing up prices, as well as increases in synthetic fibre prices. In the second quarter of 1980, prices eased as the trade and speculators began to anticipate diminution in demand because of the international recession. But in fact they rose again in the third quarter of 1980 as traders believed they had not taken account of tight supply caused by a reduction in US output in the 1980/81 season.

Inflationary pressures generally are usually insufficient to hold up raw material costs in recession, however, and 1981 was no exception. Input costs were rising, but in 1981 world cotton prices declined in dollar terms, reflecting both substantial weaknesses in the textile industries of many major cotton-importing countries and also the world recession. This applied rather less to extra-long staple cotton, where prices were more stable. In the USA, prices of competing synthetics rose due to increased raw material costs, but in the rest of the world there was still manufacturing overcapacity and synthetic fibres stayed relatively cheap, only beginning to rise in early 1981. But with textile demand weak throughout 1981, cotton prices slid by one-third, although there was a limited recovery during the first half of 1982. Prices for extra-long staples were affected by this downturn in upland cotton prices and by the continuing weakness in demand for fine yarns. Synthetic fibre prices fell during 1982 also, reducing the competitive price advantage that cotton had enjoyed during the year. Very large stocks in the market, combined with a continued recession in textile demand, led to continued price falls until early 1983, when textile demand had recovered sufficiently to balance stocks.

Price revival continued as crop prospects deteriorated in many major producing countries. With level output, cotton consumption rising to a record level high, and stocks by July 1984 fallen by 5%, prices rose until mid-1983. This was halted when crop prospects for the 1983/84 season were revised upwards, so that cotton prices levelled off until early 1984. In fact, it soon became clear that there was likely to be a large surplus in 1984/85; prices for medium staple cotton started to fall in mid-1984, declining steadily until the final quarter of 1985, at which time some grades had reached a 10 year low around the 50 cent mark. Shortages of export supplies kept extra-long staple prices higher into 1985, again showing the segmentation of the market. By mid-1985 cotton was selling at a discount to polyester.

During 1986 prices declined in line with the rise of stocks, falling to dramatically low levels in the 37 cent region for the 'A' Index. It was the poor crop prospects for the 1986/87 season which began the turnaround, and prices rose back to 86 cents/lb in August 1987, prices for extra-long staple following up at a slower rate. Polyester prices did not increase as quickly. This was cotton's version of the world bull market of the 1980s, and it undoubtedly saved many producers and possibly even trading firms from bankruptcy. Prices after the October 1987 Stock Market crash remained very volatile, as there was tremendous uncertainty about the direction the world economy was going to take. But by 1988 opinion had hardened that a recession was imminent, and cotton prices fell throughout 1988, aided by a good harvest. This marks the beginning of phase III, the pricing of cotton in a sustained recession. The 'A' Index fell to 58 cents/lb by November 1988. Once again cotton became price-competitive as ethylene glycol price increases pushed up polyester costs.

Prices recovered gradually but steadily more quickly from their August 1988 trough, following international demand increases and supply shortages, or at least fear of them. The 'A' Index increased from 56 cents in October 1988 to almost 85 cents in July 1989 and stayed at what in retrospect was an unsupportedly high level for the rest of the year. Cotton traders thought that the recession was not going to affect them; at these levels cotton was competitive. After rising steadily until July 1989, prices fell back between August 1989 and January 1990 as the full impact of continued US recession and cutbacks in the Chinese textile industry began to be felt. But soon they increased again, pushed up by speculative demand - which makes cotton prices as decisively as the fundamentals - reaching 80–85 cents/lb for the 'A' Index. Polyester prices were still higher, so cotton remained competitive. During 1990/91, cotton prices, i.e. the 'A' Index, were in the mid-80 cents range, and generally thought likely to stay there. The year saw an average price of 83 cents, 1 cent higher than in 1990. ICAC pointed out that the actual price range was very narrow, in the 79–86 cent range for the Index. From May 1991 onwards, the roller-coaster went down again, with prospects of a bumper harvest in 1991/92 combined with low consumption demand, especially from the European textile industry. The actual season average was 63 cents (compared with the ICAC forecast based on the model using relative end stocks, consumption and net exports from China of 67 cents). The average since 1973/74 has been 73 cents, excluding 1991/92 which drove the average down. During both the 1989/90 and 1990/91 seasons, the relatively favourable US situation and restrictions on US mills in importing cotton resulted in New York futures and Memphis spot prices rising at the same time as world cotton prices as measured by the Cotlook Index being stable. This divergence between the No. 2 contract (see Chapter 8) and world prices was a major impetus for the introduction of the world cotton price futures contract.

Forecasts for 1992 proved rather erroneous, however. At the beginning of the year in Bremen, the ICAC predicted that 'prices will rise from their current levels' and that 'the range will be several cents wider than in 1990/91'. At that time, in February 1992, the 'A' Index was below 70 cents. In practice it stayed below 70 cents throughout the first half of 1992 and stayed that way throughout the year, in the mid-to-low 50 cent range by December despite a mid-year rally to the 66 cent mark, although this was not to anything like the low levels of 1986. Part of the problem was undoubtedly the record world production in 1991/92, estimated by the ICAC at 20.1 million tonnes, even higher than the previous world record of 19.3 million tonnes of 1984/85 (see the section on 'Overview of world cotton production' in Chapter 4). Traders could not quite explain the downward resistance during 1992. It could be evidence of declining volatility in cotton prices as part of the New International Order, but this may be too early and too grandiose a speculation.

The received wisdom at present is that prices are determined by the intersection of demand from mills for mill consumption, and supply in terms of world cotton production. Serious economists would of course not really wish to present the price model in such stark terms. Firstly, the level of stocks themselves can change - for example, the Chinese decision to retain substantial stocks during mid-1992 undoubtedly helped prevent the price from going into a freefall dive; the ICAC was quite correct to point out in mid-1992 that 'much of the outlook for prices during 1992/93 hinges on the quantity of cotton supplied to the rest of the world by China' (*ICAC Recorder*, 1992, p. 3). In 1990/91 production and mill consumption were quite closely balanced, and there was little change in the volume, quality and geographical distribution of world stocks. This was said to have caused stable prices, although economists would probably suggest that, equally, stock stability could be a function of stable prices. However, the ICAC argument that 'if it becomes clear that China (Mainland) does not intend to export as much cotton as currently expected, world prices could rise significantly' (*ICAC Recorder*, July-Aug. 1992, p. 3), was not true: the trade continually discounted Chinese export levels throughout the year in current prices. At the end of 1992, there was a slight increase in the 'A' Index because of the market response to the lower Chinese production estimates, but central Asian cotton remained consistently the lowest quote in the Index, followed by African and Pakistani 1505. The Californian/Arizona quote was the fifth lowest - and hence part of the Index, at the end of the year, although the Memphis Territory quote remained outside it.

The declining trend in world cotton prices on the world market, measured in terms of constant US$, could be generally explained by production surpluses, strong and continuing competition from synthetic fibres and sometimes inflexible supply policies by the exporting countries. Cotton growers are not, often, able to shift to other crops. Many countries, notably those in the CIS, need cotton exports as their principal source of hard currency or its equivalent, Western chemicals, machinery or food, and bid down prices as part of a competitive strategy. A gradually increasing amount of cotton is being consumed in the producing countries themselves. In these countries prices are fixed independently of world markets through domestic government intervention. Hence cotton growers' decision to grow cotton in particular is explained by the level of this domestic price, insulating producers to some extent from the world market price. In late 1993 the ICAC said that in their view continued low cotton prices might threaten the continued production of cotton in some countries where government subsidies and inappropriate barter agreements were contributing to the low general price level. Prices were not increasing in response, for example, to considerable reductions in US crop forecasts.

It was only at the tail end of 1993 and the first few months of 1994 that both the futures and the physical market responded to the shortages of cotton available for export. Indian and Pakistani export restrictions, a 50% export tax imposed by Turkey, a lower than expected US crop, and continued logistic difficulties in the CIS, all combined to push the Liverpool 'A' Index up past the 80 cents level by March 1994. Opinion was divided as to whether this level would be maintained in the short term.

Amongst different cotton growths there is substantial variation in price, as Table 7.4 demonstrates. The variation is usually explicable in terms of relative demand for the different growths. For example, at the beginning of the 1992/93 season, quotes for Orleans/Texas Middling and Orleans/Texas SLM rose above the 'A' Index generally, reflecting the weather damage to the US crop. Another example is the shortness of Paraguayan Middling supply, resulting in a rise from 97.7% of the 'A' Index in 1991/92 to 102.4% during the first part of 1992/93, and another is the increase in Australian prices again reflecting short supply. By comparison, and continuing throughout 1992/93, quotes for Central Asian Middling are falling, as are those for Pakistani cotton, reflecting the push for high exports on the part of both governments and producers in these countries. The fact that the Chinese during 1992 were not urging substantial exports explains the relatively high price of Chinese Type 329 cotton. The relative price for extra-long staple cotton compared to the 'A' Index is approximately 1.4, with US Pima at 1.8 for the first three and a half months of the 1992/93 season.

The price differential for extra-long staple cotton is because these cottons can be spun into finer, stronger yarns than medium staples - although this is becoming less so - but they do still impart a higher lustre, strength and sheerness, as well as giving a more attractive appearance. The demand is mainly for high-quality textiles, such as high-quality sewing thread and speciality goods. As a result of this, and because even relatively small producers can affect overall supply, the price has varied between 20 and 250% of the world average.

What must be constantly borne in mind when considering fluctuations in cotton prices is that, unhedged by any successful use of the futures market, revenues and hence profits can be substantially different from one year to the next. The variations from the US crop were over 100% within the 1985–91 period. These huge variations explain the dominance of the futures markets within the cotton trade (Table 7.5).

Table 7.4 Prices of cotton cif North Europe, relative to the Cotlook 'A' Index, ratio (Cotlook 'A' Index = 1.000)

	Cotlook 'B' Index	United States					Paraguay	Central Asia		Pakistan		China	Australia
		Orleans/ Texas Middling	Orleans Texas Strict Lower Middling	Memphis Terr. Middling	Calif/ Ariz Middling	Calif/ Acala	Middling	Middling	Strict Lower Middling	Punjab	Sind/ Punjab	Type 329	MIDD
84/85	0.861	0.954	0.929	1.066	1.068	1.102	0.972				0.813	0.904	0.980
85/86	0.836	1.067	1.152	1.189	1.327	1.213	1.080	0.986	0.918		0.772	1.018	1.024
86/87	0.887	0.909	0.876	1.002	1.046	1.197	1.056	1.014	0.965	1.223	0.893	1.006	1.081
87/88	0.934	0.997	0.975	1.055	1.101	1.123	1.021	1.054	1.014	1.010	0.919	1.069	1.051
88/89	0.924	0.952	0.939	1.039	1.086	1.162	1.024	1.002	0.957	0.892	0.861	1.042	1.084
89/90	0.939	0.959	0.945	1.017	1.050	1.076	0.979	0.996	0.958	0.965	0.932	1.041	1.024
90/91	0.939	0.969	0.949	1.063	1.091	1.122	0.989	0.987	0.961	0.892	0.911	1.054	1.037
91/92	0.928	1.007	0.980	1.055	1.096	1.184	0.977	0.936	0.881	1.028	0.923	1.138	1.071
92/93[a]	0.918	1.028	1.001	1.086	1.107	1.191	1.024	0.924	0.881	0.956	0.875	1.102	1.118

[a] First 15 weeks.
Source: Cotlook Ltd.

Table 7.5 US cotton acreage, yield, production and value: the need for a hedging mechanism

Crop Year	Planted area (000 acres)	Harvested area (000 acres)	Yield per harvested acre (lb)	Production 480 lb bales (1000 bales)	Average price per pound (cents)	Value of lint production ($000)
1950	18 866	17 843	269	10 014	40.07	2 005 684
1951	29 353	26 949	269	15 149	37.88	2 868 720
1952	28 065	25 921	280	15 139	34.59	2 617 644
1953	26 872	24 341	324	16 438	33.64	2 654 683
1954	20 052	19 251	341	13 673	35.06	2 301 358
1955	17 991	16 928	417	14 698	33.72	2 379 030
1956	17 077	15 615	409	13 290	33.12	2 112 625
1957	14 310	13 558	388	10 948	30.92	1 625 119
1958	12 379	11 849	466	11 495	34.66	1 912 311
1959	15 833	15 117	461	14 527	33.04	2 304 265
1960	16 080	15 309	446	14 237	31.52	2 154 165
1961	16 588	15 634	438	14 283	34.37	2 356 309
1962	16 293	15 569	457	14 827	33.30	2 370 480
1963	14 843	14 212	517	15 294	33.64	2 469 647
1964	14 836	14 057	517	15 145	31.07	2 258 491
1965	14 152	13 615	527	14 938	29.35	2 106 088
1966	10 349	9 552	480	9 557	21.75	997 467
1967	9 448	7 997	447	7 443	26.70	953 820
1968	10 912	10 160	516	10 926	23.11	1 212 021
1969	11 882	11 055	434	9 990	22.00	1 054 981
1970	11 945	11 155	438	10 192	21.98	1 121 622

1971	12 355	11 471	438	10 477	28.23	1 419 624
1972	14 001	12 984	507	13 704	27.30	1 798 960
1973	12 480	11 970	520	12 974	44.60	2 779 504
1974	13 679	12 547	442	11 540	42.90	2 374 064
1975	9 478	8 796	453	8 302	51.30	2 043 678
1976	11 636	10 914	465	10 581	64.10	3 254 954
1977	13 695	13 275	520	14 389	52.30	3 614 938
1978	13 375	12 400	420	10 856	58.40	3 044 991
1979	13 978	12 831	547	14 629	62.50	4 391 378
1980	14 534	13 215	404	11 122	74.70	3 986 678
1981	14 330	13 841	543	15 646	54.30	4 075 403
1982	11 345	9 734	590	11 963	59.40	3 410 226
1983	7 926	7 348	508	7 771	66.40	2 478 176
1984	11 145	10 379	600	12 982	57.80	3 603 434
1985	10 685	10 229	630	13 432	56.50	3 645 410
1986	10 045	8 468	552	9 731	52.40	2 449 111
1987	10 407	10 035	706	14 760	64.30	4 555 017
1988	12 515	11 948	619	15 412	56.60	4 190 488
1989	10 587	9 538	614	12 196	66.20	3 877 888
1990	13 348	11 732	634	15 505	68.20	5 075 826
1991p	14 144	12 842	656	17 542	63.20	5 322 413

Source: USDA.

CHAPTER

8

Cotton futures

The history of futures markets

Futures markets evolved from forward markets. A forward market is similar to a cash, or spot, market in that it establishes a contractual obligation between two parties to buy or sell a specified good at an agreed price. In a spot market, delivery is immediate, whereas in a forward market, delivery is at a predetermined point in the future. Forward markets have existed for a very long time, but they became popular instruments when world agricultural trade began to expand in the eighteenth century. The reason they grew was price volatility. Without forward markets, merchants risk that the goods they bought three months ago in a far away location can be sold today at a price high enough to cover the costs of the transaction and the shipping and still make a profit. If prices fall in the three months, the merchant may actually make a loss. The more volatile prices are, the greater the risks that the merchant takes. A garment manufacturer or a spinner, for example, needs a steady supply of cotton to allow garment production all year round. The spinner could buy a year's worth of stock at one time, but would need to store all of it, and face some quality deterioration. Relying on the spot market exposes the spinner to the risk that the price of the cotton will vary from one month to the next, possibly putting him or her at a competitive disadvantage to other millers. When market prices are volatile, therefore, there are market participants with high price risk who have good cause to create a mechanism of establishing certain prices in the future.

This was what created the forward markets. A forward contract between a cotton trading company and an exporting farm or company works thus: the price, for example 100 bales of cotton, of a specified quality, for example strict middling US 1 1/32 in. is agreed on 1 January. The importer agrees to pay a price of $*x* per bale, where $*x* may vary considerably between cargoes depending on how the negotiations go and what both sides think the spot market will be like at the time that the bales are due to be delivered. The importer is spared the cost of having to keep very large stocks of cotton, and the exporting country knows the receipts from the cotton that will be received. US farmers sell between 5 and 50% (usually about a quarter) of their cotton crop as forward contracts. Most of the US forward contracts are therefore written in terms of the harvest from a specified number of acres rather than bales. Farmers agree to sell their harvest from specific fields rather than guaranteeing to supply a certain quantity of cotton. Under such conditions, high yields benefit the contract purchasers rather than the farmers, who have locked in their returns irrespective of yield. Forward contracts are also usually written in terms of a base quality. There is a schedule of discounts and premiums available from the Commodity Credit Corporation (CCC) in the USA to determine the value of cotton of different qualities. The amount contracted forward varies between different areas of the US cotton belt: in the Southwest it is low because yields on the high and rolling plains of Texas and Oklahoma are much more variable than in the Delta.

Forward contracts are usually either ‘fixed price’ or ‘call’. Fixed price contracts establish the price of the base quality in cents per lb. Call contracts fix the basis between the price received by the farmer and another quotation, most usually the futures contract. The farmer can activate the contract at any time over its specified period of life and settle on the actual price. They are effectively options on the futures market, but they are packaged as forward contracts for the farmers.

The problem with any forward market is that the risk of default is by no means removed, although in the case of the US cotton market in particular the farmer will almost inevitably produce a crop, of reasonably predictable size, and the buying merchant is probably one of a number of large cotton merchants with good credit ratings. In a forward market, by contrast, the risk of counterparty default is high. For example, the supplying country may, if the price of cotton has risen in the meantime, simply not deliver the cotton at all. Or the converse: if the price of cotton has fallen, then the trader may choose to source elsewhere - at the cost of a certain loss of reputation. One key problem is that forward contracts, being specific to two parties, do not have an active secondary market. A different trading company, with needs for a different quality of cotton delivered at different times of the year, would only buy the original exports at a substantial discount, if at all.

In a spot market, by comparison, apart from any credit risk attached to the granting of credit to the purchaser by the seller, there is no counterparty risk because the exchange of goods for money is simultaneous.

Futures markets evolved as a partial solution to the problems of forward markets. Two important features distinguish them. Firstly, futures contracts are standardised with respect to quality and quantity, delivery locations, dispute and arbitration procedures. Only the price is free to adjust to market conditions. This abstracts all the other aspects of the local buyer and seller factors such as the type of cotton required, from the transaction:

> Cotton futures are legally binding agreements, made on the trading floor of an exchange, to buy or sell cotton sometime in the future. Each contract is standard and specifies the quantity of cotton sold or brought, quality of cotton, months traded, price units, and the time of delivery or payment.
> (New York Cotton Exchange)

The second difference is that the futures market gets around counterparty risk through an institution called a clearing house. Only qualified members of the exchange, trading for commission paid by clients - known as brokers - and other members of the exchange trading for themselves, may actually buy and sell contracts. In Japan the function of the clearing house is undertaken by the exchange itself.

A futures market, such as the New York Cotton Exchange or the Tokyo Commodity Exchange (Tocom), consists of a place where traders meet to trade standardised contracts for goods. The New York Cotton Exchange is the oldest commodity exchange in New York, reflecting the importance of cotton to international trade in the nineteenth century, although since then it has diversified into frozen orange juice and financial futures contracts. A futures trading 'pit', or an electronic screen, is the place where the transactions actually occur. Once two traders, who may represent trading companies such as Volkart or Dunavant, have agreed to exchange a futures contract, that single contract, with one buyer and one seller, is registered with the clearing house or the exchange itself in Japan. It is then transformed into two contracts, one between clearing house or exchange and buyer, and one between clearing house or exchange and seller. This greatly facilitates trading, because after this registration, there is effectively no counterparty risk – the clearing house or exchange is backed by the major banks of the USA or the UK in the one case, and the entire membership of the exchange in the other. So if the seller wishes to close out his contract before the agreed expiry date, he does not need to seek out the original buyer: any buyer in the market prepared to pay any acceptable price will suffice. This new trade is also registered as a trade with the clearing house or the exchange, and the seller pays simply any difference in price between the original and the new price. The clearing house does not exist to make a profit or loss, but is funded, like the exchange, by a small fee per contract, and is backed by capital guarantees from the major banks in the host country. The NYCE has its own clearing house (the CCC), but since 1990 there has been increasing collaboration between the NYCE and the CBOT's Trade Entry and Matching (TEAM) system. The likelihood is that economies of scale will dictate that eventually the US commodity exchanges at least will combine their clearing operations, but this is not probable for some years.

How do clearing houses and exchanges protect themselves against counterparty risk? Firstly, through the initial margin, or deposit. To buy or sell a futures contract, you are required to post an initial cash deposit with your broker. He or she will call an ***initial margin/deposit*** with the clearing house or exchange to secure the contract. Initial margins are set by the exchange, in conjunction with the clearing house where one is involved, and they usually represent the equivalent of a typical day's maximum normal movement in the price of the contract. Because they are usually between 1 and 15% of the face value of the contract, initial deposits are what give futures contracts their exceptional leverage, or gearing. If the deposit is $15 on a contract bought with a face value of $100, a 15% decline in the value of the contract will wipe out all the initial investment. Conversely, a 15% rise will double the initial investment. Normally the clearing house or exchange will vary the amount of initial deposit required, depending on whether the customer is 'trade' or 'private', normally imposing a higher level of deposit on the private client. The second means is through what are called variation margins. If the contract is bought for $100, and tomorrow the price moves down to $95, then the clearing house or exchange will ask from the broker, and the broker will in turn ask from his client, the sum of $5 to cover the loss. (The broker may ask for more than the clearing house, depending on his own risk assessment of the client.) If the price then returns to $100, the money will be returned to the buyer, and if the price then moves up to $105, the clearing house or exchange will pay the buyer $5 - this money obtained, of course, from the seller. This see-saw continues for any buyer or seller until the buyer or seller "closes out' his contract by finding another counterparty. At any one time, therefore, the number of buyers and sellers holding open contracts is always the same and the net position of the clearing house or exchange is zero.

Since it should be so easy to close out contracts, because of the number of buyers and sellers in the market - at least in successful futures markets, which are called 'liquid' on account of this ease - there is little use of futures markets as actual delivery mechanisms. In the cotton futures markets that exist and have existed, physical delivery does exist as the eventual outcome of a contract that is not closed out in advance, although fewer than 10% of cotton futures contracts actually come to delivery. This is because the futures markets are used to protect against price risk and the remaining 90% of contracts are closed out for cash in advance of the delivery date. In addition, the complex procedures which the NYCE and Tocom insist must be followed before cotton can be tendered in accordance with the delivery procedures means that traders are reluctant to risk a shortage or oversupply of deliverable cotton and the price squeeze that would follow. During any one period, the stock of certifiable cotton can vary substantially, and taking cotton in and out of certification (through re-certification, which is permissible) is a major contributor to price volatility. Some futures contracts, for example those based on indices such as the world cotton contract, have a cash delivery mechanism so that no cotton ever changes hands.

Options, by comparison, are asymmetrical contracts between buyer and seller. Options have existed for centuries, as, for example, down payments for the right to occupy land at a certain rent, but they were not allowed to be traded on US exchanges until 1983. Options do now exist on futures contracts as well as on physical cargoes, in which case the counterparty risk is eliminated by using the clearing house, and there are currently exchange traded options on both the NYCE No. 2 cotton contract and the world cotton contract.

An option on a futures contract creates the right, but not the obligation, to buy or sell at a certain price at a certain time in the future. Clearly, the seller (grantor or writer) of the option has taken a much greater risk than the purchaser, whose risk is limited to the payment - the premium - that he has paid for the option. Consider a tailor-made 'over-the-counter' option between a trading company and an exporter for the exporter to buy cotton in three months' time at 67 cents per lb (a 'call' option because it involves trading the right to buy, or 'call' the goods).

In return for this right, the trading company pays the exporter, according to complex calculations of risk and value. Perhaps the price agreed is 67 cents/lb and the current price of the cotton in question is 69 cents. If the three months pass, and the current 'spot' price of cotton falls to 62 cents, the option expires worthless, and the exporter can sell the cotton elsewhere as well as having benefited from the value of the premium. The trading company buys at 62 cents. But the trading company can only lose the premium, perhaps 3 cents. If the price rises to 70 cents, the exporter must sell at 67 cents. Although more complicated option structures exist which limit this risk, in this simple option example the exporter is exposed to an unlimited price rise. Similarly, if the exporter paid the trading company a premium for an option to sell (a 'put' option), the trading company would be exposed to the full force of any price reduction, the only difference being that the maximum extent of this risk is predictable since prices will not fall below zero. The holder of the option may choose to exercise it or to expire unexercised. The choice between these two courses of action will depend on the strike price of the option and the futures price. If the strike price of a call option is below the current futures price, it is said to be 'in-the-money' and is worth exercising, and if out-of-the-money, then it is not worth exercising. Clearly for a put option, the reverse is true: the option has value if it is struck for a price above the current futures market price. A third option for the holder of an option is to close out the position by writing an exactly opposite option, and for a grantor to buy one, of exactly the same maturity and strike price. This may be a profitable exit before maturity, and means therefore that no option position need be maintained until maturity.

Clearly the complex price determination of options depends on the relative assessment of the risk of rapid price change. The price paid for an option is called the premium, and it is normally quoted in points and hundredths of a point. The premium is the cost of obtaining the right to a particular long or short futures position at a particular strike price. One option gives the right to one futures contract, so the total cost of the option in dollars equals the premium in points multiplied by 50 000 lb; see Table 8.1.

The premium is determined in the same way as futures prices, through open outcry in the New York pits. However, options' liquidity is never quite so substantial as futures', and the use of formal models to produce options' prices based on the current price and recent volatility of the futures contract (most famously the Black–Scholes

Table 8.1 Monthly volume (cotton) 1974–88

Option contract	Quoted price in points	Total dollar premium
Dec. 74 cent Call	2.50	1250
Mar. 80 cent Call	0.40	200
Aug. 76 cent Put	2.00	1000
Dec. 82 cent Put	4.00	2000

Source: NYCE.

model) is commonplace. Most theorists divide the premium into two elements: the intrinsic value, which is simply how much money the option holder would make if the option were immediately exercised, and the time value, which is the likelihood of further (or some) profit, and which diminishes over time and with movement of the option out of the money. At expiration, the option possesses only intrinsic value. The diminishing size of time value is one important reason why option holders often close out their positions before expiry. The price of an option therefore depends on the volatility of the underlying cotton contract, the time remaining before the option expires, and the relationship between the strike price and the futures price itself.

New York Cotton Exchange No.2 cotton contract specifications

Size:	50 000 lb.
Tick size:	$5.
Delivery months:	March, May, July, October, December.
Trading hours:	10.30–14.40.

The options contract trades during the same hours, and has the same size, tick size and delivery months.

Tocom cotton yarn futures contract specifications

Size:	1814.36 kg (4000 lb).
Tick size:	0.1 yen/lb.
Months:	six consecutive months.
Trading hours:	sessions at 08.50, 10.15, 12.50, 15.10.
Delivery:	warrants.

Nagoya Textile Exchange

Size:	4000 lb.
Tick size:	400 yen.
Months:	6 months.
Hours:	09.20, 10.30, 13.10, 14.20.
Delivery:	physical.

Osaka Textile Exchange

'20s contract:

Size:	2000 lb.
Tick size:	200 yen.
Months:	six months - current and following five.
Hours:	sessions at 09.45, 10.45, 13.45, 14.45.
Delivery:	physical.

'30s contract:

Size:	2000 lb.
Tick size:	200 yen.
Months:	six months.
Hours:	09.45, 13.45.
Delivery:	physical.

'40s contract

Size:	2000 lb.
Tick size:	400 yen.
Months:	six months.
Hours:	09.45, 10.45, 13.45, 14.45.
Delivery:	physical.

Uses for derivative contracts

Trading in cotton futures on the New York Cotton Exchange, and formerly the New Orleans Cotton Exchange and the Chicago Board of Trade, supplements the physical trades conducted in the spot market.

The price of a cotton futures contract fluctuates as market conditions change. For example, if more cotton is produced in Texas than expected, the NYCE cotton futures prices could rise as traders bid on the expectation of a smaller supply. The opposite can also happen. Suppose the southern United States has an excellent growing season and cotton production is up. NYCE cotton contracts could decrease in value as traders bid on the expectation of a larger supply.

The nominal objective of the New York market, which is one of the world's most active commodity futures market, is to provide a mechanism through which handlers of US cotton, including shippers in the USA, growers' co-operative associations, foreign importers and domestic and overseas spinners, can carry cotton stocks or make forward commitments without incurring large risks through unforeseen adverse price movements. Conversely, the cotton futures markets are used by handlers of cotton yarns and cloths to obtain protection against losses deriving from adverse price movements in stocks and the forward sales of their products. Because the futures market reduces risk, it assists in enabling those who make, buy or trade cotton to operate more economically, and so – in theory at least – they assist in reducing the margins that are needed between the grower and the final consumer.

This protection against adverse price changes is achieved through a mechanism known as 'hedging'. The fundamental principle of hedging in dealings with any commodity is that an opposite position is obtained in the futures market to that held in the physical market. For an unsold stock of cotton, the strategy involves selling on the futures market so that if prices fall, the futures will be able to be bought back more cheaply, thereby making a profit to offset the reduced price that the physical cotton will obtain. For a planned purchase, the strategy involves just the reverse: buying the futures so that if cotton prices rise, the futures can be sold at a profit. Hedging operations are of course based on the assumption that there is a high degree of parallelism between physical spot prices and the futures prices, which for US cotton is broadly true, although this of course depends on the allowable qualities of cotton in the New York contract. The reason that the prices on the New York Exchange closely parallel those in the spot market is that the New York market does allow for physical delivery of the cotton. Having sold on the futures market, there is no need to buy back the futures and settle in cash: the cotton itself can be tendered after trading ceases, and the buyers remaining will be obliged to receive the cotton. Futures can be used by both cotton producers who use the NYCE No. 2 contract to protect themselves against price movements in the physical market as well as cotton traders. Some examples of how this can be achieved are as follows.

Cotton grower

The biggest fear of many cotton growers is that prices will decline between the time the cotton is sown and when it is harvested and thus erode the profit margin. The cotton grower had hoped to get 70 cents for the harvest starting in mid-October, and that is the general expectation in the market, but in June it is already clear that there will be a bumper harvest and news from other producers suggests the same – prices are bound to fall. The producer now fears that the price of cotton will fall to 60 cents by October. So in June he or she sells the October futures contract for 70 cents. This is known as being 'short' of the market. Four months later in October the price of cotton has fallen but by even more than anticipated, to 55 cents. Nonetheless, he or she has to sell the cotton at 55 cents – 15 cents/lb less than hoped for. But at the same time he or she buys back the October futures contract which he or she sold in June for 70 cents, which has moved down in line with the physical market to 55 cents. Therefore he or she makes a 15 cents/lb profit on the futures transaction which offsets the 16 cents/lb loss in the physical cotton market, thus making the next revenue of the crop 70 cents/lb – just as the market originally anticipated.

Spinner

A cotton spinner had reckoned to spend about 65 cents/lb this year and has based his or her annual budget on this assumption. However, because of rains in Texas, he or she now believes that prices may rise to 70 cents. In February he or she expects to have used up the stock of low-priced cotton by May and will be forced to buy cotton at a new higher price to keep the mill going. But obviously he or she does not want to spend 70 cents if it can be helped. So still in February he or she buys May cotton futures contracts at 65 cents. This is known as being 'long' of the market. Three months later, the price of cotton has risen, but by more than expected, to 72 cents/lb. Nonetheless, the manufacturer must purchase physical supplies at the new high price, 7 cents/lb more than planned and 2 cents more than subsequently feared. However, at the same time he or she sells the May futures contracts, which have moved up in price in line with the physical market to 72 cents. On the one hand therefore, he or she makes 7 cents profit on the futures transaction, which balances out the greater than planned cost of the physical cotton, making the net cost of the cotton 65 cents, as originally planned.

Cotton trader

A cotton trader buys a 50 000 lb consignment for shipment in July, but at an advantageous price compared with the general market conditions. In order to balance the risk, the trader will sell an equal quantity of futures contracts for July as a form of price insurance. Since the quantity which the trader needs to cover is 500 000 lb, and the NYCE contract size is 50 000 lb, the trader needs to sell 10 contracts in order to hedge the risk. The trader will, in July, close his or her futures position by buying back his or her sold futures contracts. Any profit made on the futures market will offset loss on the physical deal and vice versa.

This use of the market is so important that it is worth including a fuller example.

A hedged cotton transaction handled by traders (source: Volkart Inc.)

10 July 199x. Trader buys 100 bales Mexico SM 1 1/16 in. from producer (50 000 lb). September shipping at 50 cents/lb fob (free on board) Guaymas: $25 000.

10 July 199x. Trader sells NY Exchange No. 2 Cotton Futures.

One contract for 100 bales (50 000 lb) equivalent October delivery at 48 cents/lb: $24 000.

The market risk, provided that the world cotton futures price (or the No. 2) behaves in the same way as the price of Mexican cotton, is now hedged, and the risk is reduced to a differential of 2 cents/lb 'on'.

25 August 199x. Trader sells same cotton to spinning mill, Sept. shipping at 49 cents/lb cif Bremen. Proceeds: $24 500, The trader's costs: freight, insurance, financial costs, possibly commissions, etc. – 6 cents/lb. Total cost: $3000.

Net proceeds basis fob Guaymas, 42 cents/lb: $21 500.

25 August 199x Trader buys one NYCE No. 2 contract for October delivery at 40 cents/lb: $20 000.

Result for trader:

Physical contract:	loss of $3500.
Futures contract:	profit of $4000.
Net profit:	$500.

In this bear market, the use of hedging has enabled the producer to sell at an attractive price, the mill is able to cover its needs at a competitive price when some producers may be unwilling to sell, and the trader is able to achieve a reasonable profit margin. In practice, the sums may not add up so easily, and it may be difficult to see where there is an actual profit. This may be further confused by the preferential accounting treatment afforded to hedging as opposed to speculative futures transactions.

Speculator

Some percentage, perhaps just under half, of trades carried out on the NYCE are by speculators who do not hold a physical position in cotton at all and have no intention of doing so. Their motivation in becoming involved in the market is purely to buy and sell the futures contracts at a profit. Their brokers may have discretion to use a certain amount of their money 'on call' to buy and sell, or they may telephone the client with a recommendation. In either case, the client may be an individual or a large pension fund with a certain percentage of its funds to be used in highly geared instruments such as futures.

One of the attractions of the futures market is that speculators can profit when the market goes down as well as up. They do this by 'shorting' the market - selling, and then buying back when the price falls. Another reason that speculators like futures markets is the low initial margin - returned when the speculator closes out his position, and currently standing at approximately 10% for the NYCE No. 2 contract. Finally, as futures markets generally exist only in markets that are volatile in price, the potential for rapid changes in price leading to substantial gains (and losses) always exists.

Speculators have access to a range of different trading options, of which just buying or selling (longs and shorts) are but two. For example, speculators can buy one month and sell another, thereby exposing themselves only to relative price changes between months. Another possibility is to trade a spread between the same months of different commodities. The most obvious example is to trade the spread between different delivery months, and the NYCE has introduced a special option contract to facilitate this trade. To the extent that these trades are less risky they do of course have a lesser profit potential.

Options open up another set of possibilities for both hedgers and speculators. The variety of possible options' strategies that can be used in the cotton market by other hedgers and speculators is extraordinary:

each of them has a different risk reward strategy and is appropriate to a different perception of market conditions.

Other functions of the market

One of the most important debates in futures markets is over what the futures price actually represents. Even exchange officials often disagree over this, especially with index contracts where there is no physical delivery, so it is far from surprising that international cotton traders in Europe, familiar only with the NYCE No. 2 contract and not working in proximity to other futures contracts where there is only cash settlement, are puzzled. However, with a physical commodity such as cotton that has a known storage cost and a physical delivery futures market such as the NYCE, economic analysis shows that the futures price bears a very close relationship to the spot price plus the cost of storage, or 'carry'. The USDA said in 1987:

> Spot prices should be less than futures prices, with the difference, or basis, representing the costs of storage plus delivery. As the contract delivery date approaches, the cost of storage decreases, and the basis should narrow to only the cost of delivery and of certifying that the cotton meets contract specifications. (*The US Cotton Industry*, 1987)

This certification alone is not inexpensive: it includes charges for compression, loading out, sampling, inspection and classing and amounts to about 4 cents/lb (Starbird *et al.*, 1987, p. 64). However, the other factor that affects futures prices - and this is where the debate lies - is expectations of future spot prices. One way of uniting these arguments is to say that 'As forecasts of supply, use and ending stocks change, the market signals smaller or larger rewards for the storage of cotton' (Starbird *et al.*, 1987, p. 64). Therefore a contango - a higher future than spot price - reflects expectations of high future storage costs. Conversely, if abundance is expected in the future, then the market will not reward storage. If there is a bad harvest, which leaves tight supply conditions in the US domestic market, then spot prices will be high, 'when current supplies are tight but an expected good harvest holds out the potential for rising stocks, spot prices can exceed futures prices' (Starbird *et al.*, 1987, p. 64). As harvest time comes around, if the next harvest is expected to be good, then the market will convey incentives not to store cotton, and the futures price will fall. If it actually falls below the current spot price, the market is known to be in 'backwardation', or, as is said in the USA, to have an 'inverse carrying charge'.

This relationship between the structure of futures prices and the costs of storing cotton prior to milling allow the providers of storage facilities to use the markets actively, hedging their risks. Cotton storage costs are basically the warehouse charges imposed for storing cotton bales, with insurance, the interest foregone on the cash which could have been deposited after sale of the cotton, and an element of risk premium determined by consensus expectations of future prices and storage costs. The exact storage cost for any given cotton depends on where it is harvested, milled and exported. For example, to transport Midsouth US cotton to Memphis costs less than 1 cent/lb, but to transport Arizona cotton to Galveston may be 2-3 cents/lb.

Futures prices are not, therefore, direct forecasts of future spot prices, and a comparison of futures and spot prices reveals this very much to be the case. Because there is no mechanism that affects price which is relevant only to the future, futures prices reflect storage costs rather than just expectations of future supply and demand. Poor weather during the growing season will affect the prices of new crop cotton, but it will also affect the price of the old crop because the market will need to ration existing supplies to cover the expected futures shortfall. Hence spot prices will increase; spot prices do contain an expectational element: the price of cotton now is affected by what the market expects supply and demand to be in the future. In so far as spot prices reflect expectations, so do future prices. But there is little expectational element in the futures price of a storable good.

Market discovery

The futures market provides a public source of prices - but in the case of cotton, for US cotton, which is the only cotton deliverable against the contract and the spot price of which is therefore the determinant of the futures price against the No. 2 contract on the NYCE. To some extent now other growths are also allowed in respect of the world contract. Exchanges began as meeting places where merchants would make open deals to overcome widespread suspicion that honest dealers were being shortchanged by the lack of clear public sources of prices. The prices set on the floor of the futures exchange are logged into a computer, transmitted by electronic news services such as Reuters and Manifest, and are then published in newspapers such as the *Wall Street Journal* and the *Financial Times*. Everyone can find out the price easily, and the exchange-traded prices can serve as a marker for deals done 'off-market'. The more volume of trading the futures price represents, the more credibility it has as a reflection of sentiment in the market, and in the case of cotton, NYCE futures prices have great credibility as market price signals. The heavily traded December contract, in particular, is seen as an indicator of new-crop demand and supply conditions because December is the first delivery month after most of the crop has been harvested. Almost 50% of the US crop is sold in terms of 'basis December'. The March and May crops are watched similarly for how midseason cotton demand is behaving, as the total supply is known virtually for certain by January. The price of October is influenced by the quality and amount of early harvested cotton in south Texas, by expectations about the total harvest, and also by demand.

Credit provision

Futures markets such as the NYCE help in the provision of credit to the agricultural marketing industry. An inventory of cotton that is hedged has a known value, and can be offered as collateral for a loan. Unhedged, the value of an inventory can fluctuate dramatically and is of little value as a loan.

Reducing uncertainty and volatility

Futures markets do tend to reduce uncertainty and, it is argued, price volatility. Price volatility in itself is not necessarily a bad thing, and there is academic argument as to whether futures markets do in fact stabilise cash prices. It depends, clearly, on the volume of trade in the physical market as compared with the futures market. If there is far more trade in the physical market, and a limited availability of physical supply with complex certification rules, as with cotton, then substantial speculative demand in the futures market may shift the physical market itself.

Success of the markets

The success of cotton futures can be judged by Table 8.2. Tocom's cotton yarn contract dropped to fifth in the table with volume falling over 10%. The Exchange itself says that:

> The world's heaviest volume of cotton yarn futures trading
> takes place on the NYCE. On the Tokyo Commodity

> Exchange, the representative price for cotton yarn is that for 'Cotton No. 40.' The Tokyo market serves as an important price indicator in the East Asian cotton trade. (Tocom, 1991)

In 1991 the New York contract, however, capitalised on falling volumes elsewhere to record an increase of almost 80 000 lots and thereby rose from eighth to fourth place (Table 8.3).

Cotton futures are very big business indeed; they are the major contract of the NYCE, of course, and cotton is the third largest contract of Tocom. The Brazilian exchange, BM&F launched a US$ cotton contract on 27 September 1991 but this had only traded 179 lots by the end of the year.

The international cotton futures project

The New York Cotton Exchange began to think about the idea of using the Liverpool Index as the basis for an international cotton futures contract as early as the 1960s. A number of cotton traders in particular in Europe observed that the Cotlook 'A' Index frequently diverged from the prices being quoted on the New York Exchange. The extent of this divergence can be viewed in the form of a graph, Fig. 8.1.

The divergence intensified during the three seasons after 1989 owing to shortages in US supply and restrictions in the USA against cotton imports. The effect was of course that New York futures prices rose, whilst the Cotlook 'A' Index lagged. For example, the first six months of 1991 saw NYCE prices rise by 17%, whilst world cotton prices fell by 2%. In the USA, robust export sales and domestic mill consumption which held up impressively in the USA in 1991 was actually chasing quite a limited cotton supply. As the season went on, prices rose. But internationally, freshly harvested cotton from Australia and Paraguay, and better than expected crops in Pakistan, CIS countries and in Africa, depressed world prices still further. This in turn caused overseas spinners to avoid US cotton and buy the cheaper priced world growths substitutable in quality. The NYCE was particularly aware that what was happening in the CIS states, especially their new-found right to trade independently, was having a major impact on prices. So the basis (gap) between the New York futures and physical cotton on the world market, as the NYCE itself says, 'has become unmanageable under certain market conditions'.

Table 8.2 Success of cotton futures

Position	Contract	Exchange	Country	1992	1991	1990
1 1 2	Raw Sugar	CSCE	USA	3 667 481	4 268 546	5 424 801
2 5 4	Cotton Yarn	Tocom	Japan	2 330 029	1 607 228	1 865 880
3 3 5	Coffee	CSCE	USA	2 152 383	1 772 648	1 774 050
4 4 8	Cotton	NYCE	USA	1 701 258	1 614 244	1 534 611
5 9 7	Cocoa	CSCE	USA	1 397 235	1 223 519	1 635 917
6 6 6	Cocoa	Fox[a]	UK	1 382 041	1 536 752	1 701 586
7 2 1	Raw Sugar	Tokyo Sugar Exchange	Japan	1 279 354	2 351 512	6 375 344
8 - -	Opt. Coffee	CSCE	USA	860 943	411 550	n/a
9 10 9	Coffee	Fox[a]	UK	854 746	1 195 789	1 170 176
10 7 3	Opt. raw Sugar	CSCE	USA	848 750	1 512 976	2 393 016

[a] Futures and Options Exchange, London.
Source: Futures and Options World Directories 1992 and 1993

Table 8.3 New York Cotton Exchange options volume

1974		1975		1976		1977		1978		1979		1980	
Jan	41936	Jan	37015	Jan	87505	Jan	83678	Jan	75314	Jan	142088	Jan	286058
Feb	30233	Feb	30519	Feb	65415	Feb	73973	Feb	77246	Feb	139547	Feb	278901
Mar	28415	Mar	30239	Mar	75634	Mar	97996	Mar	92494	Mar	122110	Mar	290133
Apr	29799	Apr	42701	Apr	70485	Apr	87535	Apr	75240	Apr	122110	Apr	206682
May	37526	May	42193	May	91806	May	71819	May	96764	May	131245	May	191067
June	32888	June	38620	June	106815	June	86923	June	117272	June	191763	June	128393
July	37201	July	33273	July	87990	July	65754	July	64924	July	102928	July	175619
Aug	31978	Aug	38860	Aug	73882	Aug	54936	Aug	98652	Aug	115842	Aug	188542
Sept	29918	Sept	55487	Sept	79118	Sept	49418	Sept	94758	Sept	122235	Sept	224323
Oct	36522	Oct	64298	Oct	82660	Oct	40544	Oct	120670	Oct	135924	Oct	186615
Nov	33273	Nov	52887	Nov	86984	Nov	54402	Nov	146217	Nov	207659	Nov	160939
Dec	26745	Dec	59011	Dec	61054	Dec	59717	Dec	96250	Dec	159349	Dec	173133
1981		**1982**		**1983**		**1984**		**1985**		**1986**			
Jan	156303	Jan	100240	Jan	97947	Jan	127435	Jan	61860	Jan	73576		
Feb	166292	Feb	104444	Feb	123620	Feb	133891	Feb	57034	Feb	77152		
Mar	150297	Mar	103697	Mar	139032	Mar	127448	Mar	61921	Mar	57433		
Apr	117653	Apr	106892	Apr	132447	Apr	126780	Apr	65709	Apr	74520		
May	90157	May	95597	May	139953	May	139988	May	47606	May	57479		
June	117249	June	151068	June	153932	June	111050	June	47604	June	64742		
July	99175	July	116211	July	143116	July	71522	July	41069	July	75545		
Aug	95312	Aug	99674	Aug	156680	Aug	51941	Aug	40524	Aug	94025		
Sept	96956	Sept	101119	Sept	121972	Sept	39519	Sept	42747	Sept	130989		
Oct	120301	Oct	94265	Oct	113759	Oct	83631	Oct	55046	Oct	111359		
Nov	115533	Nov	98597	Nov	137472	Nov	86729	Nov	63829	Nov	113812		
Dec	89985	Dec	83988	Dec	90187	Dec	39907	Dec	51543	Dec	84618		

Table 8.3 Continued

1987		**1988**		**1989**		**1990**		**1991**		**1992**	
Jan	110165	Jan	114672	Jan	124764	Jan	135260	Jan	133415	Jan	134531
Feb	106520	Feb	119832	Feb	135963	Feb	109852	Feb	179656	Feb	134184
Mar	88621	Mar	97533	Mar	113028	Mar	126820	Mar	148918	Mar	149711
Apr	106620	Apr	118614	Apr	118573	Apr	125045	Apr	156978	Apr	167778
May	110006	May	127880	May	135044	May	144725	May	174690	May	173128
June	132180	June	147303	June	167442	June	135040	June	122242	June	153194
July	98339	July	89907	July	91381	July	132329	July	116458	July	
Aug	127048	Aug	100447	Aug	151034	Aug	129081	Aug	115175	Aug	
Sept	131214	Sept	108285	Sept	165255	Sept	103936	Sept	107742	Sept	
Oct	124319	Oct	135221	Oct	143713	Oct	142670	Oct	125558	Oct	
Nov	126582	Nov	124221	Nov	178889	Nov	150958	Nov	150545	Nov	
Dec	107124	Dec	84334	Dec	124034	Dec	98895	Dec	82867	Dec	

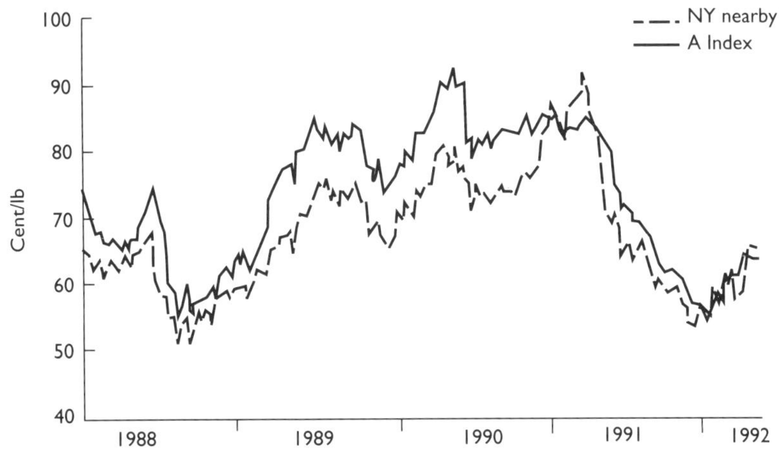

8.1 NY Nearby vs 'A' Index, cif Northern Europe, 7 January 1988–23 July 1992. Weekly prices ended Thursday (*source*: NYCE).

What this illustrates is that international traders attempting to use the New York market to hedge were sometimes caught both ways. The NYCE admitted that futures hedges on the No. 2 contract were especially difficult for the Australians in 1991, for example. Another good example of such a period of divergence was June 1992. The US market accelerated generally upwards, as domestic demand for the product held up well - indeed the US domestic mills were enjoying a boom period - and the excessive rains in Texas threatened to cause a serious reduction in the scale of the next US crop - more immediately, there was a substantial amount of decertification of the July 1992 tenderable crop (i.e. its certificates ran out of time), which raised the possibility of some squeeze on the nearby position. Meanwhile in Europe the position as reported by Cotton Outlook was anything but interesting: not even very attractive offers of Central Asian cotton were finding buyers as the market languished in the doldrums, the mills were not buying and traders were undoubtedly having a difficult time.

The logical move, therefore, was to use the Cotlook 'A' Index as the basis for a contract itself. As Joe O'Neill of the NYCE said in response to criticism of the No. 2 contract as a hedge for world cotton: 'Fortunately, there existed an index that was familiar to the trade and also possessed the confidence of world participants and the US government - the Cotlook 'A' Index' (NYCE World contract literature, 1992). In the USA, however, there were, and still are, mixed feelings about the idea of an international contract. Doubts were basically threefold:

1 *It would detract from the liquidity of the existing contract.* There is no doubt that a good percentage, perhaps up to 40%, of the business currently using the New York contract would shift across to an international contract were a liquid contract available. That would clearly reduce liquidity - the lifeblood of any contract - in the existing contract and make very large deals more difficult to achieve. Although by European commodity standards the NYCE contract is huge, trading 1 614 244 lots in 1991, up from 1 534 611 lots in 1990, this compares with over 21 million lots for crude oil, for instance, and speculative volume is quick to evaporate when liquidity starts to seep away - the fate of the London sugar contract is ample evidence of this.

2 *It would affect some US firms adversely.* Some US firms have tremendous market presence in the New York market and would not wish to see that influence dissipated by a multiplication of contracts.

3 *The Commodity Futures Trading Commission (CFTC)* had reservations, which took many months to clear up, about the permissibility of using the Cotlook 'A' Index - which Cotton Outlook freely admit is an offer index composed like the Baltic Freight Index (used for the Baltic shipping futures contract in London) as the basis for a futures contract. One trader, irritated at the delay, described the NYCE in dialogue with the CFTC as the blind leading the blind.

But in 1991 one of the Liverpool trading firms, Baumann Hinde, made an approach to the London Futures and Options Exchange (London Fox - now the London Commodity Exchange) to consider the launch of an international cotton futures contract before the Americans did. Baumann Hinde believed that a little healthy competition never hurt anyone, and rightly thought that the possibility of London Fox launching the contract would inspire the Americans to some more immediate action. Negotiations with Cotton Outlook were begun, and initial approaches produced a generally positive response from cotton traders in Liverpool and London Fox attended a conference in Geneva to explain how the contract would work to European traders who were a little sceptical of an index futures contract. They were familiar with a physical delivery contract such as New York, where the price of the futures is (to simplify an important academic debate) determined by the spot price of the physical commodity plus the cost of carry. With an index contract, settling against the value of the index at the time of delivery and in cash, the contract's value depends more on expectations in the market for the future value of the index, subject to expected net financial carry costs. This the traders found somewhat difficult to envisage, although of course in other exchanges, such as the International Petroleum Exchange, physical delivery contracts run alongside index contracts and the traders are pleased to use both. The project at London Fox was eventually scuppered by the then chief executive who insisted that other projects, involving the electronic trading of other commodities, had to have priority, and so the international cotton futures project at the London Fox (which was planned to trade in the old-fashioned style, on the floor of the Exchange) was stopped.

By this time the competition in New York, Joe O'Neill's New York Cotton Exchange, had manoeuvred, or been manoeuvred, into pressing ahead with its own version of the same scheme. Cotton Outlook reopened negotiations with New York when the London Fox decision was made. Eventually in early 1992 the NYCE made an application to the CFTC for authorisation, but with the withdrawal of the threat of competition from London Fox (by 1992 London Fox had become embroiled in internal difficulties and had disbanded its Business Development team altogether, having taken the decision not to launch any more contracts), pressure for an early launch had evaporated. By this time the NYCE had got so far with its planning, and with CFTC approval finally granted in the early autumn of 1992 after much speculation that it would take much longer, the launch date was 1 October 1992.

The international cotton contract at NYCE is cash-settled against a five-day average of the Cotlook 'A' Index, rather than on physical delivery of the cotton to any world destinations such as Bremen or Liverpool itself. There are too many such destinations and the problems associated with premiums and delivery regulations would be excessive, although the trade would undoubtedly have felt more comfortable with a physical delivery contract. The five-day average has been chosen to assist the CFTC agreeing to the contract, in that firms will find it more difficult to quote anything other than correct prices to Cotton Outlook over a relatively longer time-scale than just one day.

Cotlook World Cotton futures

Contract size: 50 000 lb (approximately 100 bales) times Cotlook World Cotton futures price.

Trading hours: 09.55 to 14.40 (NY time). Trading will cease in the spot month at 12.30 on the last trading day.

Trading months: March, May, August, October, December (seven months traded at one time).

Price quotation: cents and hundredths of a cent/lb.

Minimum price fluctuation: 1/100 cents ($5/contract).

Maximum price fluctuation: 2 cents. Higher limits may take effect under certain market conditions.

Last trading day: 17 business days from the end of the spot month.

Settlement: cash settled using a consecutive five day average of the Cotlook 'A' Index price which includes the last four trading days plus the day following the last trading day.

The advantages of the World Cotton futures and options contracts are numerous. International producers, merchants, shippers and textile mills will possess hedging opportunities to manage their international cotton price risk far more accurately than at present. The contracts will enable international textile mills to lock in a maximum purchase price for their cotton. Unlike US mills, mills internationally are able to buy from countries with lower prices. Because these lower-priced quotes (in practice over the past year, CIS quotes are certainly amongst them) are represented in the World Cotton Contract settlement price, the hedge will be viable. A caveat should be entered at present which is that CIS cotton in particular, and to a lesser extent, Pakistani cotton, is available very much cheaper than its competition, and this may make using even a Cotlook 'A' Index hedge a slightly fraught operation. When the contract actually launched, it became quickly evident that there was a lack of buyers, which remains the problem. It was pointed out that the European traders and mills, with the exception of the Italian mills, are not great futures users in the sense that they do not customarily buy and sell cotton on call. They use forward contracts instead. Volumes have not been impressive as a result, rarely being over 100 in the period since launch and showing a certain disturbing tendency to go lower. Nevertheless, open interest reached over 2500 before the first contract settlement month, for the December 1992 contract, and remained over 1500 by the end of the year. Traders were generally optimistic that once US cotton ceased to be tied via the AWP quite so closely to the world price, the No. 2 contract would decline radically in effectiveness as a hedge, and the world contract would come into its own.

The NYCE ignored these problems in promoting their new contract. They pointed out that international producers must be price-competitive in order to sell their cotton on the world market, or else they would lose market share to cheaper growths, and that 'this competitiveness in the world export market tends to draw quotes in line with each other', thereby demonstrating that the World Cotton Contract was a viable hedge for the cotton producer as well. This is only partially true, because the international market is actually more segmented than in the USA. Sellers would have been able to lock in guaranteed minimum sales prices, but that is what they would have been – minimum prices. International traders, dealing on a daily basis with the growths included in the World Cotton Contract, would undoubtedly have found that the contract was useful for them to limit their price exposure and to manage their basis risk, both of which could not be achieved through the No. 2 contract. For a contract such as cotton which is substantially used by the trade itself, there is little doubt that the World Cotton Contract would have proved highly useful.

Some examples circulated by the NYCE on how the World Contract would work illustrate the potential benefits of the contract. NYCE suggests that if an international spinner is a steady buyer of world cotton throughout the year, it currently must pay whatever price has been quoted for that growth at that particular shipment date. Actually, the price that it pays may vary considerably from the Cotlook 'A' Index quote, because the index price is offer rather than acceptance and trade. At least with the World Contract futures and options, if the world price for cotton futures was 70 cents today, and the spinner feels that world production will turn out somewhat below expectations so that there is a risk that prices will rise, he or she can buy futures contracts as far out into the future as necessary. If the World Contract price rises to 80 cents, the spinner will have to pay 10 cents more for a particular world cotton growth, but this would be offset by a 10 cent rise in the world futures price, assuming the basis remains constant. Conversely, if the spot world price moves to 60 cents, the spinner will lose the money on the hedge but pay less for the cotton on the spot market.

The producer's hedge works, as expected, in the other direction. If prices are at 70 cents, then he or she can plant the crop and sell World Contract futures at the current price, so that if prices fall to 60 cents, the producer will gain 10 cents on the futures hedge but will receive 10 cents less on the international spot market, which effectively guarantees the producer whatever the current World Futures Contract price is for the delivery month in question. Of course, the hedge can be lifted at any time, it can be a partial hedge, and the use of options can reduce the potential outlay in variation margins. The NYCE introduced options on the World Contract immediately, and perhaps would have been advised to wait until the demand built for them.

Cotlook world cotton options

Contract size: one contract World Cotton Futures Contract.

Trading hours: as futures.

Trading months: seven months of the cycle: March, May, August, October, December and the nearest two of January, February, April, June, July, September or November.

Price quotations: as futures.

Minimum price fluctuation: as futures.

Maximum price fluctuation: none.

Last trading day: as futures.

Automatic exercise: 10 points in the money.

Traders could have used the World Contract to lock into a maximum price that they were prepared to pay when they had to buy cotton, and to set minimum prices that they were prepared to achieve for selling world cotton growths. No doubt specialised products could have evolved which would have allowed financial institutions with specialised commodity departments to offer individual growth hedges to smaller trading firms, which when combined would have been hedged on the World Contract. Of necessity, since Cotton Outlook alter the selection for the 'A' Index on the basis of what is frequently traded, this would prove relatively straightforward for the financial institutions. Suppose of the five growths in the 'A' Index, a trading firm has a stock of two, and that these are the cheapest. Other firms will have stocks of the other three growths. The financial institution can issue a hedge based on the two cheapest growths if it also sells an equivalent hedge for the other three growths. Any divergence from this exact equivalence would have to be based on careful pricing of the undertaking by the financial institution to cover the basis risk.

Speculators should also have had a major interest in the World Contract but they were unwilling to risk in substantial numbers a contract that was new, and that lacked trade liquidity. Because of the considerable volatility of international cotton prices, speculators will seek to benefit from disparities between prices and perceived fundamentals of the market. Consistent news of changes in world supply and demand – for example the publication by the USDA of new world supply demand balances – news on the ratio of stocks to consumption, import and export data and policy decisions (especially by China and Pakistan), and changes in US government policy, world political economic and social events such as those occurring in Eastern Europe, and the general market mood to which all these factors contribute, all interact to determine the efficient 'world market price' for cotton. Cotton is undoubtedly a commodity where news travels fast, where there are many interested parties, and where pricing is extremely efficient.

Finally, as the NYCE pointed out, there should have been substantial arbitrage opportunities between the No. 2 and World Cotton contracts, which should have virtually guaranteed the accuracy of the price discovery mechanism of both contracts and help enhance liquidity.

It was not inconceivable that the World Cotton contract, had it been eventually successful, would have been one of the world's leading soft commodity contracts. The crucial phase of any contract, especially one traded on the floor of an exchange, is the start. It was entirely in the hands of the international cotton traders of the world as to whether the World Contract succeeded or not – they all stood aside in the early stages and expected someone else to establish prices, run risks and take positions; nothing happened and the contract failed. It was very much to be hoped that the NYCE succeeded in promoting the contract successfully and that trading companies did eventually participate in some strength. A successful new soft commodity futures contract would have inspired the futures industry worldwide at a time of doubt and recession, but little comfort in the short term can be taken from the performance of the World Cotton Contract. Trading fell almost to zero by the beginning of 1994, although prices were still quoted.

The regulation of futures trading

Like other financial market activities, futures trading is regulated by the authorities. In fact, the majority of regulation is conducted by the exchanges themselves, since they determine the day-to-day 'rules of the game' and have established mechanisms for the settlement of disputes. There are many disputes between traders, for example over the incorrect registration of trades, that are quickly resolved through these procedures. Regulation in the more general sense of involvement by public authorities backed by law is, however, very common and extensive.

The economic justification for regulation is that unregulated markets cause a socially undesirable transfer of resources and bring the markets themselves into disrepute. In financial markets, there are two kinds of market failure that can justify intervention by elected or appointed authorities. First is held to be asymmetric information: trading companies and brokers know a lot more about the workings of future markets than do members of the investing public and the public has a right to be protected from unscrupulous people posing as investment advisers. It is often felt that this danger is greater in the futures markets because the investment is highly leveraged - high profit opportunities are combined with equal opportunities for losses. Thus investors need to be protected, it is suggested, by disinterested third parties, which in turn will encourage investors to participate in the markets, giving them much-needed liquidity. To the extent that the economy as a whole has an interest in seeing that there are free and fair 'level playing-fields' to determine prices, regulation, it is said, cannot be left to the market itself.

The second problem is that of maintaining the market. Financial markets lie at the centre of the economy, and distress in the financial systems can cause extremely serious problems in the real economies of the world, as the events of 1987 demonstrated.

On the other hand, financial regulation tends to be self-justifying, self-accelerating and stifling of the very initiatives that protect the system itself. Futures markets have been banned at various times - index contracts in particular were banned in the USA until 1983, and in Japan (from 1868 until after World War II) and Germany (until 1986). But there is an equal danger if regulations are not enforced, or are too lax: this provides an environment where fraud can flourish and the public loses confidence in market institutions.

Regulations in the USA and Japan, the two countries in which there are cotton futures markets, are different in structure. In the USA, futures market regulation is provided for by the CFTC, which in 1975 replaced the older Commodity Exchange Authority. The former body was under the authority of the Department of Agriculture, since futures trading was primarily in agricultural goods. The growth in financial futures trading made a move from Department of Agriculture control logical. The CFTC has two main purposes, and has a wide range of legal powers to enforce its views. Firstly, the CFTC aims to ensure that futures trading meets its main purpose, so it has power to approve the operating rules of an Exchange and to approve any new contract before it begins trading. The CFTC also monitors positions held by traders and can impose limits on positions held. This is to prevent market manipulation. Secondly, the CFTC regulates the dealings of the futures industry with the general public. Thus it imposes minimum standards of training of brokerage staff, has capital adequacy standards for brokers and demands the segregation of clients' from brokers' funds. The CFTC is far too small to monitor all US exchanges all the time, but it does frequently intervene in the markets. Often its decisions arc challenged by the exchanges in the courts: it operates within a clear legal framework.

There has been a long-running dispute between the CFTC and the Securities and Exchange Commission (SEC), the body responsible for the regulation of securities markets - including equity options - in the USA. It has long been thought anomalous that trading in stocks should be regulated by one body whereas trading in stock index futures should be regulated by another. This sentiment was particularly marked in the October 1987 stock market crash, when there was debate over the role played by futures markets in the market decline. It seems likely that some redefinition of responsibilities between the two bodies will be inevitable, although the Chicago exchanges, including the CBOT, whose lobbying power in Washington is formidable, have expressed a strong preference to remain under the authority of the CFTC, which they regard as more sympathetic than the SEC.

In Japan, there is no one central organisation responsible for the regulation of all futures and options contracts. The choice of regulatory body depends rather on the particular instrument or commodity underlying the contract. Cotton futures contracts, being agricultural commodities, are under the Commodities Exchange Act and are governed by the Ministry of Agriculture, Fisheries and Food (MAFF). In Japan, protection for futures traders is achieved in a similar fashion: deposits of cash and securities are made with commodity commission merchants (CCMs) for margins. In order to protect clients against the potential bankruptcy of their CCM, in 1967 the Commodity Exchange Law was revised and the 'guarantee money' system was established. A CCM deposits monies with the Exchange - which in Japan acts as its own clearing house, partly accounting for the pressures which resulted in the revisions made to the Commodities Exchange Act in 1990, towards merger, first internationally and now within Japan. Of course, little has yet happened because of territorial rivalries between exchanges. The amount of the guarantee monies is determined by the Law. In July 1975, the 'Association of Compensation Fund' was established to supplement the guarantee money with which it is interdependent. Should a CCM conclude a compensation agreement with the Association, then the guarantee money requirements are halved: if the CCM becomes insolvent, and unable to discharge his liability to his clients, then the Fund will compensate the client up to the amount settled by the contract if the liability exceeds the guarantee money level.

For the further security of clients' monies, there is also client segregation from house accounts and rules governing the use of clients' funds. In 1988 cash settled index futures and options became legal, and in 1991, the Japanese Diet approved new legislation, the Commodity Investment Act, aimed at promoting and regulating the offering of domestic commodity funds in Japan, allocating supervisory responsibility amongst the various ministries; the guidelines for operation of this new law were completed in 1992. As a result of this legislation, a self-regulatory organisation, the Japanese Commodity Fund Association, has been established, aiming to promote the business development of such funds and to enforce existing consumer protection measures - there are already over 40 members. Further reform of the Securities and Exchange Law to expand the definition of security is envisaged, but no major reforms of the current system or regulation for commodities futures has been promulgated.

Forecasting the market

An assessment of likely price movements is a central component of decision making by cotton traders, governments, hedgers and speculators on futures markets, and it is the lifeblood of the trading companies such as Cargill and Meredith Jones. Speculators will only take up a trading position - thereby providing valuable liquidity to the market - or trading companies purchase or sell (including sell short, i.e. sell what they have yet to buy) stocks of cotton - if they have a firm view of what is going to happen to prices. Although hedgers are risk-averse, they may choose not to hedge substantial components of their portfolio if they believe that prices are set to rise substantially (if they are holding stocks of cotton).

Most cotton traders rely on their own experience to trade cotton. Experienced cotton traders made a great deal of money, for instance, when US cotton prices rose substantially during June 1992 when the rains occurred in Texas and prices increased. In the cotton trade an appreciation of not just the direction of price movement, but its extent and likely duration is knowledge which is at a great commercial premium, as in fact there are relatively few experienced physical cotton traders with the necessary judgement, contacts, experience and courage to take a position as to, for example, whether to enter into a forward contract for Central Asian cotton which could be worth up to US$1 million.

Assessing likely price movements is a central element of decision making for both hedgers and speculators in the New York and Tokyo cotton futures markets. Speculators will only adopt a trading position if they take a view of what prices are likely to do, and although hedgers are risk-averse, they frequently decide not to hedge a position or to hedge only part of it, if they have a strong view of prices. Much trading and judging of prices is experience, especially when balancing risks and rewards and deciding the precise time of entering a trade and closing it out. Apart from this experience and personal contacts, there are two basic methods that are used, especially in the futures markets where there is an available 'track record' of prices: fundamental analysis and technical analysis. These methods are quite important for cotton futures because, owing to the volume of transactions, both the New York and the Tokyo markets are relatively liquid - price movements are reasonably reliable estimates of opinion and change.

Fundamental analysis

Fundamental analysis relates price movements to changes in supply and demand conditions: it is therefore the kind of price analysis with which economists are familiar. Because long-term prices always relate to supply and demand conditions, fundamental analysis is always advisable for participants in futures markets - and the cash market. For example, cotton traders will always pay close attention to the weather conditions in the US Delta, because this may determine the availability and quality of the next year's US crop. Demand tends to fluctuate throughout the year, but there are also longer-term considerations, such as the attractiveness of synthetic substitutes for cotton, growth in real incomes and consumer tastes, which have an impact on cotton prices. Traders always pay close attention to all weather factors, although to the extent that changes in supply follow already anticipated patterns, the price will not change. Carryover stocks are also important contributors to price levels, because they affect import demand and hence price levels. The location of stocks between countries and organisations also affects price levels.

International organisation and national policies, especially of the US government, is a major determinant of cotton prices in the USA and hence New York futures prices. This explains in large measure why cotton traders are very concerned to be close to the USDA and the US Foreign Agriculture Service. US government decisions can have an effect in both the short and long terms. For example, traders restrained the size of their positions in July 1992 in anticipation of the first survey results of the USDA on the 1992/93 crop, which were published on 12 August 1992.

All cotton traders study the supply and demand characteristics for cotton worldwide. They are familiar with drought conditions in China if they exist - as these will affect whether China needs to import or can export. They will study monsoon times and insect pests. Their analysis of the supply and demand conditions for cotton is comprehensive and thorough, though usually unaccompanied by any formal econometric models. Part of the reason for not adopting complex econometric models is this very difficulty of incorporating climatic factors. Another problem would be the need to construct models for every different national market - and where does one market stop and another begin? There is little doubt that fundamental analysis is worthwhile: droughts, late monsoons, gradually increasing populations, relative prices of substitute goods such as synthetics and other factors such as the income elasticity of demand for cotton products are what determine cotton prices in the long run.

Inflation and the growth in real incomes, together with the steady increase in world population and the relatively fixed areas available for cotton cultivation, foreign exchange reserves, currency values and other economic variables have all had their effect on long-term cotton prices. They will continue to do so (Tables 8.4 and 8.5).

ICAC's forecasts, unlike those of Cotton Outlook, RER or other more 'hands on' forecasting companies, are based on very serious econometric models. They take their cotton consumption figures as a function of economic growth and international demand, and these figures in turn are taken from such organisations as the World Bank

Table 8.4 Textile fibre consumption per capita: results of logarithmic regressions

Country group/period	Constant	GDP per capita	Textile price index ($t \div 1$)	Structural income weight	Structural price weight	Autocor-relation	*R*2	Durbin Watson statistic
World	−4.748	0.478	−0.067	0.903	0.131	0.807	0.97	1.14
1963–1990	(2.504)	(20.449)	(2.772)	(2.496)	(2.378)	(4.960)		
Eastern Europe and USSR	1.720	0.252	−0.082	-	-	0.471	0.82	1.68
1963–1987	(6.205)	(2.915)	(1.488)			(2.925)		
Industrial countries		0.763	−0.119	-	-	0.837 941	0.87	2.03
1963–1987		(8.923)	(1.454)			(6.549)		
Developing countries	−2.193	0.774	−0.024	−0.007	-		0.96	1.33
1963–1987	(5.153)	(9.622)	(1.642)	(1.718)				
Developing:								
Africa	−1.905	0.466	0.121	0.016	-	0.588	0.85	2.24
1963–1986	(1.433)	(1.627)	(5.120)	(3.487)		(3.265)		
Asia	−2.307	0.843	−0.092	-	-	0.530	0.95	1.89
1963–1986	(2.126)	(5.815)	(0.927)			(2.571)		
Europe	1.185	0.360	−0.248	-	-		0.53	1.70
1963–1986	(4.403)	(4.357)	(4.332)					
Middle East		0.649	−0.175	0.010	-	0.811	0.91	1.56
1963–1986		(5.355)	(1.642)	(1.682)		(5.483)		

Table 8.4 Continued

Country group/period	Constant	GDP per capita	Textile price index ($t \div 1$)	Structural income weight	Structural price weight	Autocor-relation	$R2$	Durbin Watson statistic
Latin America and Caribbean 1963–1986		0.382 (27.860)	−0.018 (1.787)	−0.009 (2.723)	-	0.669 (4.090)	0.89	1.79

* The dependent variable is textile fibre consumption per capita for each country group. All equations were regressed in logarithmic form. The numbers in parentheses are t-statistics. $t-1$ indicates a time lag of one year. The autocorrelation coefficient is an estimation defined by a Cochran–Orcutt process. The textile price index is the relative textile fibre price index for the appropriate region, except in Eastern European countries. Africa and the Middle East, where, due to multicollinearity, an absolute textile price index was used. Income and price weight variables are the GDP per capita and fibre price index multiplied by a dummy variable that distinguishes between years of normal consumption growth and years where per capita consumption declined. See: ICAC, *World Textile Demand*, October 1990 for details of the model.

Table 8.5 World cotton and non-cotton consumption per capita: results of logarithmic regressions[a]

Country group/period	Constant	GDP per capita	Price of cotton/other textile fibres price index ($t \div 1$)	Autocor-relation	*R*2	Durbin Watson statistic
Cotton						
World 1975–1989	−4.231 (2.817)	1.167 (3.622)	−0.073 (2.088)	0.618 (4.487)	0.91	1.26
Non-cotton						
World 1975–1989	−3.825 (5.360)	1.097 (6.428)	−0.070[b] (1.587)		0.84	1.59

[a] The dependent variable is end-use cotton consumption per capita for cotton and end-use non-cotton textile consumption per capita. Both equations were regressed in logarithmic form. The numbers in parentheses are *t*-statistics. $t-1$ indicates a time lag of one year.

[b] The price variable is the relative price of non-cotton textiles. See: ICAC, *World Textile Demand*, October 1990 for details of the model.

and the IMF. The World Bank figures state that world growth in GDP slowed from 3% in 1989 to 2.1% in 1990, and 1.5% in 1991. In fact, world economic growth was almost certainly negative in 1991, 1992 and 1993: small increases in developing countries were overshadowed by large falls in the first world, for example in the USA and the UK, and of course also by the collapse of the former Soviet and East European economies. The responsibility for the over-optimism of early 1992, therefore, rests with the agencies that predicted macroeconomic revival in 1992, not with the ICAC's connecting equations which are quite reasonable. ICAC believed in 1991 that the year's cotton consumption figure worldwide would be 38.9 million tonnes, but it did not reach that level. ICAC believes that fibre consumption generally follows economic performance in the world's regions, and also maintains that world economic growth of 2% minimum is required for international fibre consumption to rise; in other words that the income elasticity of demand for cotton is sharply kinked at 2%. That certainly explains why fibre consumption did not rise in 1991 or 1992, but it overlooks the demand picture when broken down by country.

There are other shorter-term fundamental factors. The most noticeable is seasonality, because of the new harvests in both hemispheres which affect prices. There are many dozens of types of cotton with different characteristics and harvest times - November/December in the northern hemisphere and April/May in the southern. Consequently there are two major seasons when cotton initially comes to market, although it can be stored for several seasons. The main harvest takes place in the northern hemisphere between October and January and prices are likely to follow seasonal patterns: high prior to harvest and low just following the harvest - not so much because of the availability of the cotton itself as because of the increased certainty of its availability. The production and stocks in the largest producing/consuming countries, China, the USA, India, Pakistan and the CIS countries may alter the seasonal pattern. To the extent that these seasonal patterns conform to market expectations, there will be no substantial market response.

Stock levels are also extremely important. The stock/consumption level plays an important role in determining prices. The ICAC state that their price forecasts are based 'on the statistical relationship in the last 19 seasons between the Cotlook "A" Index and the ratio of ending stocks to consumption.' The equation used also takes into consideration changes which are expected in, for example, government stocks of cotton held by numerous countries - including most importantly for world trade prices, China. All national stocks have a much larger effect than the smaller stocks held by private companies.

Finally, and extremely importantly, the weather. The amount of rain that falls during the main growing season is of course crucial in determining the annual crop level. Cotton-growing areas particularly in Asia frequently experience periods of drought or floods, although this can also happen in the USA. Even after the main crop has been harvested, weather continues to play an important role.

Cotton analysts and traders are sensitive to both national and international policy too. Cotton is an important commodity internationally and access to political decisions is crucial. There are several key policies internationally which cotton traders watch. First and foremost, the policy of the US government: how much cotton does it want planted, and how much exported? Cotton traders know that the US agro-industrial complex is so interrelated with government that what happens in Washington is every bit as important for determining cotton prices as events in Arkansas or Texas on the cotton plantations themselves. Secondly, Chinese export policy and production must be considered. Thirdly, the EC agricultural policy and the CAP: does the EC intend to bring down its intervention price barrier to imports, and if so, over what time period and to what extent? Governments may take unexpected actions that drastically affect the price of cotton in the open market. Imports or exports may be restricted by means of bans and/or taxes or encouraged by means of subsidies. For example, China does not reveal its crop production figures until relatively late, partially so that other countries will not know the size of its import demand or export surplus (and partially because the information collection and analysis is not yet adequate to the huge task). The accuracy of information collected through the analysis of these economic and political factors is essential for the prediction of the future direction of prices and a successful trading strategy.

The problem with fundamental analysis is that of translating the general principle that demand and supply conditions are the determinants of prices into short-term trading instructions, in a market such as cotton where the accuracy of the futures market in interpreting short-term trading signals is famous. Cotton traders are still sceptical of the statistical techniques that can be used to quantify the impact of administrative changes, but they tend to work better in predicting long-term changes - indeed, for long-term forecasting, there is no alternative to statistical methods. But long-term modelling is no use if short-term margins - in a highly geared instrument such as futures - work against the trader; or if the specific deal the trader intends in a physical market does not work because of loading problems, quality failure on inspection, or worst of all, failure to pay.

For short-term analysis it is virtually impossible to build a model that is sensitive to daily changes in supply and demand conditions that depend on market sensitivities often on the other side of the world. Cotton traders use econometric models, if at all, to provide background analysis of long-term trends. Given that variation margins are taken daily, traders operating on fundamental analysis stand to suffer serious cash-flow losses if short-term trends move against them on the New York futures - especially if the short-term factors are peculiar to the New York market. In the short term, it is clear, expectations change so rapidly that the construction of statistical models incorporating them for cotton prices has not been attempted. All traders are sceptical, anyway, of economic forecasting - although cotton traders, having seen prices rise and fall schematically for so many years, perhaps less than many others.

Technical analysis

The alternative, which is certainly used by some participants on the New York markets and is closely followed by cotton analysts in the major US cotton trading companies, though not apparently by any in the European physical cotton markets, is 'technical' analysis, based on the evaluation of cotton futures prices themselves.

This type of analysis is more commonly used for short-term trading; there are a large number of different systems in operation. Arguably, markets can be observed performing distinctive patterns in the evolution of their prices: the market price is revealed by the pattern of prior price movements. Technical analysts analyse whatever trends can be detected in order to try to establish what is likely to happen next. It is also the case that technical analysis can be combined with cyclical theory - at its most extreme, the long-term economic cycles that are currently affecting all markets including commodities. Sometimes technical patterns can be constructed through the interaction of cycles in price series. Technical analysis further assumes that prices follow patterns that do not require external economic considerations to understand and predict, at least in the short term, but technical traders also use other data from the markets, such as trading volume, open interest, price spreads between contract months and the equivalent data from options contracts. The existence of a liquid cotton options market creates greater technical trading interest than there would otherwise be.

The key difficulty is always to establish that the patterns can be validated, which is why experienced technical traders usually put in stops (sell or buy orders cancelling out their original positions) at crucial price points which would invalidate their reasoning. Technical traders contend that all fundamental factors have already been incorporated into the current price of the security - for example, a cotton future - and that changes are instantly discounted. The high liquidity of the cotton futures market in New York is an incentive for technical traders, as it provides an incentive for speculators to become involved, and technical analysts place great store in liquidity itself, as this smoothes out the price distortions caused by individual orders which may be irrational. So far as the NYCE cotton market is concerned, there is little doubt that the cotton price interacts with the spot market price. New York prices respond to all three sets of stimuli: fundamentals impacting the physical market; what one might like to call 'quasi-fundamental' factors - reports, analyses and comments on the market which quickly become incorporated into short-term market psychology; and technical factors (monitored by the NYCE just as closely). The NYCE publishes market volume and open interest (number of contracts yet to be closed out) statistics on a daily basis. Although there are in theory an infinite number of potential traders in a market, in practice the data provide some evidence of why prices are evolving in the way that they are. For example, if both price and volume figures rise together, this is probably because of fresh long positions in the market, so that once volume begins to fall, price is very likely to follow suit. Again, the higher the open interest, the more vulnerable a market becomes to sudden changes, especially given that because of the very high volumes involved in the market, individual contract months on the NYCE can become subject to 'squeezes' where there appears to be insufficient cotton to meet the necessary deliveries.

Other technical indicators are essentially ways of following trends and changes in trends. Identifying a trend can be simply a matter of looking at a bar, point or figure chart and relying on visual inspection. A more complex way is to calculate moving averages (averaging the sum of a predetermined number of daily closing prices - as a new closing price is added to the total, the first price is subtracted), which reduces the extent of variation caused by a single day's trading; these can be taken across different time periods, for example over 9 or 15 days, with the problem that the longer the period taken, the less distortion but also the less information that is likely to emerge from the analysis. Similarly, the momentum indicators can also be used. Chart analysis is perhaps the most commonly used method, however. This entails graphing prices on a bar chart, where each day's highs and lows are written, with a line connecting them and a horizontal line showing the day's closing price. These charts can show the evolution of prices over time very clearly, and changes in trend - identifying support and resistance zones - are often thought to be signalled by characteristic chart patterns. One of the most famous is the 'head and shoulders' formation, in which prices form a curve, fall back, rise again above the top of the first curve to form the 'head', and then repeat to form the second shoulder. Technical analysts see this as a bearish signal. Other patterns that are looked for are 'downtrend channels', 'double tops', 'wedges', 'flags' and 'pennants', 'island tops' and 'island bottoms', and 'key reversals'. Once prices emerge from a depressed state they tend to trade in a channel, where declines towards the uptrend line are buying opportunities and stops can be placed outside the trend line. Charts that go back decades are studied to establish significant trends in the market; most traders study long-term charts on a monthly and weekly basis, so they are able to spot potential longer-term price reversals before analysts looking only at daily prices do. Considerable scepticism attends all these contentions, mainly from those who believe that the markets follow, albeit imperfectly, the basic fundamentals of the underlying cash market, and that the technical patterns simply reflect a process of adjustment by the market to changes in outside conditions. That is not to say, however, that individual traders cannot make substantial profits out of this lack of correlation: they can.

The problem for technical analysis is how to tell the difference between a change in price that is due to daily fluctuations, and one that is due to a trend. The many aspects of technical analysis can be combined into a trading system, i.e. a system of rules which are used by traders and fund managers. There are many successful trading systems, but they are not publicly available, and the suspicion must be that a measure of competent intuition is allowed to disrupt rigorous technical instructions from time to time. Comparing technical and fundamental analysis as trading tools, it is clear that fundamental analysis is very limited in advising the initiation of trades and is weak in being able to predict the effect of sudden shocks. By comparison, technical analysis can reveal something of market psychology, but can generate a host of conflicting signals.

How successful has forecasting been?

The answer has to be: not very, and the more complex the forecasting method, the worse the results.

Price forecasts for cotton do change frequently and dramatically.

Table 8.6 Cotlook 'A' Index prices

	1988	1989	1990	1991
Q1	68.72	64.25	77.03	84.15
Q2	66.70	76.64	89.60	83.74
Q3	59.31	82.48	84.45	74.57
Q4	59.17	80.63	82.72	63.20
Year	63.48	76.00	82.70	76.42

The actual prices as measured by the Cotlook 'A' Index are as given in Table 8.6. In June 1992, the EIU forecast the Cotlook 'A' Index at 58 for the second quarter of 1992, 59 for the third, and 59 for the fourth, compared with 63, 65, 67 and 68 in February 1992. Predictions for 1993 were 60, 62, 64 and 65 for each quarter, compared with 70 and 72 for the first two quarters in February 1992. This particular forecast appears to react very sensitively to current short-term trends.

Another example can be seen when cotton consumption figures of the present season have very much failed to live up to the relatively optimistic estimates of February 1992 – 19.3 million tonnes. Evidence from 1990/91, when consumption had actually fallen slightly (see the section of 'Trends in consumption and stocks' in Chapter 5) and the prominent continuation of the world recession might have suggested otherwise. However, to be fair to the ICAC, there was a slight return of optimism amongst economic pundits in the early part of 1992, until a grimmer realisation set in during the spring.

The ICAC produced some results of its forecasting exercises in September 1991. They said that the ICAC textile demand model forecasts produced in 1990/91 had been adjusted within 1% of current estimates: the same year forecast of world textile consumption for 1989 prepared for the 48th Plenary Meeting in Scottsdale, Arizona, in October 1989, was 0.2% below the most current estimate for that year. The forecast prepared for 1990's Plenary Meeting in Montpelier, France, was 0.7% above the most current estimate for 1990. However, the consumption forecast for 1990 carried out in 1989 was 2% over-optimistic. The importance of the accuracy of these forecasts for cotton companies internationally cannot be over-emphasised: a cumulative 2% over-optimism over 10 years produces a result for 2000 of over 26% inaccuracy: eating well into investment decision profits through under-utilised machinery, apart from the personnel implications. The effect of the optimistic 1992 ICAC consumption forecast was an optimistic price forecast, owing to the reasons stated above. The ICAC said in February 1992 that they expected only a moderate fall in prices in 1991/92, down to a season average of 78 cents for the 'A' Index, followed by another fall of 3 cents to 75 cents in 1992/93. (This should be compared with the reality described in the section on 'Trends in consumption and stocks' in Chapter 5.) Less optimism was expressed by Cotton Outlook, unencumbered by econometrics and over-optimistic inputs from international bodies, who looked only at the figures presented to them by traders themselves and concluded that there was no especial reason for expecting prices to rise above 70 cents during either season - a view that with great justification they continued to hold. Cotton Outlook forecast only consumption and production, not prices. To be fair to the ICAC, in February 1992 they did point out that in the past their forecasts had been accurate to within about 4 cents.

It would be churlish to dwell too greatly on the inaccuracy of the 1991 ICAC forecasts, but the point about cotton prices is too important to ignore. In 1990, the ICAC suggested that cotton prices would remain in the mid 80 cents range, and they averaged 83 cents, which was an accurate forecast. The ICAC was also correct in its forecast that the variation would be narrow - it was between 79 and 86 cents. But in 1991/92, although the ICAC correctly predicted price falls, they underestimated the extent of the fall. Rather than trading around 78 cents, and rising to that level from the 70 cents achieved at the end of 1991, the 'A' Index fell throughout 1992 almost consistently, reaching levels in the 50 cent mark by the end of the year. What happened to produce the differences from the ICAC projection was the increase in production combined with a failure of the mills internationally to increase consumption.

The US Foreign Agricultural Service published an analysis of the reliability of its own projections in its document WASDE-277-30, for a number of commodities. The results for cotton were as shown in Table 8.7.

What all this hindsight also demonstrates is that forecasts based on long economic series are extremely vulnerable to 'shock' - fundamental changes in the relationships between the variables. What happened in 1992 is that the long-term economic growth determinants did not 'bounce back' in the way that most forecasts suggested they might. Green considerations and the long-term effect of the worst recession since the 1930s in the West, together with the constant availability of low-cost CIS cotton, may mean that previous assumptions of cotton demand growth being highly correlated with economic growth statistics may be erroneous. More than that, if the ideas of Professor David Pearce and others are seriously adopted, following ideas put forward as early as 1970 by Kenneth Boulding (in

Table 8.7 Differences between the projected (in April) and final estimates for cotton, 1981/82–1991/92[a]

Region	Average (%)	Average (million 480 lb bales)	Difference (million 480 lb bales)	Difference (million 480 lb bales)	Below final (years)	Above final (years)
Production						
World	0.8	0.6	−3.0	0.4	8	2
US	0.2	0.0	−0.1	0.1	4	2
World excluding US	0.9	0.6	−3.0	0.4	7	3
Exports						
World	3.2	0.7	−2.3	0.9	6	5
US	2.3	0.1	−0.5	0.3	1	8
World excluding US	4.1	0.7	−2.3	0.9	6	5
Mill use						
World	1.5	1.2	−2.4	1.0	6	5
US	3.2	0.2	−0.6	0.2	8	1
World excluding US	1.6	1.1	−2.0	1.2	6	5
Ending stocks						
World	6.6	2.0	−3.9	3.3	7	4
US	5.9	0.3	−0.3	1.3	4	7
World excluding US	6.8	1.8	−3.8	2.7	7	4

[a] Final estimate for 1981/82–1990/91 is defined as the first November estimate following the marketing year and for 1991/92 last month's estimate.

Helfrich, pp. 157–70), the very measures of GNP that we currently use will be phased out to be replaced by measures that actively take into account the environmental cost of production – and cotton remains a high consumer of fertilisers.

CHAPTER

9

The international cotton trade

The pattern of cotton trading after World War II

After World War II, the number of exporting countries increased, as well as the volumes traded. Production after 1945 grew more rapidly than consumption: productivity grew and many smaller countries increased the amount of their land allocated to cotton production.

It took time for wartime controls to be lifted: currency restrictions and price support mechanisms (for example the Raw Cotton Commission in the UK between 1947 and 1954) became part of the international trading regime well into the 1950s. The UK Labour Government's decision, against fierce international opposition, not to allow the free determination of cotton prices in Liverpool at the end of World War II was decisive in preventing the re-emergence of the Liverpool cotton market. It was only in 1954 that the market was resumed, and it failed to succeed, closing down in the 1960s. Several factors were responsible: special US export programmes weakened the link between the US markets and Liverpool with the long-term effect of reducing the relative significance of Liverpool. Speculation in the markets declined, as did stockholding. The USA produced more than it consumed before and after World War II. The majority of the

US crop has been exported since 1945. Presently, the bulk of the cotton produced in the western portion of the belt, and part of that produced in the central portion, is exported. US government programmes have affected international cotton trading in respect of the volume of US exports, the stabilisation of world prices and the protection of the US garment industry. The whole process began as early as 1931 when a Grain Stabilisation Loan was extended to the Chinese government; the Export–Import Bank and its predecessor agencies made several such loans to China and Europe in the pre-war years, and PL-320 and straight barter were used in the 1940s. Import quotas were established in the 1930s and have continued unamended since 1950 at 1 1/8 in. staple or more, 95 188 bales/yr (first come, first served, in terms of importing countries) and 30 240 bales total for cotton of less than 1 1/8 in. staple, with country quotas such that Mexico has the largest quota, of 18 507 bales. The import of 5.5 million lb of cotton waste is also allowed. The USA has attempted to use both the GATT and the Multi-Fibre Agreement (MFA) to restrict the import of textiles and clothing to the USA, although there have been measures to allow imports of 21 days' mill use if the monthly average spot price exceeds the previous 36 month average by 130% or more.

Between 1946 and 1954, the US loaned cotton to China, Japan and Europe to assist them in re-establishing their textile industries. It was commented in 1958 by the Liverpool Cotton Association that

> the task of the cotton merchant is not that of his pre-war counterpart. Mills have become merchants and US shippers selling on Liverpool spot terms ex quay have taken the place of the merchants. The cotton merchant does not feel that it is worth the risk to trade unless the price of the futures has been higher than the price for imported cotton plus the differential for storing and tendering on the Liverpool market.

In the late 1960s, world trade in cotton was about 3.8 million tonnes annually. The USA, Mexico and Latin America were the largest exporters, and the largest importers were the EC and Japan. The USA continued to provide S32 government funded subsidies until 1970. Soft currency sales, long-term dollar credits and barter were all assisted by PL-480 until the mid-1970s. The cotton trade was trans-Atlantic and trans-Pacific, from the New World to the Old and to the Far East. Cotton exported from the USA used to go primarily to the UK, to other European countries, to Japan, to China and to Canada, in that order. US cotton competed only very indirectly with other cotton in the US market, because transportation costs made its import into the

USA uneconomic, but competition with other cottons internationally has been intense since 1945. The value of exports in the late 1960s averaged some $2.4 billion.

By the mid-1970s, world trade had increased only slightly to 3.9 million tonnes, worth $4.4 billion at current prices, showing the effect of inflation over the decade but little else. The USA was still the largest exporter but the Middle East now followed it very closely and exports from the centrally planned economies had also increased - this was the height of the attempt to create an 'alternative world economy'. largely based on barter, amongst centrally planned economies within Comecon. In the mid-1970s, the energy crisis and its adverse affect on economic growth and trade worldwide had the effect of cropping the growth in the international trade of raw cotton, which had expanded in 1973 to 4.6 million tonnes and then contracted by 10–15% in 1974. After 1974 the US no longer supported its prices above world prices, and the need for export assistance was much reduced. Limited use of PL-480 continued and the Export–Import Bank loans were still used to finance some cotton shipments, but CCC credits and credit guarantees became the main form of assistance to cotton exports. Cotton goods output fell, stocks in importing countries were drawn upon, and importers such as India and China increased their own production. The EC in particular reduced its imports, although other countries such as the Mediterranean countries, Japan and those Far Eastern countries aiming to maintain their developing textile export businesses continued their imports: Hong Kong, South Korea, Thailand, the Philippines, Malaysia, Indonesia and Bangladesh should be counted amongst them. But the recession badly affected some exporting countries such as Turkey, Brazil, Sudan and Pakistan. The USA started to subsidise its exports, although in general US government restrictions limited the amount of cotton grown. The objective everywhere was to fight to retain market share.

After the slight recovery in 1975 from the recession of the previous year, the volume of world trade fell again in 1976 by 2%. A number of different factors caused this fall, all connected with the continued recession and the structure of cotton prices. The market began to change: between 1970 and 1986, developing countries increased their participation in the international cotton trade dramatically. The level of exports increased from 4.2 million to 13.2 million bales, up over 200%. In value terms this increase was far greater: up from US$571 million in 1970 to US$2.831 billion in 1986, an increase of 70% in real terms (Tables 9.1–9.4).

Table 9.1 Exports (million tonnes)

	1966–70	1971	1972	1973	1974	1975	1976	1977	1978	1979	1980	1981
Developed												
USA	0.764	0.936	0.701	1.246	1.197	0.871	0.748	0.973	1.279	1.527	1.823	1.269
Others	0.115	0.124	0.093	0.128	0.105	0.093	0.138	0.111	0.117	0.143	0.164	0.177
Developing												
Latin America	0.967	0.666	0.831	0.823	0.652	0.764	0.605	0.704	0.898	0.748	0.641	0.604
Near East	0.943	1.104	1.046	1.076	0.768	0.832	1.034	0.747	0.803	0.723	0.377	0.530
Pakistan	0.133	0.153	0.292	0.165	0.110							
Other Far East	0.039	0.041	0.081	0.062	0.056	0.248[1]	0.261[1]	0.138[1]	0.161[1]	0.156[1]	0.377[1]	0.467[1]
Africa	0.346	0.393	0.365	0.351	0.292	0.278	0.341	0.275	0.275	0.336	0.324	0.313
China												
CPEs	0.525	0.548	0.653	0.729	0.800	0.801	0.880[2]	0.975[2]	0.859[2]	0.794[2]	0.863[2]	0.921[2]
Total	3.833	3.965	4.062	4.581	3.980	3.887	4.007	3.923	4.392	4.792	4.808	4.281

	1982	1983	1984	1985	1986	1987	1988	1989	1990	1991	1992	1993	1994
Developed													
USA	1.392	1.205	1.497	1.095	0.657	1.433	1.34	1.68	1.70	1.45	1.13	1.31	1.50
Others	0.227	0.275	0.251	1.095	0.657	1.433	1.34	1.68	1.70	1.45	1.30	1.31	1.50
Developing													
Latin America	0.598	0.539	0.487	0.640	0.373	0.537	0.64						
Near East	0.583	0.623	0.601	0.482	0.611	0.331	0.45						
Pakistan								0.30	0.27	0.45	0.30	0.41	0.28
Other Far East	0.327[1]	0.510[1]	0.401[1]	0.649[1]	1.379[3]	1.209[3]	1.25						
Africa	0.297	0.325	0.343	0.400	0.491	0.540	0.65	0.45[6]	0.50[6]	0.51[6]	0.52[6]	0.47[6]	0.46[6]
China				0.261	0.558	0.508	0.36	0.19	0.20	0.13	0.15	0.15	0.20
CPEs	0.949[2]	0.824[2]	0.651[2]	0.698[2]	0.731[2]	0.803[2]	0.75[4]	0.73[5]	0.33[5]	1.81[5]	1.78[5]	2.00[5]	2.06[5]
Total	4.839	4.301	4.231	4.259	4.646	5.182	5.70	5.27	5.05	6.08	5.54	5.99	6.08

Source: USDA and ICAC
1 Including Pakistan 2 Excluding China 3 Asia and the Pacific 4 USSR only 5 Central Asia 6 Francophone Africa only

Table 9.2 Imports, 1966–74 (million tonnes)

	1966–70	1971	1972	1973	1974
Developed					
W Europe	1.444	1.264	1.287	1.511	1.241
Japan	0.742	0.755	0.802	0.855	0.814
Other developed	0.146	0.134	0.143	0.140	0.130
Developing					
Far East	0.520	0.628	0.557	0.717	0.695
Others	0.115	0.119	0.128	0.118	0.105
CPEs	0.923	1.069	1.067	1.230	1.000
Total	3.840	3.969	3.979	4.584	3.985

Source: USDA

Table 9.3 Imports, 1989–94 (million tonnes)

	1989	1990	1991	1992	1993	1994
Developed						
W Europe	1.15[1]	1.04[1]	0.99[1]	0.90[1]	0.96[1]	0.98[1]
Developing						
Far East	1.65	1.68	1.56	1.36	1.30	1.31
CPEs	0.76[2]	0.47[2]	1.52[2]	1.26[2]	1.20[2]	1.26[2]
China	0.41	0.48	0.36	0.05	0.08	0.08
Total	5.53	5.32	6.34	5.88	5.99	6.08

Source: ICAC
1 EC only 2 E Europe/CIS only

The international cotton trade: 1980–90

The cycle is clear and repeated: anxiety over the nature of the northern hemisphere crop reaches a peak in the spring/early summer of any year, and beings to dissipate as the crop is first harvested, then ginned.

Until 1985 relatively high international cotton prices sustained a virtuous circle: peasant farmers received adequate incomes, domestic tax takes were high, fertilisers could be supplied profitably, and the cotton companies internationally made satisfactory profits. Trade grew steadily; see Fig. 9.1.

This virtuous circle was abruptly broken in the mid-1980s, creating a cotton depression which despite 'false dawns' has basically continued into the 1990s. In May 1984 a kilogram of fibre sold at

Table 9.4 Stocks (million tonnes)

	1966–70	1971	1972	1973	1974	1975	1976	1977	1978	1979	1980	1981
USA												
Total net exporters	2.086	0.922	0.721	0.892	0.846	1.221	0.800	0.650	1.193	0.846	0.575	1.444
Other net exporters	0.024	0.022	0.024	0.046	0.041	0.083	0.085	0.081	0.125	0.142	0.144	0.204
EC	0.294	0.262	0.238	0.311	0.244							
Other net exporters	0.394	0.389	0.401	0.497	0.486	0.713[3]	0.673	0.539	0.626	0.592	0.535	0.703
Developing net exporters	1.098	1.148	1.291	1.241	1.432	2.148	1.349	1.352	1.540	1.314	1.344	1.054
Developing net importers	0.705	0.590	0.741	0.826	0.749	0.792	0.668	0.629	0.816	0.877	0.671	0.848
CPEs	0.945	1.108	1.195	1.161	1.535	1.496	0.627[4]	0.475[4]	0.529[5]	0.512[4]	0.735[4]	0.474[4]
Afloat (estimated)	0.024	0.043	0.087	0.163	0.087	0.108	0.108	0.043	0.022	0.022	0.022	
China							0.716	0.876	0.647	0.487	0.663	0.944
Total	5.62[1]	4.484[1]	4.698[1]	5.137[1]	5.420[1]	6.561	5.026[2]	4.645[2]	5.498[2]	4.792[2]	4.694[2]	5.671[2]

	1982	1983	1984	1985	1986	1987	1988	1989	1990	1991	1992	1993	1994
USA	1.728	1.728	0.605	2.338	2.035			1.55	0.65	0.51	0.81	1.02	1.11
Total net exporters						5.996	5.42	5.73	4.86	5.47	8.10	7.65	7.08
Other net exporters	0.150	0.131	0.131	0.219	0.277								
EC													
Other net importers	0.637	0.717	0.705	1.069	0.629	1.653	1.45	1.58	1.44	1.34	1.30	1.17	1.12
Developing net exporters	1.030	1.079	1.135	1.772	1.748								
Developing net importers	0.888	0.857	0.721	0.689	1.231								
CPEs	0.443[4]	0.428[4]	0.594[4]	0.536[4]	0.590[4]								
Afloat (estimated)													
China	0.801	0.801	1.662	3.978	3.902	2.169	1.43	1.13	0.99	1.55	3.21	3.00	2.53
Total	5.677[2]	5.741[2]	5.558[2]	9.157[3]	10.412[6]	7.649[6]	6.87	7.30	6.30	6.82	9.36	8.83	8.19

Source: USDA and ICAC
1 Beginning of year 2 At 1 August (closing stocks) 3 Including EC 4 USSR only 5 USSR and E Europe only 6 Opening stocks 7 Total net importers

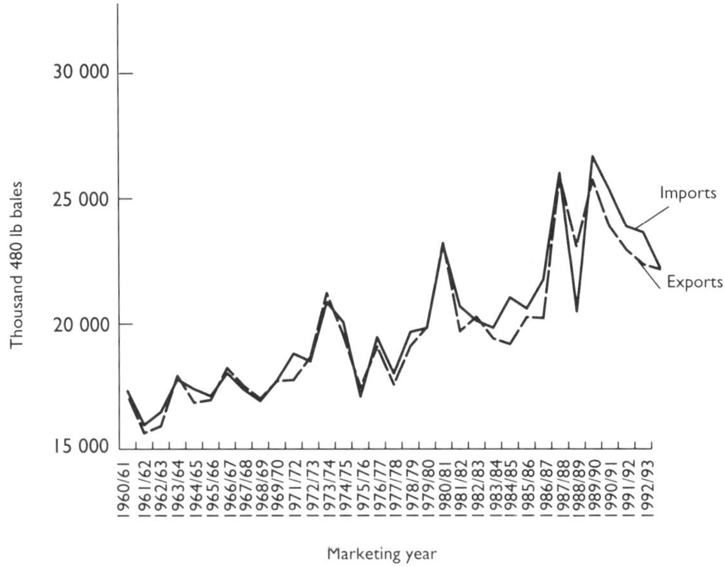

9.1 World cotton imports and exports (thousand 480 lb bales).

US$3.18, 15 months later the price had dropped to US$0.94. The spectacular harvest and the sale of highly competitively priced Chinese cotton were the causes. In West Africa the effect on national budgets and the national cotton companies alike was extreme and deleterious. In one year alone losses of 100 billion francs were announced - too great for national treasuries conceivably to absorb.

The market did improve towards the end of the 1980s, but the improvement was shortlived.

1988

Export earnings from cotton in 1988 were US$7.5 billion (FAO) and were expected to rise further in 1989, reflecting higher prices and larger shipments, combined with lower production. This optimism encompassed a belief that stocks would fall further from their level of 7.1 million tonnes (FAO) in August 1989 - the third year in succession that they had fallen. World trade volume in 1988/89 increased by 9% to nearly 5.7 million tonnes (FAO), which was the second highest recorded. The USA's export surplus was reduced, owing to impressive domestic demand and consumer textile spending which was accompanied by a build-up in stocks. China also reduced its exports, diverting cotton to meet the need for lint of its own textile industry;

Chinese stocks also fell, down to three months' mill requirements, half the level of three seasons earlier. The USSR and Pakistan compensated, the latter exporting a record 0.8 million tonnes (FAO). Australia, Africa and Latin America all continued their export growth, Brazil excepted.

1989

The US cotton marketing loan was designed to move US cotton to the market while providing some income protection to US farmers. In fact, a comparison of activity prior to the cotton marketing loan and since the passing of the 1985 Farm Act shows a 40% increase in average annual US cotton exports. The average annual farm value of those exports has risen over 40% since the beginning of the loan. Under the 1981 farm bill, covering the years 1982–85, offtake was between 8.4 and 12.7 million bales, averaging 10.9 million bales. The average farm value, excluding government payments, was $3.1 billion. Between 1986 and 1991, offtake ranged from 13.9 to 16.5 million bales, averaging 15.3 million with an annual average value of $4.4 billion: a 42% rise in annual average farm value. In the past, the US cotton industry had been the residual world supplier in the face of growing world stocks and low world cotton prices. In the following year, the US cotton industry would absorb the majority of the adjustments to correct any world cotton market disqualification.

1990

In 1990/91 stages one and two of the US government cotton marketing loan were activated. The objective was to prevent the USA becoming the residual world supplier when there are surplus world stocks, such as since 1990. Indeed, US exports have traditionally followed the world stock/use ratio, but in 1992 US exports kept up despite the fall in use relative to stocks. The objective has been to avoid the accumulation of US stocks – which can only be disposed of in a following year of underproduction – through the combination of user certificates and loan deficiency payments (LDPs). LDPs increase cotton sales and lower loan redemption costs. Non-US cotton producers therefore bear an increased share of the adjustment costs as international market conditions change.

The international cotton trade in the 1990s

The 1990s in the international cotton trade can be characterised as the period of the three conditions:

1 The effect of the international recession. It was clear by 1989 that international demand for textiles was falling because of the decrease in demand in the industrialised countries, combined with the collapse of the formerly Soviet cotton procurement policy.

2 The increasing perception of lengthy and intense recession throughout the world, not likely to return to growth conditions until 1994 at the earliest.

3 A gradually increasing perception that previous relationships between the textile industry and economic growth in general may have been superseded by a new reality, of less correlation and perhaps a sustained 'underperformance' of the textile sector by comparison to the overall statistics. Condition 3 is a hypothesis and is unproven: it will take many years to demonstrate its accuracy – but elements of the underlying logic are already present in writing on cotton in many journals.

By June 1990, a relatively buoyant mood had arrived in the market and prices moved up to the 85 cents/lb mark. It was said that 'the traders are attempting to eke out stocks before the major harvesting and shipping period, and there are traditional uncertainties about the 1990/91 northern hemisphere crop' (Economist Intelligence Unit, World Commodity Forecast June 1990) – which is where the overwhelming majority of the world's cotton crop is grown. A further impact was made by the US 1990/91 farm policy, whereby a larger area of cotton qualified for price support than previously. A final impact on the market was the belief that there would be supply constraints, for example caused by the continuing poor weather which hampered sowing in Uzbekistan. This view continued throughout the summer of 1990, with dark references in contemporary analyses of alleged transport and processing constraints on the southern crop, and adverse weather and a poor seed crop having a deleterious effect on the Indian crop.

These uncertainties took some months to evaporate, so that by October 1990 stable prices of approximately 80 cents/lb had been reached as the harvesting of the northern hemisphere crop dissipated uncertainties. World production for 1990/91 was now estimated at 86 million bales, which would have been a 7.8% increase over 1989/90 and 0.5 million bales higher than the production estimates current in the spring. World production estimates do have an important influence on trade conditions because they affect international perceptions of supply tightness or otherwise. By December 1990 the bulk of the northern hemisphere crop had been harvested and market uncertainty was definitely evaporating. This process was reinforced by the relative closeness between the projected and the actual harvest for the 1990/91 season, which resulted in little change in raw cotton prices towards the tail end of 1990 - around the 82 cents/lb level. The USDA published their estimate of the US crop showing a 22.6% increase from the previous season, which also had some influence on calming the market. Traders in Europe, seeing the decline in mill take from the former Warsaw Pact countries and Romania, recognised as early as 1990 that the direction of Central Asian cotton exports was likely to be a crucial aspect of the cotton trade of the 1990s, and they intensified their efforts to establish offices there and become involved in the export of this cotton.

1990 was another year in which the differentiation between textile trade and raw cotton trade became visibly apparent. It was another year in which textile trade did not reflect the international downturn in trade following the worldwide recession. In 1990, despite the international recession, the volume of textiles manufactured in ITMF member countries increased by about 7.5%. Contrary to this trend, however, yarn traded expanded by more than trade in fabrics - 9.5% compared with 5.5%. These rates have declined a little from 1989 but they are still impressive compared to growth rates in the developed economies ranging from 5% in Germany and Japan to 1% in the USA, Canada and the UK. Geographically, the 1990s were immediately marked by an 'equalisation' in favour of the newly developed countries, and in favour of the Far East in general. Yarn imports increased 21% for Taiwan, but fell by 14% for the USA, for example, which nevertheless continued to run a deficit in textile and apparel trade. Yarn exports increased by 56% for India, 32% for Pakistan, 33% for Spain, and 20% for Taiwan, but down by 8% for Japan and by 24% for Turkey. Fabric imports (these figures are, of course, for all fabrics, not just cotton) increased by 15% for Belgium, 14% for the Netherlands, 134% for Turkey; but they decreased by 21% for Japan, which was in the grip of a serious textile recession, achieving in this a reduction in its textile and apparel trade deficit in 1990 of 11.1%, to US$6.5 billion. Imports increased by 35% for Spain. Fabric exports increased by 11% for Belgium, 17% for both India and Japan, and by 49% for the USA. The textile and apparel deficit of the largest four countries in the EC (France, Germany, Italy and the UK) was $11.1 billion, up by 42.7% over 1989.

During early 1991 the raw cotton market continued relatively calmly despite the outbreak of hostilities in the Gulf. To be fair, so did the oil and other commodity markets. There were ideal growing conditions reported in Brazil and Australia, and demand was expected to fall due to Condition 2: the perception of the slow recovery of the market. Condition 3 was also in evidence in the market perception that 'Taiwanese investment was shifting away from the textile sector' (World Commodity Review, February 1991). In Japan, stocks of cotton were increasing and it was not thought likely that substantial imports would be needed.

1991 saw the beginning of the major depression in the international cotton trade that continued relentlessly into 1992 and showed no immediate sign of recovery during 1993. By April 1991 the EIU was describing European mill profitability as 'far from satisfactory', and although prices continued to hold at 85 cents/lb, there was a widespread feeling that the fundamentals did not justify this, and that prices would be falling. The relationship between price and fundamental demand does seem to have been relatively confused at this juncture. The southern hemisphere crop was being harvested, whilst in the northern hemisphere, land preparation and sowing was beginning. The US acreage reduction programme was lowered from 12.5% to 5%. The general perception at that time was that stocks were relatively low in comparison with existing and expected future world consumption, which was one of the principal reasons underlying the continued relatively high price levels. This was commenting on a stock level of about 29 million bales at the beginning of August, thought to be likely to rise slightly in 1991/92 to about 31 million bales.

Prices continued to stay at a relatively high level throughout the summer of 1991, mainly because of doubts about the US cotton crop owing to the slow start to planting, especially in the Delta – there was poor weather in Mississippi and the Far West; by mid-May only 26% of the crop was sown compared with 37% a year before. An increase in the Chinese crop was expected, because of an increase in acreage of some 9%: and China was granted Most Favoured Nation status by the USA which provided a boost to the market – this optimism was replaced later in the year by a realisation that US/Chinese relations were once again on the slide and additional trade was unlikely. Although there were floods in China in the summer of 1991 the market was relatively unaffected, and growing conditions in the USA improved, so supply was now deemed adequate. When the USDA production estimates were issued in October 1991, at 17.5 million bales, prices began to slide, more than outweighing the effect of the perceived reduction in output from the former Soviet Union; but the Economist Intelligence Unit still believed that 'With tentative estimates for the 1992/93 season indicating a closing of the gap between production and consumption prices are unlikely to weaken further and to stay in the 70/75 cents/lb band.'

It was upward revisions in the US crop estimates and above-average yields throughout the northern hemisphere during the later part of 1991 that began to send prices downwards - the November 1991 estimate of the US crop from the USDA was 18.2 million bales, including 0.4 million bales of American Pima, which was the highest since the record 1937/38 harvest. The estimated average yield of 649 lb/acre implied almost perfect growing conditions and minimal losses during the harvest.

Recent trends in the international cotton trade

Trade in 1992/93 was expected throughout 1992 to hold near the 1991/92 record of 23 million bales. The reason cited by the USDA was that consumption growth was concentrated in the major producing countries of China, the USA, Pakistan and India. World exports of cotton fell in 1992 to 5.54 million tonnes, down from 6.08 million tonnes in 1991, but rose to 5.99 million tonnes in 1993 and are expected to stabilise at another 6.08 million tonnes in 1994. The world market, the ICAC expect, will probably be in approximate balance in 1993 and 1994 - imports and exports approximately at the same levels.

The ICAC is clear that the cotton market is segmented and that different grades face different long- and short-term demand curves. Lower production led to a decrease in the amount of extra-fine cotton - used in higher yarn counts - exported in 1990/91, and there was no revival of this trade in 1991/92. The Egyptian authorities claimed that production bottlenecks were responsible for the decline in exports of Egyptian styles in the 1980s. However, in the view of the ICAC the main reason for this decline is the installation of more advanced spinning equipment capable of producing expensive textile products using cotton in the fine and high-medium categories. As a result of these factors, the total level of exports of extra-fine cotton fell below 165 000 tonnes in 1991/92, the lowest level for several decades. The decline is expected to continue in 1992/93 and 1993/94 to significantly less than 150 000 tonnes. A similar picture has prevailed with respect to fine cotton.

This leaves the broad division between high-medium and medium cotton on the one hand, accounting for just over half of world supply, and coarse cotton on the other, where production is decreasing and stocks are rising, which has caused lower prices for coarse cotton. What the market has seen over the past few years is a trend towards concentration: a trend towards standardisation of the high-medium categories and an increased importance for electronic quality testing. The nature of the cotton market internationally has not changed dramatically - but one reason for that is that the international cotton trade is already far in advance of most other commodity trades in almost every aspect, from price reporting to industrial concentration.

The USA began the 1991/92 marketing year with only 0.81 million tonnes of cotton stocks. Such tight supplies contributed to reduced export markctings in the beginning of the crop year. US cotton exports of 7.8 million bales in 1991 marked the first time in over 50 years that exports had exceeded 7.7 million bales twice in succession. Export commitments were above-average for 1992/93, probably reflecting the impact of the competitiveness provisions of the US cotton loan programme. US cotton did particularly well in South-East Asia, where 1.7 million bales were exported, becoming the second largest export area. This conceals a geographical change within Asia: there are lower sales to China, Indonesia, Thailand and the Philippines, but higher sales to Korea and Hong Kong. Cotton Incorporated maintains an office in Singapore and has moved its European office from London to Basle in Switzerland, increasing focus on Eastern Europe. Import commitments for 1992/93 are higher than 1991/92 in all EC countries except Portugal, with Italy and Spain accounting for 60% of the increase in US sales to Europe. The cotton marketing loan provisions have ensured the competitive position of US cotton in the world market. Despite a lack of price competitiveness, US cotton exports reached 1.45 million tonnes in 1991. Net exporting and importing figures are given in Tables 9.5 and 9.6.

Projected world exports of cotton, the USDA said in April 1993, were 22.2 million bales, a level that the USDA had maintained throughout early 1993. The USDA in August 1992 estimated that US exports would continue at 6.7 million bales in 1992/93, although this estimate fell to 5.8-6.1 million bales by mid-1993, reflecting precisely this lack of price competitiveness causing sluggish foreign demand and shipments in spring 1993. US shipments are likely to fall, the USDA said in April 1993, by 13% to the lowest level since 1985/86. In practice, US exports have remained at relatively low levels - 1.3

Table 9.5 Net exporting countries: cotton supply and utilisation, 1992/93[a] (thousand 480 lb bales)

Country	Beginning stocks	Production	Imports	Total supply	Use	Loss	Exports	Ending stocks
Western hemisphere								
Argentina	753	900	5	1 658	550	9	320	779
Brazil	1 799	3 000	460	5 259	3 200	0	400	1 659
Colombia	282	320	200	802	540	3	90	169
El Salvador	43	17	75	135	85	0	5	45
Guatemala	50	80	45	175	140	0	18	17
Honduras	4	10	0	14	10	0	0	4
Mexico	307	150	600	1 057	735	25	40	257
Nicaragua	7	9	0	16	5	0	9	2
Paraguay	470	900	0	1370	70	0	800	500
Peru	393	260	35	688	220	0	100	368
USA	3 704	16 259	2	19 965	9 700	−135	6 000	4 400
Total	7 812	21 905	1 422	31 139	15 255	−98	7 782	8 200
Asia/Oceania								
Afghanistan	33	100	0	133	80	0	30	23
Australia	1 100	1 600	2	2 702	135	0	1 700	867
China	14 018	21 000	400	35 418	21 000	0	1 100	13 318
India	2 642	10 000	0	12 642	9 050	0	550	3042
Iran	49	525	0	574	500	0	25	49
Israel	64	150	50	264	100	0	100	64
Pakistan	3 083	9 000	10	12 093	6 800	100	2 000	3 193
Syria	230	985	0	1 215	325	0	620	270
Turkey	410	2 780	250	3 440	2 660	0	300	480
Yemen	4	65	0	69	25	0	40	4
FSU-12	3 681	9 400	100	13 181	6 250	250	4 000	2 681
Total	25 314	55 605	812	81 731	46 925	350	10 465	23 991

Africa								
Angola	7	20	0	27	15	0	5	7
Benin	42	300	0	342	20	0	275	47
Burkina Faso	44	300	0	344	20	0	280	44
Cameroon	20	230	30	280	50	0	200	30
Cent. Af. Rep.	14	50	0	64	15	0	35	14
Chad	17	250	0	267	20	0	225	22
Cote d'Ivoire	211	620	0	831	65	0	435	331
Egypt	183	1 500	250	1 933	1 550	20	160	203
Malawi	13	30	10	53	35	0	10	8
Mali	34	530	0	564	50	0	450	64
Mozambique	60	60	5	125	50	0	25	50
Senegal	31	75	0	106	25	0	40	41
Sudan	103	400	0	503	60	0	350	93
Tanzania	161	320	0	481	60	15	260	146
Togo	30	160	0	190	15	0	150	25
Uganda	45	45	0	90	20	0	35	35
Zambia	100	90	0	190	50	0	60	80
Zimbabwe	89	400	50	539	300	13	75	151
Total	1 204	5 380	345	6 929	2 420	48	3 070	1 391
Total exporting	34 330	82 890	2 579	119 799	64 600	300	21 317	33 582
Total foreign	36 365	69 666	22 345	128 376	76 165	245	16 415	35 551
Total world	40 069	85 925	22 347	148 341	85 865	110	22 415	39 951

[a] Season beginning August 1.
Source: Prepared or estimated on the basis of official foreign government statistics, other foreign sources, reports of US foreign service.

Table 9.6 Net importing countries: cotton supply and utilisation, 1992/93[a] (thousand 480 lb bales)

Country	Beginning stocks	Production	Imports	Total supply	Use	Loss	Exports	Ending stocks
Western hemisphere								
Bolivia	10	60	25	95	75	0	0	20
Canada	14	0	200	214	185	0	0	29
Chile	29	0	90	119	90	0	0	29
Costa Rica	11	4	10	25	13	0	0	12
Cuba	60	5	145	210	150	0	0	60
Dominican Rep.	3	10	0	13	8	0	2	3
Ecuador	49	50	75	174	120	0	0	54
Haiti	8	10	0	18	10	0	0	8
Panama	1	0	5	6	5	0	0	1
Uruguay	5	0	10	15	10	0	0	5
Venezuela	38	70	100	208	150	0	30	28
Total	228	209	660	1 097	816	0	32	249
European Community								
Belgium	30	0	180	210	175	0	10	25
Denmark	3	0	10	13	10	0	0	3
France	106	0	550	656	460	0	70	126
Germany	202	0	1 000	1 202	850	0	115	237
Greece	153	1 260	115	1 528	800	0	500	228
Ireland, Rep.	37	0	10	137	100	0	0	37
Italy	507	1	1 450	1 958	1 400	0	0	558
Netherlands	7	0	40	47	40	0	2	5
Portugal	200	0	640	840	600	0	0	240

Spain	182	335	365	882	600	2	130	150
United Kingdom	120	0	80	200	85	0	0	115
Total	1 547	1 596	4 530	7 673	5 120	2	827	1 724
Other Western Europe								
Austria	5	0	140	145	140	0	0	5
Finland	23	0	50	73	50	0	0	23
Norway	8	0	15	23	15	0	0	8
Sweden	10	0	25	35	25	0	0	10
Switzerland	190	0	300	490	290	0	0	200
Total	236	0	530	766	520	0	0	246
Eastern Europe								
Albania	7	20	30	57	50	0	0	7
Baltics	33	0	0	33	250	−250	0	33
Bulgaria	50	78	75	203	150	0	0	53
Czechoslovakia	53	0	300	353	300	0	0	53
Hungary	51	0	150	201	150	0	0	51
Poland	113	0	300	413	300	0	0	113
Romania	86	0	250	336	250	0	0	86
Yugoslavia	45	1	230	276	230	0	0	46
Total	438	99	1 335	1 872	1 680	−250	0	442
Asia								
Bangladesh	133	75	450	658	515	5	0	138
Myanmar	71	110	0	181	85	3	25	68
Cambodia	1	0	5	6	5	0	0	1
Cyprus	1	1	1	3	2	0	0	1
Hong Kong	283	0	1 100	1 383	860	0	165	358

Table 9.6 Continued

Country	Beginning stocks	Production	Imports	Total supply	Use	Loss	Exports	Ending stocks
Indonesia	143	14	1 600	1 757	1 500	50	0	207
Iraq	12	65	50	127	100	0	0	27
Japan	575	0	2 475	3 050	2 450	0	0	600
Lebanon	3	0	7	10	7	0	0	3
Malaysia	26	0	200	226	190	0	0	36
North Korea	40	5	150	195	170	0	0	25
Philippines	122	30	235	387	240	0	0	147
Singapore	63	0	125	188	125	0	15	48
South Korea	566	1	1 900	2 467	1 800	0	0	667
Sri Lanka	6	5	40	51	45	0	0	6
Taiwan	480	0	1 500	1 980	1 470	0	0	510
Thailand	322	150	1 700	2 172	1 800	0	30	342
Vietnam	106	5	250	361	250	0	0	111
Total	2 953	461	11 788	15 202	11 614	58	235	3 295
Africa								
Algeria	6	0	225	231	200	0	0	31
Ethiopia	8	46	75	129	120	0	0	9
Ghana	13	13	40	66	50	0	0	16
Kenya	15	20	35	70	55	0	0	15
Madagascar	15	60	0	75	60	0	0	15
Morocco	86	10	150	246	160	0	2	84
Niger	2	5	5	12	8	0	2	2
Nigeria	128	290	95	513	375	0	0	138

Somalia	1	7	5	13	12	0	0	1
South Africa	50	200	160	410	320	0	0	90
Tunisia	9	3	115	127	120	0	0	7
Zaire	4	16	20	40	35	0	0	5
Total	337	670	925	1 932	1 515	0	4	413
Total importing	5 739	3 035	19 768	28 542	21 265	−190	1 098	6 369

[a] Season beginning August 1.
Source: Prepared or estimated on the basis of official foreign government statistics, other foreign sources, reports of US foreign service officers and other information. Foreign Agricultural Service, Tobacco, Cotton and Seeds Division December 1992.

million tonnes in 1992 and 1.31 million tonnes in 1993, although the ICAC expects a revival to about 1.5 million tonnes in 1994. Changes in forecasts suggest that Pakistani exports will also be reduced by more than the increase in Paraguayan exports, bad weather and better yields being the two causes. The textile industry is also placing pressure on the US government to curtail exports. The US government has been increasing, at regular intervals, the minimum export price which has eliminated profit margins to raw cotton exporters.

The importance of China to the international cotton trade can hardly be underestimated. The world's largest cotton producer, its consumption and trade operations can have a major effect on international prices and the distribution of trade. China is so important that its net exports are the major determinant of world prices in the ICAC cotton price model, along with stocks, consumption and production. In 1992/93 China did shift to being a new exporter after three seasons as an importer, but this was brief. Imports for China are estimated to have been 180 000 tonnes more than exports in 1991/92, but in mid-1992 the ICAC believed that in 1992/93 Chinese exports would be approximately 530 000 tonnes. This would have produced 710 000 tonnes more cotton for the world market in 1992/93 and would have toppled prices dramatically. In fact, the Chinese chose to increase their stocks during the year, which will allow them comfortably to maintain mill production during 1993/94 despite their reduced crop production in the next season. The ICAC expects China to export 0.2 million tonnes in 1994.

These statistics for cotton export must be seen against an important human and economic dimension for cotton-producing countries, especially in developing countries. As cotton provides the most important source of cash income for many developing countries' farmers, it is often, as Morris says, 'seen as the major platform for both rural and industrial economic development' (1988, p. 12). At a national level, the importance of cotton as a hard currency earner for many developing countries should not be underestimated. For example, cotton is the principal source of hard currency by comparison with other agricultural commodities for Egypt, Burkina Faso, Mali, Nicaragua, Pakistan, Paraguay, the Sudan, Syria and Zambia. In many developing countries cotton is *the* major foreign exchange earner: Paraguay, for example, the Sudan, Burkina Faso and Nicaragua. In Egypt, cotton is amongst the 10 largest sources of foreign exchange revenue. Very importantly, it is a vital source of hard currency for many developing countries which the UN classes as least developed. The overwhelming majority - 95% - of these exports is cotton lint.

By February 1992 the 1991/92 crop had been harvested and the

1992/93 had yet to be planted in the northern hemisphere. World production in 1991/92 was thought to have reached a record 91.7 million bales. This was accounted for by upward revisions for China and Pakistan: in particular, Chinese yields in the northern provinces of Hebei, Henan and Shandong were higher than expected, and there were good growing conditions in the three largest southern hemisphere growing countries of Brazil, Australia and Paraguay. By contrast, the market saw longer than expected harvests in the Texas High Plains, North India and Zimbabwe where there had been a poor rainy season. Adverse weather led to resowing in some areas of Argentina, and it was believed that a lack of moisture would reduce yields in South Africa. The overall picture was increasingly depressing for the traders, because of downward revisions in world consumption estimates. In particular, textile manufacturing in Russia was declining rapidly - Russia had previously taken the majority of CIS cotton - and was expected to continue to decline into 1992/93. In the USA, under the 1992/93 Acreage Reduction Programme, upland growers were required to reduce their plantings from their Crop Acreage Base by 10%, compared with 5% in 1991/92 to be eligible for CCC loan support and target price protection.

Prices continued to slide in the months following. Condition 2 certainly applied, and the out-turn for Chinese production was higher than expected. A 'scissors' was evident in the international market, with US sales beginning to pick up. The trade began to doubt whether the US market was representative of the world trade in general - or indeed whether Condition 3 simply would not apply to the USA as the world came out of recession. US apparel sales actually rose during January and February, and there was a strong continuing demand for textile products. There was also some revival in demand in Hong Kong for textiles where a number of spinners concluded yarn agreements in mid-1992. However, in general, East Asian markets have been sluggish, and the rise in import share in Japan shows the importance of the cotton and textile import question in Japanese textile politics that has already occurred.

In Europe mills continued their policy of buying only for immediate needs. Europe as a whole was a declining market for textiles of all types, except silk, in the early 1990s. For example in 1988/89, Europe had a demand of 2 million tonnes. This fell to 1.6 million tonnes in 1990/91 and to 1.3 million tonnes in 1991/92. The ICAC expects EU demand for cotton to remain under 1 million tonnes for the rest of the century.

In 1992, even the very low-priced CIS cotton was not finding a home and warehouses were stocked full of it. The Uzbek export infrastructure improved its efficiency throughout 1992 as merchants worked hard to smooth the export process, overcoming the bottlenecks that had been noted in Black Sea ports during 1990/91 and 1991/92, and the CIS could continue indefinitely to be highly competitive on price because it is such a bonus for it to receive any hard currency at all. The USDA estimated in July 1992 that the countries of the former Soviet Union would export 9.2 million bales in 1992/93, compared with 9.343 million in 1991/92, of which 3.5 million and 3.2 million respectively represent exports out of the former Soviet Union, and the remainder inter-republic trade. (These figures are based on those available for 1988 which are the last available from Soviet sources.) Exports from Central Asia to major European countries are still increasing, especially from Uzbekistan and Turkmenistan, and the ICAC estimated total CIS exports in 1992/93 at 780 000 tonnes to countries outside the CIS and 2 million tonnes in total. About 50% has already been committed, mainly to barter deals of one sort or another, sometimes through intermediary companies in Turkey and Pakistan. Even if these figures are overestimates, the scale of CIS cotton exports is still impressive and an important aspect of the international cotton trade in the 1990s.

CIS export aggressiveness in turn is constraining the new leadership at Chinatex (the Chinese textile organisation) and its regional outlets and contributory organisations – no doubt they would like to sell, but any price cutting on their part would mean registering losses since their procurement price is still 60 cents/lb.

Chinese production was some 26 million bales according to the final production figures issued by the State Statistics Bureau (one of the most reliable Chinese statistical sources) in March 1992, which was over a quarter higher than the previous year's crop and 3 million bales above the best previous forecast. The trade refused to cut prices drastically in anticipation of a large Chinese export volume, considering, rightly, that several factors would inhibit the Chinese from exporting.

In May–June 1992 cotton prices stabilised for the 'A' Index above 60 cents/lb for 1991/92 cotton and 2.5 cents higher for 1992/93. The EIU said that 'the lack of any discernible price trend is an indication that many mills are waiting to see the outturn of the 1992/93 northern hemisphere crop before making any purchasing decision beyond immediate needs'. (EIU, WCF, June 1992) Production estimates for 1991/92, at 95.5 million bales, were an undoubted record, and it was expected that the level for 1992/93 would be some 5.8% lower at 90 million bales. The US harvest was being estimated at 17 million bales, 600 000 bales less than the 1991/92 out-turn. By mid-May some 57% of the crop had been planted, compared with 30% in 1991/92 and a five-year average of 45%. Commentators believed that Chinese and Central Asian crop prospects were even more uncertain to predict than in previous years: areas planted were higher than in 1991/92, which could, it was thought, produce an output even greater than 26 million bales and well above the five-year average of 24 million bales. Forecasters continued to believe that the Chinese would not try to compete with the CIS and dump thousands of bales of cotton on the world market, but that if they did, prices would spiral downwards quickly towards the 50 cent level. But even small downward adjustments in forecast output from the CIS due to weather-caused resowing was totally insufficient to revive the market.

Traders were holding off simply because ever more interest surrounded the release in August of the new USDA numbers for the projected crop of 1992/93 and the world supply and demand balance. USDA pointed out that this was the first estimate based on survey data - and the survey *did* take account of the Texas rains; it provides an estimate of acreage and yields. However, USDA pointed out also that planting was late on the High Plains and that therefore there is additional uncertainty there. Even usually it is not until November/December that real clarity is established about the crop, and the previous weather will exacerbate this. Traders knew, however, that the record of the August forecast is impressive, though: over 10 years it has never been more than 2 million bales out from a harvest that has ranged from 7.8 to 16.6 million bales. The average error has been 700 000 bales and although the forecasts have tended to be on the low side (one was on; seven low; and two high) the error on average has been only 5.2%. The July estimate for the crop was 16 million bales. Traders waited on the 12 August number; the New York Exchange therefore moved little within a 1 cent band; the dollar fell; European mills waited on further clarification of trends. Only rumours of an Australian drought that began to circulate seemed at all likely to bring any further excitement into the market and such a drought did not occur. Positive price signals therefore only began when it became apparent that the Chinese crop was not to be as substantial as had been thought.

As the year continued it became clear that the uncertainties surrounding world economic recovery would continue to persuade mills not to buy for anything more than their immediate requirements. IMF/World Bank expectations for revived world economic growth in 1992 were dashed, and as a result supplies rose relative to use. International cotton buying has, as the ICAC has pointed out, declined. Total world cotton imports have fallen from 5.7 million tonnes in 1988/89 to 5 million tonnes in 1991/92 and will not be much more in

1992/93. Export commitments recorded by the ICAC at the end of 1992 demonstrated a relatively high level of commitments to expected exports, as usual, but US exports are showing a trend away from traditional Far East markets. Estimates of 1992/93 production (see the section on 'Overview of world cotton production' in Chapter 4) in China, Pakistan and India were decreased at the end of 1992, but this did not substantially revive the market although the stock: consumption ratio outside China is now falling towards the levels seen in the late 1980s. As the cotton trade entered 1993, world stocks were still at historically high levels.

US exports during 1992/93 were expected to be 6 million bales, down 646 000 bales from the season before, facing intensified competition especially from CIS countries. This cut the US share of world exports to 26% from 30% the year before. The main US markets continued to be Korea, Indonesia, Japan, Mexico, Canada, Hong Kong, Italy, Taiwan and Thailand. The rest of the world, the USDA said in December 1992, exported 16.4 million bales in 1992/93, as prices fell and the CIS and Pakistani exports in particular maintained their market share. All these figures are subject to revisions of the order of 5% either way when final figures are available. Import and export figures are given in Fig. 9.2 and 9.3 and Table 9.7.

In 1993 US cotton production shifted towards export types. There is an increased supply of coarse cotton in both Texas and Oklahoma. Declines in production in the Memphis area have affected domestic consumption. Mills are paying more, due to the shortage, and are therefore seeking alternatives.

US shipments to the Western hemisphere are likely to rise in 1993/94, according to the ICAC. This is partly due to the relative ease of penetration of US cotton into the rapidly growing Brazilian market, and partly due to the continuation of US export subsidies. The

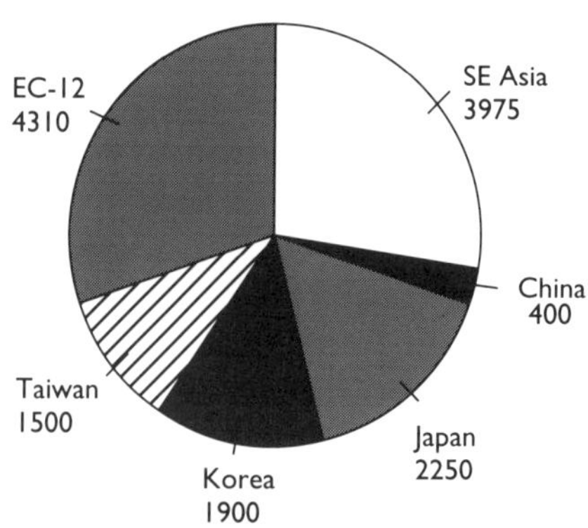

9.2 World cotton exports: 1992/93 estimate (*source*: USDA/FAS).

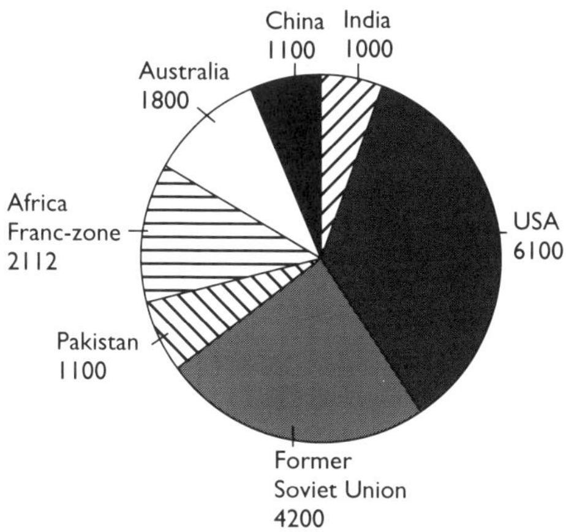

9.3 World cotton exports: 1992/93 estimate (*source:* USDA/FAS).

Table 9.7 World cotton imports and exports (thousand 480 lb bales)

	1987/88	1988/89	1989/90	1990/91	1991/92 Estimate	1992/93 Estimate
Imports						
World total	24 096	26 691	25 368	23 892	23 650	22 244
China	86	1 448	1 873	2 205	1 662	400
Japan	3 431	3 491	3 165	2 949	2 705	2 250
Korea	1 957	2 145	2 040	2 052	1 798	1 900
Taiwan	1 608	1 781	1 301	1 479	1 484	1 500
EC - 12	6 040	5 677	5 629	4 848	4 499	4 310
SE Asia[a]	2 519	3 268	3 303	3 855	4 275	3 975
Exports						
World total	23 106	25 757	23 893	22 959	22 372	22 175
United States	6 582	6 148	7 694	7 793	6 646	6 100
FSU - 12	3 457	3 656	3 330	2 000	3 500	4 200
Pakistan	2 358	3 780	1 371	1 357	1 923	1 100
Franc-Zone Africa[b]	1 680	1 982	2 093	2 077	2 292	2 112
Australia	818	1 319	1 319	1 372	2 067	1 800
China	2 322	1 636	865	928	600	1 100
India	34	78	1 070	929	60	1 000

[a] Includes Indonesia, Malaysia, Philippines, Singapore, Thailand and Vietnam.
[b] Includes Benin, Burkina, Cameroon, CAR, Chad, Côte d'Ivoire, Mali, Niger, Senegal and Togo.
Source: USDA/FAS.

medium-term trend that is behind these figures is that a higher percentage of raw cotton is being milled nationally before export as yarn, or increasingly as finished product, rather than exported for milling in Western countries; the total percentage of milled cotton is now about 25%, compared to over 30% even in the 1980s. The trend is most noticeable in India and Pakistan and the USA, where mill use has increased by 1 million tonnes against a stagnant world background, although US cotton is now being exported for milling and garment manufacture in developing countries and in NICs such as Thailand and Taiwan. Against this trend, India's Agricultural Ministry, traders and the cotton production lobby are all continuing to press for an increase in the cotton export quota. This is currently (1993) set at 1.5 million bales (170 kg bale). But because of slow domestic mill demand and the fact that domestic prices are 13–16 cents less than world cotton prices, there is pressure to increase exports. Opposition to the exports comes from the Indian Textile Ministry which observes that if exports are increased, mill prices would be pushed down even further as a result of higher raw material prices. The Textile Ministry has therefore refused export permits for additional exports on the grounds that domestic prices are too high, and already above support levels. Some 500 000 bales were released for export in 1993. There was a considerable increase in textile exports after liberalisation in early 1993. The release of cotton for export was intended to bolster India's reputation for reliable supplies.

Government price support schemes in each of those countries have been partly responsible for the curtailment of exports. Estimates suggest that up to half of the change may be accounted for by these institutional factors. Trade between China and Japan is evolving, whereby Japan exports fabrics to China, which then re-exports the finished clothing back to Japan on a commission basis. The ratio of fabrics exported to China and imported back to Japan in the form of clothing is estimated to be more than three-quarters of cotton fabrics. Japanese trading houses have started joint ventures specialising in clothing production in Indonesia to make blouses, skirts and trousers, and in Thailand, a blouse and suit plant began manufacturing. Japan will continue to manufacture abroad, especially in China, Vietnam, Thailand and Indonesia.

Probably the most important recent trend in the international cotton trade, and proof that commodities can still spring surprises in the 1990s, has been the rise in cotton imports into countries such as Mexico and Brazil. Both countries have become new importers of cotton owing to production falls (for example in Brazil from 3.445 million in 1991/92 to 2.6 million bales in 1992/93); imports are also expected in Central American countries, and not just from traditional South American exports such as Paraguay, but from the USA and West Africa too. Other previous exporters have become cotton importers because of bad weather. Most notable are the countries of southern Africa, where drought had severe adverse effects on production which will continue unless there is better weather. Even Turkey is now set to become a net importer of cotton on a continuing basis.

Existing international arrangements: export marketing and physical contracts

The international cotton community is well supplied for information. There are better, more regularly produced statistics for cotton production, consumption and trade than for many other commodities, although there is of course still great secrecy surrounding specific deals. Statistics emerge from five principal sources: the NYCE, Cotton Outlook, the USDA, the United Nations FAO and the ICAC, which has its headquarters in Washington. The International Federation of Master Cotton Spinners and Manufacturers' Associations, with headquarters in Manchester in the UK, provided statistics for many years, but the ITMF has taken over this role.

Probably the main reason for the efficient distribution of information is the success of the futures markets. The NYCE (and when it was in existence, the New Orleans Cotton Market) compiles and publishes annual statistical reports on the supply and distribution of both US and other cottons. Various cotton trade organisations, notably the cotton exchanges at Alexandria in Egypt and Bombay in India and many other government agencies used to issue important reports relating to the cotton trade in their respective countries.

Above all, however, there is the Cotton Outlook organisation in Liverpool, which has become the central information exchange of the cotton world. Their sources of information are extremely numerous and usually very reliable. Cotton Outlook publishes daily, not just price information, but on a wide range of cotton topics, often influencing opinion as much as reflecting it. The same applies to the USDA which also publishes on market trends, although much less frequently.

Spot markets

Among the leading spot markets for cotton are Dallas, Houston, Memphis and New Orleans in the USA, Liverpool, Le Havre, Bremen and Milan in Europe, Bombay and Karachi in India, Shanghai in China, and São Paulo in Brazil. There are many other spot markets, and thousands of farmers' markets in producing countries and many other spinners' markets in consuming countries.

Current terms of trade

Recent price levels have accentuated the problems of developing countries relying on commodities such as cotton for their primary export earnings. The current selling price for cotton – throughout the early 1990s – means that the African cotton companies are no longer in a position to assist the farmers financially. Incomes are falling, investment is squeezed, and much of the good work done in the 1980s is likely to be undone. Price levels have been unsatisfactory since 1985/86 and there have been economies; it is accepted that the costs of farmer training and rural development can no longer be met by cotton production on its own. Subsidies for chemicals have been eliminated; the village associations have gradually taken over marketing, and producer prices have fallen sharply. Most West African countries have fixed a nominal minimum of approximately 80–100 FCFA for each season, with uplift. Some African farmers are now questioning cotton's viability and areas are falling in Africa and in the southern hemisphere. What has happened is that as the market has become more international, freight costs have been honed and a greater demand has been set up in the Far East close to China. Economies of scale are dominating the market and the number of producers looks set to fall. At the ICAC meeting in New Delhi towards the end of 1993, the organisation made a number of recommendations regarding the production and marketing policies and cotton imports and decided to work together to resolve issues concerning raw cotton supplies – including the issue of non-standard bale sizes, honeydew contamination and country damaged bales. The possibility of re-establishing a cotton research institute to aid the developing countries was also raised.

The Multi-Fibre Agreement

The Multi-Fibre Agreement (MFA) is an agreement between textile importing and exporting countries providing for acceptable protectionist measures and establishing a framework for bilateral negotiations. The USA's first efforts to restrict textile imports were in the early 1950s when the Japanese first provided competition. These early negotiations were conducted under GATT auspices. During the 1960s the number of countries participating in these negotiations increased, and the MFA was agreed in 1974. The trade of Japan and the developing countries has always been expected and agreed to rise, although the USA and Western Europe have consistently reserved the right to control the level and rate of growth of their imports of textiles. By mutual agreement, the USA, Canada, Western Europe, Australia and New Zealand do not restrict trade amongst themselves. The MFA requires countries to notify mutually changes in quantitative import restrictions and to receive approval of these changes from the Textile Surveillance Body of the MFA. Such approval is only given when the restrictions are limited to products from a particular country or countries that are disrupting a specific domestic market, i.e. anti-dumping operations. When the potential damage is imminent and serious, the importing country can request immediate bilateral negotiations with the exporter to restrict the exports. Over 30 countries have had quotas imposed for their exports to the USA under these regulations, for example. China has not signed the MFA, but the USA negotiates with China as if it had.

The USA and other developed countries also collect tariffs on imported cotton and textiles – 11% on cotton and an average 20% on clothing. Such tariffs are collected on imports from all sources. There are two different US regimes for export marketing of cotton, one for ELS cotton and the other, much larger, programme for upland cotton.

Adjustments to the AWP are permitted when the AWP is less than 115% of the base loan rate, and when also US cotton price quotes are higher than the 'A' Index. 'Step 2' payments are currently made to US farmers when the US northern Europe price exceeds the 'A' Index by more than 1.25 cents/lb for four successive weeks. 'Step 2' payments during 1992 were payable to both US exporters and domestic mills for spot cotton and also to exporters for forward (post 30 September 1992) deliveries. The amendments to the Cotton Research and Promotion Act voted on by US producers and importers in the July 1991 referendum created new mandates: US upland cotton producers can no longer receive refunds for assessments paid, and imported cotton and products are now assessed at the same bale rate from July 1992 onwards. Payments made by the USDA under the first handler and the user marketing certificate after April 1992 are in cash. But user marketing certificate payments are not now made in any week after four weeks in which the AWP exceeds 130% of the current crop year base quality loan rate, nor in any week after a 10 week consecutive period in which the US price is more than 1.25 cents/lb higher than the 'A' Index. The Federal Register publishes the assessment rate for cotton.

The gap between US and international cotton prices both for spot and forward shipment (determining the issue of 'Step 2' certificates) during mid-1992 declined during April and May but widened in late June 1992. The AWP rose throughout mid-1992, and in June the loan repayment rate was equal to the AWP adjusted for specific quality and location, plus applicable interest and storage charges. The assessment rate for cotton from July 1992 onwards was $1 plus ½% per bale.

Cotton trading companies

Most of the trading companies have their head offices in the larger interior markets of the USA, in Liverpool, in Geneva, in Memphis or elsewhere. They are usually organised as unincorporated private firms, although some, such as Baumann Hinde, are part of larger organisations (in their case, Lonrho). They perform the function of intermediary between the mills in the USA, Europe and elsewhere in the world and the growers. Trading companies buy cotton, often in lots of mixed quantities, from growers and exporters, classify the cotton in the mixed lots and sell it in lots of uniform quality to domestic and overseas spinners and overseas importers. The trading companies have facilities to ensure that the cotton is compressed, from the relatively large, loosely packed gin bales into smaller, tightly packed bales for economical transportation, and they ensure that the cotton is moved overland and overseas to the specified delivery ports. Trading companies compete - US companies tend to sell in competition with Central Asian, Indian and Pakistani cotton being bought and sold by local firms there as well as the European traders; US companies, of course, buy much of the world's cotton themselves. Many of these sales are on a forward basis, often for periodical deliveries running a year or more ahead from crops not yet actually grown when the deal is struck. Consequently all these trading firms make extensive use of futures markets to protect themselves against adverse price movements on both unsold cotton and uncovered forward commitments.

Cotton trading firms differ enormously in size, style, scope of operations, organisation and funding. Some firms trade as little as 5000 bales per annum, whilst the biggest trade over 3 million bales annually. The very largest US firms, such as Volkart and Dunavant, buy cotton from all over the USA and the world - Dunavant has a notable presence in Central Asia, for instance - and sell in all parts of the world. These firms have become very much more international since World War II. The firms that have been squeezed, and in many cases that have disappeared, are the more moderate sized companies that originally bought in only part of the cotton belt - perhaps a few states - and sold to a few recognised markets such as Liverpool. The bulk of the US crop, some 80% of it, and a sizeable percentage of other countries' crops, perhaps 60% in all, is normally sold by the growers and farmers' markets to the international trading companies, which used to be known as shippers. In the 1950s there were about 300 such firms; now the number is under 100 and of sizeable firms perhaps 50, but they are of course larger, at least in the quantities that they deal with if not in the total number of their staff. A survey after World War II in the USA showed that about 20% of US cotton used by mills was bought through mill buyers, agents or other subsidiaries, 64% was obtained through central markets, 10% through ginners and other local first buyers, 4% through the CCC, 1% through co-operatives and 1% directly from growers. In the 1990s, these percentages have not greatly changed except that more cotton is obtained through government sources.

Currently, the international cotton companies exist 'because spinning mills require a service which, in most cases, would be too costly and time-consuming to provide for themselves; and because producers need help to market their cotton for the same reason' (Nick Earlam, Plexus Cotton). In the international cotton industry, perhaps more than in any other international commodity market, the relationship between the merchant, the producer and the spinning mills is one of partnership.

The USDA says that 'the critical link between cotton producers and final domestic and export markets is provided by various types of cotton marketing firms' (USDA, 1986 p. 88). These firms operate in both local farm markets and in the central markets, and these firms are called 'merchant shippers' in the USA. Elsewhere in the world, and increasingly in the USA as well, they are referred to more generally as trading companies.

Volkart (*Company History*, 1991) suggests that

> a commodity trading company has two basic functions. It sells specific grades of a commodity for delivery at a specified time to customers/consumers should they wish to purchase them, and buys specific grades or entire harvests from supplier/producers should the latter wish to sell them.

Volkart point out that back-to-back trades are relatively rare: 'Spinning mills often submit detailed orders for specific grades of cotton before the producer has even decided whether he wants to plant cotton or an alternative crop'. It is undoubtedly at this point that the cotton trading company is most crucial, and this gap between producer and mill is what the trading company bridges (Table 9.8). The trading company takes on the risk but benefits from the producer's added value. This added value results either from the company's own processing or cotton ginning operations in the producing country, by its selection procedures, or from quality control, but also through the trading company's good relations with shippers, insurance brokers and financial services. The company's operations bear small margins, so constant calculations of the

Table 9.8 Traders as bridge between producer and spinning mill

Service to producers	Bridging of	Service to spinning mills
Up-to-date market information	Languages	Up-to-date market information
Competitive prices	Space	Competitive prices
Buy, when producers want to sell	Time	Sell, when spinning mill wants to buy
Buy all qualities	Financing	Delivery of correct specified quality as required
Buy at local terms	Exchange rate risks	Sell at local terms
Guaranteed off-take contract perform	Market price risks	Guarantee correct contract
Minimum price guarantee	Terminal market options/hedging	Maximum price guarantee
Unfixed purchase	Quality needs/option	Unfixed sale/buyer's option to fix
Prompt payment	Countertrade	Shipment/delivery on time
Prefinancing support	Cultural areas	Broker services, training, technology

contributory factors to profit must be carried out, together with a review of what hedging operations in the futures markets are required.

The firms vary in size considerably, from enormous companies such as Dunavant and Cargill (which owns Ralli and Hohenberg) to small firms; there are also large selling co-operatives. Such co-operatives rebate the equity they receive from back-to-back sales to the producers: approximately 25% of the US crop is sold through this method each year. The remainder of the crop is bought from the growers or other first-hand merchants by mills in the cotton-growing areas. There are four major co-operatives: Calcot Ltd, Plains Cotton Co-operative Association, Southwestern Irrigated Growers and Staplcotn. These together in 1971 formed Amcot, which is internationally represented including in the UK.

The world's major cotton companies can be listed, but the important division is fourfold. Firstly there are the major US companies, foremost amongst which is Dunavant; US companies together trade approximately 50% of the world's cotton trade, bearing in mind that a large percentage of the world's cotton is traded directly with governments, as part of barter deals and generally out of the private sector: for example by ChinaTex for Chinese cotton (and formerly Novaexport for Soviet cotton). The percentages quoted below are of the total amount of cotton that at one time or another reaches the free world market. Dunavant, the world's most important cotton trading company, trades itself one-sixth of the world's cotton trade and therefore deserves close examination.

The company grew out of T.J. White & Co and since 1961 William B. Dunavant Jnr has taken the company from 100 000 bales annually to its present position, that of the 'world's premier cotton merchant'. It trades over 3 million bales - possibly closer to 4 million - annually with an annual sales volume of over $1.2 billion - and this claim would appear authentic. The company's headquarters are in Memphis, Tennessee, the world's largest spot cotton market. There are sales offices in the USA, Belgium, Hong Kong and Japan, merchandising offices in Argentina, Switzerland and Guatemala (and formerly in Germany), buying offices in the USA and agents all over the world; the global sweep of the company is undeniable. Dunavant buys cotton from all the world's cotton-producing areas, and their company literature argues that

> the price they get is solid, based on the kind of up-to-the-minute information and flexibility that only a marketing network as vast as Dunavant's can offer. The company has designed, engineered and now also markets its own proprietary software systems revolving around Honeywell Bull DPS-8/49 hardware and Honeywell Bull DPS-6 mini-computers. The use of HVI classing further places Dunavant on the cutting edge of industry technology.

The company's warehouses can accommodate over 1.5 million bales and there is an on-line computer link to all warehouses. There is a truck brokerage operation based in Memphis, and the company says that 'through its relationship with carriers, Dunavant commands respect in scheduling domestic and international transportation'. Dunavant is also the largest cotton ginning operator in the world, ginning over 600 000 bales annually. It owns a new four-stand gin in Moree, Australia. The company bought Geo. H. McFadden & Bros, thereby acquiring gins in Mexico and Brazil, and Producers Cotton Oil Company to increase its dominant cotton merchandising position in Arizona and California. The Dunavant Commodity Corporation deals in cotton futures and options in New York, whilst the holding companies have diversified into real estate and professional sport.

The company's sheer size is both its greatest advantage and biggest weakness. Although it can move prices, both physical and spot, it can be seen coming both geographically and temporally, and does not therefore obtain as many individual high profit trades. Nevertheless its position now and into the twenty-first century under a new generation of Dunavants remains assured, with low overall debt and substantial equity reserves.

The remaining one-third of world trade is taken up by the major US companies of Amcot, Conticotton, Allenberg, Hohenberg, Esteve, Ralli Bros (owned by Cargill) and Weil Bros. Very approximately these companies can be described as '1 million bale' sized companies, although Amcot, for example, trades much more than this amount every year. Conticotton in particular trades a very substantial part of non-US growths and has a major trading operation out of Geneva which is handling a substantial portion of CIS cotton exports currently. In all, Conticotton trades perhaps 2 million bales annually.

The second group of companies are those based in the UK. In the UK the cotton industry has contracted painfully and continues to diminish, with Courtaulds recently announcing the closure of another two mills, for instance, although one was taken over and now operates under another company. In the 1950s, there were several different types of UK firm. The cost insurance freight (cif) agents were intermediaries between the exporting shippers in the USA or elsewhere and the UK buyers, and operated on commission. The importing merchants, except those that had branch offices in the USA, bought US cotton and sold it in the UK and continental Europe. It is estimated that one-sixth of the world's cotton is still traded by companies based in the UK (Liverpool). That is far from saying that one-sixth of the world's cotton is ever landed in the UK. However, companies such as Meredith Jones, Alexander Eccles and Stern (now fused since 1991 with Weil Bros in the USA), Baumann Hinde, Ralli (part of Cargill), A. Hannay and Plexus are far from small; it is estimated that Ralli trades perhaps 350 000 bales annually. The Cargill Group's involvement in the cotton industry extends far further than 'just' cotton trading. Three companies, Cargill Technical Services Ltd (CTS), the British Cotton Growing Association (BCGA) and Minster Agriculture, are also part of the group. CTS was established in 1987 and co-ordinates the Group's provision of services worldwide; the BCGA was established by Royal Charter in 1904 and has been involved thereafter worldwide in the growing, processing and marketing of cotton, being acquired by Cargill PLC in 1981; Minster was established in 1975 and has undertaken over 400 projects worldwide, being acquired together with its Thai subsidiary in 1990. The group employs over 50 specialists with experience in cotton agronomy, agricultural economies, processing, marketing and training. The group has equity holdings in a number of cotton companies worldwide: 51% of Cotton Ginners Ltd, which gins 90% of the Malawian crop, 40% of Cotton & Agricultural Processors Ltd, which gins 95% of the Nigerian crop, and

48.5% of BCGA (Punjab) Ltd, which has a 600 acre cottonseed breeding and multiplication farm.

Meredith Jones trades approximately 250 000 bales annually, A. Hannay and Baumann Hinde perhaps the same, and Plexus rising probably towards 200 000 bales. None of these companies is totally open about their trading volumes, and no one should expect them to be. They are even less open about whether they have any particular speciality in any one area: but they do. For example, it is well known that Baumann Hinde has a particular speciality in exotic, especially West African, cottons; Meredith Jones has been making inroads into the Far East along with a number of other importers. Plexus is a new company based in Liverpool with an aggressive attitude to its market share. In Liverpool today there is a quiet optimism about both the city's future in cotton trading and the likely future success of companies based there. Liverpool will almost certainly never function again as a major importing port (perhaps its only hope is the Channel Tunnel traffic), but it remains an extremely important concentration of cotton trading companies, second only to Memphis, and above Geneva. Lunch in a small Liverpool restaurant has found the analysts of Cotton Outlook at one table and the Chairman of Ralli at another; and they frequently meet together.

The third group of companies are those that trade into Japan and that are predominantly Japanese companies. These are the companies which along with Pakistani exporters, Korean mill owners and Japanese mill owners constitute the key users of the Osaka and Tokyo futures markets: the Japanese trading companies almost all trade cotton; they include Mitsui and Marubeni as well as C. Itoh and Toyamenka. These companies together trade another one-sixth of world trade and they have both the financial strength and the global connections to mount a challenge - in their own good time - to Western companies' domination of the global cotton market. Current interest in exporting CIS cotton from Vladivostock is not unconnected with this possibility.

Finally there are all the other companies, mainly trading out of continental Europe. These range enormously in size, from large international trading operations specialising in importing Russian, Turkish and Greek cotton into Europe, such as Bunge, now owned by Louis Dreyfus, through CDI, Cogecot, Compagnie Cotonière, Cotimport, Cottip, Deca, Polycoton, Reinhardt, Rheinschelde, Stahel Hardmeyermer and Volkart, which estimated in 1985 that it had a 4% share of the world's cotton business. Built by the Swiss Reinhart family, the cotton trading department of the company has been based since 1989 in Dallas although it maintains a powerful trading presence in Europe.

The company acquired Western Cotton Services (Anderson Clayton) in Phoenix as part of its strategy to integrate vertically backwards into cotton ginning, warehousing and milling. In 1991 the company acquired Sethos AG, a yarn trading company based in Winterthur, Switzerland. The company intends to increase its presence steadily in the textile trading and processing sectors, which it states as desirable particularly because of the increased industrialisation of the cotton-producing countries - and hence the rising importance of trading yarn rather than raw cotton. As with many trading companies, the company is also interested in financial services, purchasing shares in several holding companies and banks.

Cotton trading is characterised by a much larger number of active trading firms than for most other commodities, as a glance through the membership list of the Liverpool Cotton Association demonstrates, and by an absence of brokers. Almost all the companies continue to be in private hands, notably the US giants Dunavant and Conti, but also Bunge (now Dreyfus-owned) and Cargill. There is also price disclosure through the active futures market and the daily published Cotlook indices, which makes for a relatively level playing-field between traders. No cotton company, not even Dunavant, can maintain prices for even one growth at a beneficial level for very long, if at all: the companies are competitive, although they avoid 'dumping' cotton stocks and have an interest in an orderly world market for cotton. There are also a few cotton shipping companies. There are also opportunities for new companies, as Plexus has demonstrated. It is thus a much 'healthier' commodity in terms of international trade than many others, and the main concern that the cotton companies have is that the international demand picks up, i.e. the continued health of the primary textile industry, or else they will be trading yarn. However, there has been a steady tendency to merger over the past few years: most recently there has been the Volkart/Reinhardt merger; the Weil Bros and Stern merger; Lonrho took over Baumann Hinde, and Ralli and Cargill joined forces. There is also the powerful dominance of Dunavant to consider.

The cotton freight market

Dozens of companies internationally are constantly competing for cotton freight business, and each of the international cotton merchants has established relations with at least a dozen such companies, with occasional use of up to 30, and the intermediate action of shipping brokers. Different companies tend to specialise in the freight of cotton out of specific destinations, for example Latin America or West Africa, and are most competitive for those routes. Cotton is normally quoted by the shipping companies across specific routes per tonne over a fixed period, for example three or four months.

The contracts agreed between the shipping companies and the merchants have over time become honed to a point of simplicity and straightforwardness - a single Telex, received by the shipper, serves to confirm the order, and the conditions of 'full liner terms' are normally specified as well as the agreed rate (Fig. 9.4). Should problems arise, for example deterioration in quality due to the cotton being stowed on deck (which should not occur, but sometimes does), the merchants, shippers and receivers tend to negotiate them away, rather than use complex and often lengthy insurance procedures. The shippers themselves insure their cargoes, as they have done for centuries, often via Lloyds of London and sometimes there are claims against them by their clients.

92-12-23 14:34
Msg 329 Title:

TO: SHIPPING CO./AGENT
FROM: SHIPPER

WE CONFIRM OUR TELEPHONE CONVERSATION TODAY WHEN YOU ACCEPTED OUR FIRM BOOKING FOR SHIPMENT OF ABOUT 500 METRIC TONNES RAW COTTON IN BALES FROM DOUALA TO KEELUNG PER M/V (NAME) ETA DOUALA (DATE) AT (RATE PER M/T) FULL LINER TERMS.

PLEASE CONFIRM BOOKING

THANKS + REGARDS
++++

9.4 A typical Telex contract between a shipping company and a trading company.

Ocean shipping rates for cotton took effect on 1 November 1992. Cotton shipping rates for TransPacific Westbound Rate Agreement (TWRA), a consortium of leading freight carriers to Asian destinations, ranged from $72 to $123/tonne and the new rates ranged from $73 to

$131/tonne. TWRA includes some of the largest ocean freight companies such as Sea-Land Service, American President Lines Ltd and OOCL (USA) Inc. This group also serves the Asian market, the main purchasers of US cotton. The system has stayed this way for upwards of 20 years: the divisions of responsibility appear to work well, and there is little incentive to change the way it works.

The future of international trade in cotton

For the short term, the ICAC textile demand model forecast suggested in 1991 that total textile fibre consumption would have increased to 39.7 million tonnes in 1992 and 40.6 million tonnes in 1993. These forecasts can now be seen to be based on an over-optimistic assumption about the rate of economic recovery. The ICAC believed that fibre consumption in both industrial and developing countries would be boosted by lower fibre prices and increased availability, but in fact this has yet to occur, and moreover the decline in consumption in the former Soviet Union countries and in Eastern Europe can be expected to continue for at least several years. If the 1991 projections for 2000 are achieved, it will be cutting across the line from below as a result of a growth surge in the later 1990s. Cotton's market share amongst textile products, however, can be expected to be at least maintained and if it increases, then an overall weak textile industry can still support steady growth in cotton production and mill consumption. Evidence from the USA in 1992 in particular suggests that for largely cultural reasons, cotton in the 1990s is a good business to be in. Both the analysts at Cotton Incorporated and the US traders are relatively optimistic.

Taking a longer perspective, it is clear that since the 1960s textile consumption worldwide has increased dramatically. Some 40% of this increase can be attributed to population growth, and the UN forecasts suggest that such an increase can again be expected over the next 30 years (Fig. 9.5).

Production forecasts are co-determinant with prices and consumption beyond one year. At present, reforms being enacted by the Chinese government suggest that production will increase next year and thereafter; and Chinese stocks are already high, at 13.3 million bales. There is clearly an optimal ratio of stocks to production. Planned acreage in the states of the former Soviet Union is likely to fall, and it is strongly to be expected that continued political unrest

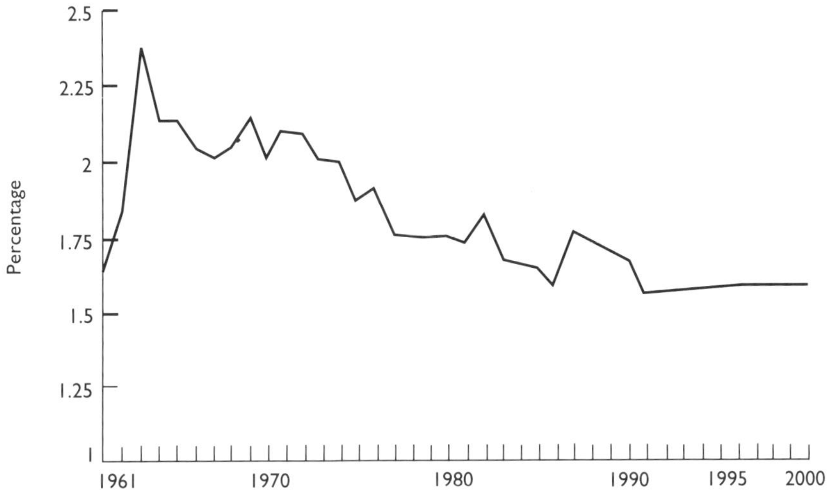

9.5 World population growth projection to 2000 (*source*: UN).

will affect production adversely in the key regions of Tajiskistan and Azerbaijan in the period to 2000, despite gradual improvements in these countries' cotton distribution systems. They still face dreadful infrastructure demanding huge investments that can neither be afforded nor politically justified in the present state of unrest. Pakistan has now started to reduce cotton production, although this is not likely to be a sustained or substantial reduction because of continued yield growth and high and rising domestic consumption. African cotton production is unprofitable, likely to remain so, and may well decline in the period to 2000. Finally the USA's production, which on the justified assumption of continued US farm support programmes can be expected to continue at existing high levels, makes up the balance of any increase in world production required to meet demand.

The remaining 60% of projected consumption increase can be attributed to economic growth, textile real prices falling and increased competition amongst fibres which has made consumer preferences more sophisticated. As demonstrated above, however (see the section on 'Regional cotton consumption' in Chapter 5), this pace of growth has slowed (Fig. 9.6).

Cotton is likely to continue to benefit from a preference for 'natural' products, and will therefore sustain a higher rate of demand growth than non-natural fibres. The ICAC textile demand model in 1991 projected textile consumption and the share of cotton in 1995 and 2000. Operating the model in its original form, without short-term income and price adjustment variables drastically reduces the

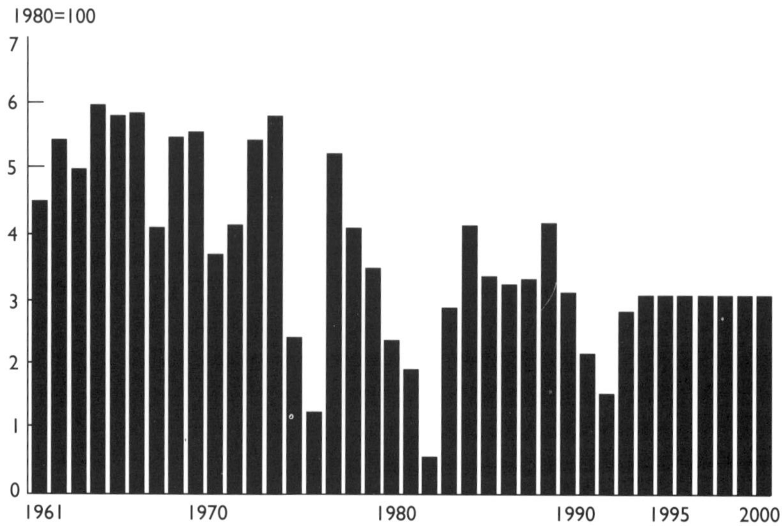

9.6 World GDP growth projection to 2000 (*source*: ICAC and World Bank).

accuracy of the results, because the assumption set included a 3% annual growth rate - borrowed from the 1980s - and UN population projections; on the negative side, the model assumed a continuation of the ratio between cotton and its competitors in terms of price, and textile prices rising in line with inflation - both of which are assumptions that can now be challenged. The results of the 1991 'False Dawn' analysis suggested that textile consumption would grow at an average annual rate of 2.3%, reaching 43 million tonnes in 1995 and 48 million tonnes in 2000. Per capita consumption was forecast to rise from 7.3 kg in 1991 to 7.9 kg in 2000.

Cotton consumption was projected to grow at an average annual rate of 2.5%, which produced the often-cited headline of these very words in February 1992's *Textile Month* magazine, whilst non-cotton textile consumption was projected to grow at 2.2%. The faster rate of cotton consumption with respect to other textiles according to the model allowed cotton's market share to reach 49.7% (Fig. 9.7). These projections were not perceived as especially optimistic at the time they were made: the rate of growth of demand was 2.8% in the 1980s, but only 1.7% in the 1970s and 1.6% in the 1960s.

In the short term, the ICAC pointed out that textile consumption growth had been very strong immediately after a recession, for example in 1983 there was a 4.5% increase worldwide. But the 1990 decline did not of necessity bring revival in 1991, and in particular the continued decline in the former Soviet Union countries and in Eastern Europe created the poor consumption figures. The former East

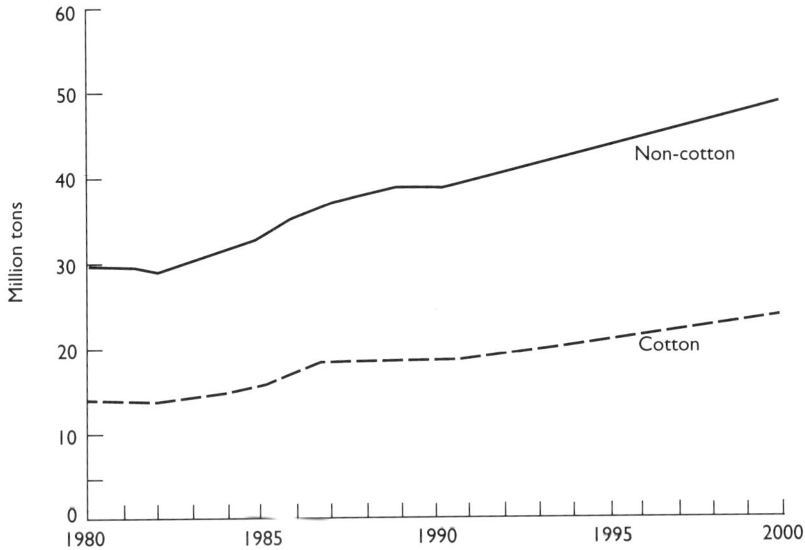

9.7 Cotton's share of the market projection to 2000.

German textile industry, along with those of the former Soviet Union and the other East European countries, proved largely inadequate to face international competition and has shrunk significantly, with the result that demand for yarn continues to fall there and may be expected to do so for several years yet, even though niche markets may be created throughout the region. As the ITMF Director noted in 1991:

> We now have to ride out a 5–7 year long cyclical roller-coaster of global dimensions. We must, at the same time, face up to a rising mountain of surplus capacity, now in spinning and tomorrow also in weaving. It is this latter structural aspect which has changed the name of the game and its consequences will in future affect all players, irrespective of their location. (ITMF *Annual Report*, 1991, p. 58)

The ICAC are the first to admit that in the light of what has actually happened in 1992 their 1991 forecasts must be drastically revised. The RER forecast for 1995 and 2000, based on a heuristic interpretation of Cotton Outlook prices throughout 1992, and available data on consumption, production, stocks and trade, suggests that with an average economic growth rate much closer to 2% than 3%, with UN population projections proving approximately accurate, and with cotton prices falling relative to other textiles, textile consumption could be increasing at an annual rate of just under 2% in the 1990s. Cotton consumption should increase slightly faster, at

2.15%, due both to low prices relative to competitive synthetic fibres and also to increased consumer preferences for cotton internationally. However, the rate of growth can be expected to remain extremely low in the period to 1995, and the overall 2% is based on the idea - widespread in 1992 - that economic recovery will occur, but later in the decade. Consequently textile consumption should be approximately 38 million tonnes in 1995 and 43 million tonnes in 2000. Cotton's share can be expected to increase in line with ICAC's expectations, however, to reach about 48.5% by 1992/93 and 49.5% by 2000. Consumer preferences are expected by many analysts, for example RER, to shift further in favour of cotton in countries such as Japan and Taiwan, following what has already happened in Europe and the USA - but this process may well be slow.

Intensive and successful marketing produced the rise in cotton consumption from 18.8 million tonnes in 1990 to 21.1 million tonnes in 1993 and 24.1 million tonnes in 2000, which compares with world supply even in 1992/93 of 29.3 million tonnes (production plus stocks). Geographically, the ICAC believed that the increase between 1990–2000 would come as shown in Table 9.9. This forecast is now open to question in respect of Eastern Europe and the former Soviet Union, where in practice there has already been a very substantial fall in production and in both mill use and end consumption. If both return to their 1990 levels by 2000, domestic leaders in the successor states to the Soviet Union and in Eastern Europe will be pleased enough. In the industrial countries, the end-use projections, given a fair economic revival by 2000, are plausible, but there seems much less justification for an increase in mill use, when in practice mills were closing throughout Europe in the early 1990s. Finally, there are the developing countries. Here the forecasts seem most accurate, based on the rapid rates of growth in mill consumption seen in both importing countries such as the Philippines and Indonesia as well as in exporting countries such as Zimbabwe and Pakistan. Even these forecasts of consumption growth do not allow for substantial or sustained price increases or for prolonged stock reductions. The

Table 9.9 Increase in consumption (million tonnes)

	In mill use	In end use
Eastern Europe/former USSR	+0.6	+0.6
Industrial countries	+0.4	+0.2
Developing countries	+4.3	+2.7

overall picture remains one of constrained growth, combined with historically high stocks and highly competitive markets.

The outlook for prices can be sharply divided into two: the short term and the rest. In the short term there is general consensus and it is very difficult to envisage sufficient factors combining to alter it. The ICAC says that the magnitude of the fall in prices over the 1991/92 and 1992/93 seasons was greater than previous correlations with supply and demand suggested. The ICAC suggested in October 1992 that the average 'A' Index figure for 1992/93 would be 69 cents, which compares with the NYCE World Cotton Contract indicating a figure of less than 55 cents. Other trade estimates suggest that the 55 cent level is likely to be far closer to the mark for the next two seasons. However, the experience of the world cotton market suggests that the outlook for prices in the longer term is probably more bullish. It is not possible to say for certain which shocks are likely to push prices up again, but some changes will almost certainly do so in the longer term.

Prices may be expected to remain commensurately low, if not in the doldrums, for several years. The USDA commented that 'Buyers expect prices to remain depressed until fundamental changes occur in cotton production and consumption markets'. There is too much productive capacity for savage price rises to be likely in the period to 2000, at least for them to be sustained; nor is it likely that production will fall to allow a sharp rise in prices: pressure on other commodity prices is just as severe. The sheer size of the physical market makes it impossible for there to be a physical squeeze; the reliability of the 'A' Index will always be quite sufficient for it to reflect any over-supply, and therefore the only price that could conceivably leap is the US domestic price. At that time the discretionary powers of the US Farm Act would probably be used to subsidise exports and the world contract in New York would enjoy great success as a hedging mechanism. What does seem likely, therefore, as predicted by the 1991 ICAC forecast, is that 'Cotton prices are assumed to continue to fall somewhat relative to other textile fibre prices' (ICAC, *1991 Conference Report*, p. 51; Fig. 9.8). Consequently, in this relatively flat market, less price volatility than experienced in the 1980s may tentatively be expected, and also less variation from high to low during a season.

There is also the question of how the relationship between coarse cotton and medium count will evolve and it is generally thought that existing trends are likely to continue. In 1991 the discount averaged approximately 2.5 cents, about 4% of the 'A' Index. Improvements in the quality of the CIS crops and the increase in production and stocks of coarse cotton mean that the discount has widened and is likely to

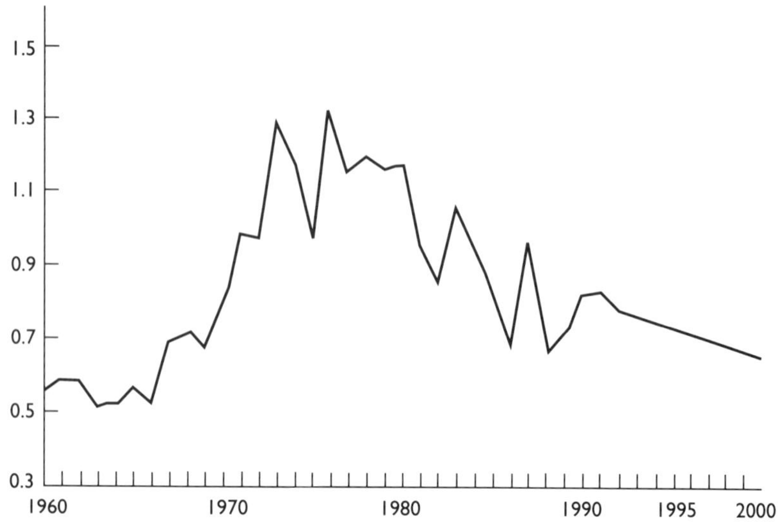

9.8 Ratio of cotton prices to non-cotton textile price index 1960–2000.

stay at these wider levels for several more years. Exports of fine cotton, where the US San Joaquin Valley styles are the most popular, amounting to some two-thirds of world exports, are expected to continue their fall, started in 1989/90. Production increases of extra-fine cotton are expected by the ICAC to be in countries that do not usually export, so there is little likelihood of a substantial trade revival and the high price premiums enjoyed by these grades is contributing.

What forecasts of this type suggest is that world trade in cotton and apparel is likely to increase by about 50% in the period to 2000. It is most likely, however, having seen developments between 1990 and 1993, that trade in raw cotton is by no means as likely to rise as fast as in the finished products. What pays newly industrialised countries is to produce the cotton, spin it and produce garments themselves to export to industrialised countries.

For raw cotton, ICAC estimates suggested that US exports will be about 1.4 million tonnes in 1992/93, explained by relatively high US prices and the limited supplies available for export sales: exports to the Far East are expected to bear the brunt of any decline, as the USA faces both decreased demand and probably also Chinese competition. Concentration amongst the five largest sellers is expected to continue, surpassing 70% by 1992/93 and continuing to rise for some years. Pakistani sales can be expected to experience State curbs but as the CIS gradually reorganises, its exports can be expected to continue their current high levels of competitiveness. It is expected that owing to the need to grow food there, and also some concern at what the

ICAC describe as 'the environmental effects of the cotton monoculture', cotton production will stabilise at about 2.4 million tonnes. Domestic use will take perhaps a decade to build up again with the construction of new mills. For some years, it can be expected that exports will be about 2 million tonnes, divided between other former Soviet Union states (about 1.2 million tonnes or a little less) and the rest of the world, principally Europe for the remainder. Production in Australia can be expected to continue to be profitable, but there are strains on the EC Stabex scheme in 1993/94 as cotton production in Francophone and other parts of Africa is fast becoming once again uneconomic as prices continue to slide. This production already requires substantial subsidy, but will of course continue so long as that money is forthcoming.

EC rules and regulations currently encourage cotton production in both Greece and Spain as part of the general subsidy to EC agriculture. Greece has become an important exporter in 1992/93 and will continue to be so provided that these substantial (50 cents/lb) subsidies are continued. However, they may not be continued indefinitely. The effect of reducing them may not be as great on the international cotton trade itself by the time that they go as it would be now, because of the switch away from raw cotton trading into Europe in favour of straight apparel imports. However, viewed globally, take away the quantitive restrictions on the textile trade which have been around since the GATT agreements of the early 1960s, and the export of textiles could leap, to the detriment of European mills in particular. Even if the present Uruguay Round just fizzles out without there ever having been concluded a new GATT agreement, within 10–15 years it is highly likely that virtually all textile trade will be liberalised and that quota restrictions will have disappeared. The current hostility to GATT in the European countries overlooks what the UK Minister of Agriculture, John Gummer, said to the Conservative Party Bow Group in December 1992: 'If GATT is not concluded, I don't know where the money to pay the subsidies is going to come from.' Textile trade would then be conducted under the general rules of GATT as is already happening throughout the Americas; the MFA would be finished, and there would not be an international authority governing textile trade. Although the ITMF in 1991 believed that under these circumstances, what the industry internationally would need would be better and faster information, better communication and a readiness to invest, these are precisely what would be missing under such highly competitive conditions. The only likelihood will be greater cross-border collaboration, probably taking the form of joint ventures, with European, US and Japanese companies providing the capital, the know-how, for some years still the machinery, and the marketing, and the newly developed countries providing the land, the labour and the production expertise.

From the available data, it is true that if cotton consumption in the developing countries is to reach the same levels as those prevalent in the developed world, then there is still a vast expansion required in the international cotton trade to match the demand. That is the most positive thing that can be said at the annual dinners of the Liverpool Cotton Association. What cannot be predicted with certainty, except in the negative, is when this will occur. It seems unlikely that cotton consumption in developing countries will reach levels achieved in the developed countries by 2000. However, cotton's gradually increasing share of world textile consumption despite a fall in 1992/93 will result in a need for more trade, since many of the increased consumption countries will not necessarily produce and spin all they need. There is an argument, promulgated by analysts rather than traders, and which has been circulating since the 1960s, that raw material producers, especially China, India and Pakistan, will produce and export increasing quantities of the finished product rather than raw cotton. Analysts say that recent figures, which show Asian countries' demand increasing - countries which are aiming to process greater amounts of cotton internally - back up this contention. Hence, the ICAC argues, international trade volumes can no longer be calculated with any degree of certainty from production or consumption figures.

Even if the ICAC is wrong in the short to medium term, there must be some force to the argument that the liberalisation of China's trading system, devolution to the provinces and eventually, no doubt, to the private sector, will encourage the production and export of higher value added cotton goods. This in turn must diminish the opportunities for export from West Africa, the US, India and Pakistan to mills in countries with higher labour costs.

A similar long-term threat is likely to emerge from the CIS states when production of cotton goods is finally revived there on a capitalist basis. CIS wages are set to remain much lower (up to 50% on a comparative purchase cost basis) than in Western Europe, well into the next century, experts on European labour developments forecast. The relative structure of labour costs is therefore likely to be the major determinant of cotton trade volumes, as it already is for other textiles such as silk.

Some traders do not entirely concur with this hypothesis. They say that the international trade has already proved remarkably resilient and question the accuracy of current official consumption figures. They also believe that the international textile industry is still very firmly in the grip of recession with the revival not expected until 1994. That, they say, explains better the flat international trading figures.

This implies that the international trade statistics, and the information garnered in Liverpool and Memphis on the volume of trade being done by the major companies, are more accurate than the national production and consumption figures - especially of some developing countries.

However, this argument concedes that some of the larger cotton trading companies may face a long-term threat from the development of domestic mills. Among these one can single out Dreyfus, Continental and Cargill - all of whom have been accustomed to servicing large mill demand, especially in Europe. Their own cost bases are relatively large and they need to trade, each of them, some 2 million bales annually to some 24-25 million bales. This is not difficult but some mergers can be expected over the next decade in response to competitive cost pressures.

It is not surprising that the traders arguing this particular point strongest are those in Europe with niche markets, for example servicing the new product lines that Courtaulds is busy introducing at present. For these companies, trading of the order of 2-500 000 bales annually, there is a continued market at keen prices throughout Europe, Asia and especially South America. Traders have been forced to adopt geographically mobile policies in the last year - more so than ever before - in seeking good markets for their cotton. This trend can only continue.

Relative price stability and widespread availability of price information makes knowledge of demand the crucial factor.

All the firms are absolutely united in refusing to enter the market for finished products. They do not see their future as wholesale providers of finished products or even grey goods to the retail market. They do not believe their expertise is well suited to this task, and in many cases their vertical diversification is too well entrenched towards the actual field production of the crop, harvesting, storage and transport - for example Dunavant in the US.

A good example was the sudden demand for imported cotton by Brazil in 1992/93 which was being forecast to be 200 000 tonnes short of domestic demand in 1992/93 and is likely to continue for a few years. Mexican demand is another example. This hypothesis is perfectly consistent with the argument that in the short term countries will mill a higher percentage of their production than they have in the past and therefore trade will ease off perhaps for some years. The crucial fact, on which all agree, is that a trade revival in the medium term therefore depends on rising consumption worldwide, and especially in cotton importing countries. In the long term, technical and even climatic changes allowing these countries to grow cotton themselves may alter the balance again.

The GATT, which far from being signed at the end of 1992 now lies on the table as a high initial priority for the Clinton Administration, does include an outline of what should happen to agriculture: the Dunkel proposal. For cotton, however, non-GATT signatories have an important role, although this has not been considered in depth yet. The GATT project involves, as part of the international first step towards free trade, converting all non-tariff barriers into straight tariffs. The USDA proposes (December 1992) a tariff of 7.5 cents/lb for raw cotton imports into the USA. Similarly the North American Free Trade Agreement (NAFTA) will probably adopt the 'fibre forward rule of origin' which would ensure that gains from trade are confined only to signatories. The US cotton lobby is extremely anxious to avoid substantial imports and to maintain existing levels of support: not just for raw cotton production but also for the US textile industry, where the US National Cotton Council has estimated that the adoption of the Dunkel proposal would result in a reduction of domestic textile production by two-thirds. Domestic pressure within the USA is for 'at least' a 15 year phase-out of the MFA to assist the US industry during the transition period. Given the protectionist mood of the new US Administration and within the EC, radical free trade - permitting imports into the USA - in raw cotton and textiles is unlikely, certainly within the next 5 years and probably within 10. The experience of the past 30 years, however, and the weight of economic experience, suggests that only lower costs will produce long-term productive capacity. Otherwise both the US cotton belt and the US and remaining European mill capacity will be placed under threat. Negotiations that are eventually concluded between the USA, Canada and Mexico on the creation of a NAFTA would also of necessity create a readjustment of the location of textile industries in North America owing to the difference in wage costs between the USA and Mexico in particular. The NAFTA agreement, which participants wish to see concluded in 1993/94, aims at benefiting producers within the NAFTA area. The USA is negotiating for a fibre forward rule for yarn and a yarn forward rule for fabrics and textiles. Yarn exports within NAFTA will be permissible with NAFTA trade preferences if domestically produced cotton is used; the yarn forward rule will allow internal status to be conferred on fabrics woven or knitted from internal yarns. Little is expected to change once this agreement comes into force, except to increase Mexican textile exports to the USA. The USDA stresses that the cotton industry would benefit from an extension of NAFTA to Latin America and the Caribbean.

So the outlook for the international cotton trade in overall volume terms can be described as static in the short to medium term, and then increasing towards 2000, probably reaching not less than 10% more in volume terms than the 1992/93 season by then. This compares with a growth of 50% from 1993/94 compared with 1983/84, and no change from 1983/84 and 1973/74. What is particularly remarkable is that most of the increase in world exports has taken place precisely during the international recession - rising from 5.2 million tonnes in 1989/90 to 6.4 million tonnes in 1993/94. Production similarly has risen dramatically and now stands worldwide at consistently over 20 million tonnes. Moreover, in this rising volume of world exports there is ample room for trading companies, taking progressively more of world exports as the position of direct government and agency trading declines. At the same time it must be actively questioned, although not definitively answered, as to whether small trading companies can profitably compete in the 21st century with Dunavant, Conticotton and their few immensely powerful rivals. The 21st century is surely likely to see a further rationalisation of the numbers of active cotton trading companies, most likely through international merger between US, UK and even Japanese trading companies, towards an optimum number of cotton trading giants. It is not at all possible to say with assurance, however, what that optimum number is, nor how many small niche companies may profitably operate in their midst.

Indian productivity gains in the period to 2000 may provide the opportunity for India to become a regular and significant exporter of cotton and cotton textile products. In 1992/93 India exported more cotton than in any year since 1983, accounting for 6% of world trade. Export restrictions would have to be eased, however, for this to happen.

The future to 2000

Technically, quality control and demands are bound to escalate worldwide, fuelled by the progressively more widespread and cheaper availability of increasingly accurate and sophisticated HVI technology. Open-end spinning may be applicable to all fibres. The use of glandless cotton will spread. But the negative aspects of the current international recession will themselves restrain the growth in consumption, trade and production for some years, it seems. In many countries yields can only be increased through the additional application of fertiliser, especially in densely populated areas where leaving land fallow is not economically viable. Farmers cannot and will not be able to afford this additional quantity of fertiliser: moreover the international environmental lobby is certain to demand more 'naturally produced' cotton; some garments are already stamped 'The cotton used in this product was organically grown' and such cotton will attract a premium in the market.

Research is therefore being carried out to try to find a less expensive way to raise yields, perhaps a closer link between agricultural and pastoral systems. Some adjustments in fertiliser application, in the use of insecticides, biotechnical production control and further advances in IPM can all be expected. In particular, biocontrol is a strategy that fits in well with small-scale production and it can be expected that regional and even national biotechnical industries will develop, improving the resistance of local varieties of cotton to specific weeds, diseases and pests. Genetic lines with resistance to broad-spectrum herbicides are almost certain to be commercially available; cultivars carrying the *Bt* toxin genes are similarly very likely to become commercially available in the USA by the end of the century, although in the long term the insects will develop resistance to these bacteria as much as they already have to certain chemicals. The upshot of this process is likely to be the establishment of a similar type of relationship between producers and the biotechnical companies as already exists between the producers and the chemical companies. The further use of insect toxins and the pests' own genes to engineer plant resistance can be expected. Sophistication may come through the use of plant gene promoters which will control when certain proteins are actually needed, e.g. when the plant is under attack.

Biotechnology may also play a key role in the promotion of hybrid cotton production through the easing of reproductive problems. Indeed, there are likely to be important contributions throughout the whole cycle provided by biotechnology, most importantly on pest resistance and fibre quality. The period to 2000 may be a little slower than expected in the 1980s because of the commercial unviability of many projects; but as soon as the economic demand is apparent, the projects will be rejuvenated. It is likely that in order to achieve the levels of textile consumption throughout the world already experienced in developed countries, this type of innovation will be required in general use throughout cotton-growing countries.

Production forecasts are co-dependent on prices and consumption beyond one year. At present, reforms being enacted by the Chinese government suggest that production will increase in 1994 and thereafter; and Chinese stocks are already high, at 13.3 million bales. Production in the former Soviet Union will continue to fall in area, and it is strongly to be expected that political unrest will continue in the cotton-producing regions of Tajiskistan and Azerbaijan for some years at least, despite gradual improvements in the cotton distribution system. Pakistan has now started to reduce cotton production, although this is not likely to be a sustained or substantial reduction because of high yields and still rising domestic consumption. African cotton production is unprofitable, likely to remain so, and may well diminish over the next few years. Finally the USA's production, which on the assumption of continued US farm support programmes along current lines can be expected to continue at existing high levels, makes up the balance of a continued rise in world production and hence stocks.

For the cotton textile industry, there will in European countries be increased stress on the environmental implications of production. This is already of great significance in the Netherlands. ITMF recognises that the treatment of polluted effluent requires future investments equivalent to a typical annual new machinery investment. The EC may press for this investment to be made, but the effect if implemented will be to drive the industry further to countries with lower environmental standards which must merely be concerned with a shortage of effective labour resources such as Taiwan.

The economies of textile production are likely to be affected, as ITMF rightly surmises, by the tremendous - perhaps 35% - overcapacity which will remain in force for the world until 2000. Inventories are likely to fall during the 1990s, and capacity utilisation will remain less than satisfactory for several years yet. Textile manufacturers did not entirely believe the ICAC's forecast that world fibre consumption would grow by 2% in 1992 and 3.6% in 1993. RER's estimates, 0.5% and 1.2%. would appear closer to the likely outcome even with the Clinton victory in the US elections. The results of this continued excess of capacity over use is that the average age of machinery will age throughout the 1990s. Those textile machinery manufacturers that remain by 2000 may have substantial catch-up business to do as well as exploiting an Eastern European market that by 2000, at least according to the European Bank for Reconstruction and Development's 1992 estimates, should be in a position seriously to start buying textile and other manufacturing machinery.

It is perhaps too speculative an endeavour to forecast textile production in the medium term worldwide. But on existing trends, it is almost inevitable that by 2000, European mill use will have actually fallen, perhaps to around less than 1988 levels but with consumption rising roughly in accord with the limited economic growth expected in the 1990s, and concentrated in the less developed countries of Europe, including by 2000 the former Comecon countries registering growth in textile consumption. This would produce a considerable growth in textile trade over the period, mainly in finished products but also to some extent in yarn as well. The risk to this scenario is not so much whether it will occur, but the time period over which it is likely to happen. The 1990s are proving difficult years for the expansion of trade and the diffusion of effective consumer markets which in many Western countries are actually contracting.

In the measurement of cotton prices, it seems unlikely that the present system of collecting reported quotations from the trading companies by Cotton Outlook will change in the period to 2000. Nevertheless, the use of the 'A' Index for the NYCE world cotton contract has focused attention on it as never before, and a Cotlook East index would be a useful innovation, for example - a cif Japan equivalent has already been tried but there may be another attempt; there is already an office in Memphis. Moreover, Cotton Outlook's link to Liverpool is just a question of history and preference. For a transaction price index and weighted by transaction volumes, based perhaps in Geneva, the next century may be too soon to expect it - if ever.

APPENDIX

I

Cotton supply 1992/93

Table AI.1 Cotton supply, use and trade by country, 1992/93: 1000 tonnes

	Beginning stocks	Production	Imports	Total supply	Consumption	Loss[a]	Exports	Ending stocks
Afghanistan	7	22	0	29	17	0	7	5
Albania	15	0	4	20	5	0	0	14
Algeria	1	0	49	50	44	0	0	7
Angola	2	4	0	6	3	0	1	2
Argentina	164	174	11	349	120	2	70	158
Australia	239	348	0	588	29	0	392	167
Austria	1	0	30	32	30	0	0	1
Baltic States	7	0	0	7	54	−54	0	7
Bangladesh	29	16	98	143	112	1	0	30
Belgium	7	0	39	46	38	0	2	5
Benin	9	65	0	74	4	0	60	10
Bolivia	2	13	5	21	16	0	0	4
Brazil	453	566	201	1 220	729	0	29	461
Bulgaria	11	17	16	44	33	0	0	12
Burkina Faso	10	70	0	79	4	0	65	10
Burma	15	24	0	40	19	1	5	15
Cambodia	0	0	1	1	1	0	0	0
Cameroon	4	45	7	56	11	0	40	4
Canada	3	0	46	49	44	0	0	5
Central Afr. Rep.	3	5	0	8	3	0	3	2
Chad	4	50	0	54	4	0	45	5
Chile	6	0	20	26	20	0	0	6
China	3 052	4 529	87	7 668	4 572	0	239	2 856
Colombia	39	67	25	131	109	1	11	11
Costa Rica	2	1	2	5	3	0	0	3

Côte d'Ivoire	46	135	0	181	14	0	95	72
Cuba	13	1	32	46	33	0	0	13
Cyprus	0	0	0	1	0	0	0	0
Czech. & Slvk Reps	12	0	65	77	65	0	0	12
Denmark	1	0	2	3	2	0	0	1
Dominican Rep.	1	2	0	3	2	0	0	1
Ecuador	11	11	16	38	26	0	0	12
Egypt	40	348	44	432	337	4	35	55
El Salvador	9	4	16	29	19	0	1	10
Ethiopia	2	10	16	28	26	0	0	2
Finland	5	0	11	16	11	0	0	5
France	23	0	120	143	100	0	15	27
FSU–12	801	2 090	11	2 902	1 252	54	914	682
Germany	44	0	196	240	163	0	25	52
Ghana	3	3	9	14	11	0	0	3
Greece	14	275	25	314	174	0	109	31
Guatemala	11	17	10	38	30	0	4	4
Haiti	2	2	0	4	2	0	0	2
Honduras	1	2	0	3	2	0	0	1
Hong Kong	62	0	239	301	187	0	36	78
Hungary	11	0	33	44	33	0	0	11
India	595	2 134	0	2 729	1 921	0	218	590
Indonesia	31	3	348	383	327	11	0	45
Iran	11	114	0	125	109	0	5	11
Iraq	3	14	11	28	22	0	0	6
Ireland	8	0	22	30	22	0	0	8
Israel	14	33	11	57	22	0	22	14
Italy	113	0	290	403	310	0	0	93
Japan	125	0	490	615	501	0	0	114

Table AI.1 Continued

	Beginning stocks	Production	Imports	Total supply	Consumption	Loss[a]	Exports	Ending stocks
Kenya	3	4	8	15	12	0	0	3
Korea, North	9	1	33	42	37	0	0	5
Korea, South	123	0	414	537	392	0	0	145
Lebanon	1	0	2	2	2	0	0	1
Madagascar	3	13	0	16	13	0	0	3
Malawi	3	7	2	12	8	0	2	2
Malaysia	6	0	44	49	41	0	0	8
Mali	8	120	0	127	11	0	103	13
Mexico	67	33	152	252	174	5	9	64
Morocco	19	2	33	54	35	0	0	18
Mozambique	13	13	1	27	11	0	5	11
Netherlands	2	0	9	10	9	0	0	1
Nicaragua	2	2	0	3	1	0	2	0
Niger	0	1	1	3	2	0	0	0
Nigeria	28	63	11	102	71	0	0	31
Norway	2	0	3	5	3	0	0	2
Pakistan	677	1 568	2	2 247	1 481	22	239	505
Panama	0	0	1	1	1	0	0	0
Paraguay	102	196	0	298	15	0	174	109
Peru	86	35	8	128	48	0	22	58
Philippines	35	7	44	85	52	0	0	32
Poland	25	0	65	90	65	0	0	25
Portugal	44	0	139	183	131	0	0	52
Romania	19	0	54	73	54	0	0	19
Senegal	7	20	0	26	5	0	12	9
Singapore	14	0	27	41	27	0	3	10

Somalia	0	2	1	3	3	0	0	0
South Africa	10	20	44	73	60	0	0	13
Spain	40	73	79	192	131	0	28	33
Sri Lanka	1	1	9	11	10	0	0	1
Sudan	22	87	0	110	13	0	76	20
Sweden	2	0	5	8	5	0	0	2
Switzerland	41	0	65	107	63	0	0	44
Syria	50	214	0	265	71	0	142	52
Taiwan	105	0	327	431	320	0	0	111
Tanzania	35	70	0	105	13	3	57	32
Thailand	101	27	348	477	392	0	7	78
Togo	7	40	0	47	3	0	36	8
Tunisia	2	1	25	28	26	0	0	2
Turkey	123	606	71	799	579	0	87	133
Uganda	10	10	0	20	4	0	8	8
United Kingdom	26	0	17	44	19	0	0	25
United States	806	3 540	0	4 347	2 134	−29	1 328	914
Uruguay	1	0	2	3	2	0	0	1
Venezuela	8	15	22	45	33	0	7	6
Vietnam	23	1	54	79	54	0	0	24
Yemen	1	14	0	15	5	0	9	1
Yugoslavia	10	0	44	54	44	0	0	10
Zaire	1	3	4	9	8	0	0	1
Zambia	12	20	0	31	11	0	11	9
Zimbabwe	25	60	13	98	54	3	11	30
Totals	8 853	18 104	4 843	31 799	18 508	24	4 828	8 440

[a] For FSU and the Baltics, adjusts for trade between them; for other foreign countries, reflects cotton lost or destroyed in the marketing channel; for the USA, reflects difference between Bureau of Census data for ending stocks and implicit stocks based on supply plus total use. A negative 'loss' is a positive number.

Source: USDA/FAS

Table AI.2 Cotton supply, use and trade by country, 1992/93: 1000 480 lb bales

	Beginning stocks	Production	Imports	Total supply	Consumption	Loss[a]	Exports	Ending stocks
Afghanistan	33	100	0	133	80	0	30	23
Albania	69	1	20	90	25	0	0	65
Algeria	6	0	225	231	200	0	0	31
Angola	7	20	0	27	15	0	5	7
Argentina	753	800	50	1 603	550	9	320	724
Australia	1 100	1 600	2	2 702	135	0	1 800	767
Austria	5	0	140	145	140	0	0	5
Baltic States	33	0	0	33	250	−250	0	33
Bangladesh	133	75	450	658	515	5	0	138
Belgium	30	0	180	210	175	0	10	25
Benin	42	300	0	342	20	0	275	47
Bolivia	10	60	25	95	75	0	0	20
Brazil	2 079	2 600	925	5 604	3 350	0	135	2 119
Bulgaria	50	78	75	203	150	0	0	53
Burkina Faso	44	320	0	364	20	0	300	44
Burma	71	110	1	182	85	3	25	69
Cambodia	1	0	5	6	5	0	0	1
Cameroon	20	205	30	255	50	0	185	20
Canada	14	0	210	224	200	0	0	24
Central Afr. Rep.	14	25	0	39	15	0	15	9
Chad	17	230	0	247	20	0	205	22
Chile	29	0	90	119	90	0	0	29
China	14 018	20 800	400	35 218	21 000	0	1 100	13 118
Colombia	178	310	115	603	500	3	50	50
Costa Rica	11	4	10	25	13	0	0	12

Côte d'Ivoire	211	620	0	831	65	0	435	331
Cuba	60	5	145	210	150	0	0	60
Cyprus	1	1	1	3	2	0	0	1
Czech. & Slvk Reps	53	0	300	353	300	0	0	53
Denmark	3	0	10	13	10	0	0	3
Dominican Rep.	3	10	0	13	8	0	2	3
Ecuador	49	50	75	174	120	0	0	54
Egypt	183	1 600	200	1 983	1 550	20	160	253
El Salvador	43	17	75	135	85	0	5	45
Ethiopia	8	46	75	129	120	0	0	9
Finland	23	0	50	73	50	0	0	23
France	106	0	550	656	460	0	70	126
FSU-12	3 681	9 600	50	13 331	5 750	250	4 200	3 131
Germany	202	0	900	1 102	750	0	115	237
Ghana	13	13	40	66	50	0	0	16
Greece	66	1 261	115	1 442	800	0	500	142
Guatemala	50	80	45	175	140	0	18	17
Haiti	8	10	0	18	10	0	0	8
Honduras	4	10	1	15	10	0	0	5
Hong Kong	283	0	1 100	1 383	860	0	165	358
Hungary	51	0	150	201	150	0	0	51
India	2 734	9 800	0	12 534	8 825	0	1 000	2 709
Indonesia	143	14	1 600	1 757	1 500	50	0	207
Iran	49	525	0	574	500	0	25	49
Iraq	12	65	50	127	100	0	0	27
Ireland	37	0	100	137	100	0	0	37
Israel	64	150	50	264	100	0	100	64
Italy	520	1	1 330	1 851	1 425	0	0	426
Japan	575	0	2 250	2 825	2 300	0	0	525
Kenya	15	20	35	70	55	0	0	15

Table AI.2 Continued

	Beginning stocks	Production	Imports	Total supply	Consumption	Loss[a]	Exports	Ending stocks
Korea, North	40	5	150	195	170	0	0	25
Korea, South	566	1	1 900	2 467	1 800	0	0	667
Lebanon	3	0	7	10	7	0	0	3
Madagascar	15	60	0	75	60	0	0	15
Malawi	13	30	10	53	35	0	10	8
Malaysia	26	0	200	226	190	0	0	36
Mali	35	550	0	585	50	0	475	60
Mexico	307	150	700	1 157	800	25	40	292
Morocco	86	10	150	246	160	0	2	84
Mozambique	60	60	5	125	50	0	25	50
Netherlands	7	0	40	47	40	0	2	5
Nicaragua	7	9	0	16	5	0	9	2
Niger	2	5	5	12	8	0	2	2
Nigeria	128	290	50	468	325	0	0	143
Norway	8	0	15	23	15	0	0	8
Pakistan	3 111	7 200	10	10 321	6 800	100	1 100	2 321
Panama	1	0	5	6	5	0	0	1
Paraguay	470	900	0	1 370	70	0	800	500
Peru	393	160	35	588	220	0	100	268
Philippines	159	30	200	389	240	0	0	149
Poland	113	0	300	413	300	0	0	113
Portugal	200	0	640	840	600	0	0	240
Romania	86	0	250	336	250	0	0	86
Senegal	31	90	0	121	25	0	55	41
Singapore	63	0	125	188	125	0	15	48
Somalia	1	7	5	13	12	0	0	1

South Africa	44	90	200	334	275	0	0	59
Spain	182	335	365	882	600	2	130	150
Sri Lanka	6	5	40	51	45	0	0	6
Sudan	103	400	0	503	60	0	350	93
Sweden	10	0	25	35	25	0	0	10
Switzerland	190	0	300	490	290	0	0	200
Syria	230	985	0	1 215	325	0	650	240
Taiwan	480	0	1 500	1 980	1 470	0	0	510
Tanzania	161	320	0	481	60	15	260	146
Thailand	465	125	1 600	2 190	1 800	0	30	360
Togo	30	185	0	215	15	0	165	35
Tunisia	9	3	115	127	120	0	0	7
Turkey	564	2 782	325	3 671	2 660	0	400	611
Uganda	45	45	0	90	20	0	35	35
United Kingdom	120	0	80	200	85	0	0	115
United States	3 704	16 260	2	19 966	9 800	−134	6 100	4 200
Uruguay	5	0	10	15	10	0	0	5
Venezuela	38	70	100	208	150	0	30	28
Vietnam	106	5	250	361	250	0	0	111
Yemen	4	65	0	69	25	0	40	4
Yugoslavia	45	1	200	246	200	0	0	46
Zaire	4	16	20	40	35	0	0	5
Zambia	53	90	0	143	50	0	50	43
Zimbabwe	114	275	60	449	250	13	50	136
Totals	40 660	83 150	22 244	146 504	85 005	111	22 175	38 763

[a] For FSU and the Baltics, adjusts for trade between them; for other foreign countries, reflects cotton lost or destroyed in the marketing channel; for the USA, reflects difference between Bureau of Census data for ending stocks and implicit stocks based on supply plus total use. A negative 'loss' is a positive number.

Source: USDA/FAS (1993).

Table AI.3 Area, yield and production: world and selected countries and regions[b]

Region and country	1991/92 1000 ha	1992/93 1000 ha	1991/92 kg/ha	1992/93 kg/ha	1991/92 1000 bales[c]	1992/93 1000 bales[c]
Western hemisphere						
USA	5 245	4 514	731	784	17 614	16 260
Brazil	1 969	1 580	381	358	3 445	2 600
Mexico	250	45	724	726	831	150
Argentina	580	400	431	435	1 148	800
Colombia	275	115	574	587	725	310
Paraguay	480	350	318	560	700	900
Peru	96	60	528	581	233	160
Guatemala	39	17	1 050	1 025	188	80
Nicaragua	36	2	665	980	110	9
Venezuela	32	35	476	435	70	70
Others	81	94	419	384	156	166
Total[a]	9 083	7 212	605	649	25 220	21 505
Europe						
Greece	230	318	940	863	993	1 261
Spain	78	70	1 024	1 042	367	335
Others	18	18	1 016	980	84	81
Total[a]	326	406	964	899	1 444	1 677
Africa						
Egypt	358	353	814	987	1 338	1 600
Sudan	185	188	494	463	420	400
Zimbabwe	236	275	88	218	95	275
South Africa	78	80	262	245	94	90
Tanzania	450	430	180	162	372	320

Cameroon	100	90	501	496	230	205
Nigeria	430	510	140	124	276	290
Chad	210	180	321	278	310	230
Others	1 386	1 409	365	388	2 322	2 512
Total[a]	3 433	3 515	346	367	5 457	5 922
Asia and Oceania						
China	6 539	6 750	869	671	26 100	20 800
FSU - 12	3 010	2 850	814	733	11 250	9 600
India	7 684	7 480	267	285	9 430	9 800
Pakistan	2 881	2 460	756	637	10 000	7 200
Turkey	599	633	937	957	2 578	2 782
Australia	282	251	1 770	1 388	2 292	1 600
Syria	174	212	1 104	1 012	882	985
Israel	12	20	1 633	1 633	90	150
Burma	165	170	133	141	101	110
Thailand	91	67	471	406	197	125
Afghanistan	60	60	363	363	100	100
Others	409	399	429	433	806	794
Total[a]	21 906	21 352	634	551	63 826	54 046
Foreign total[a]	29 503	27 971	578	521	78 333	66 890
World total[a]	34 748	32 485	601	557	95 947	83 150

[a] Totals may not add because of rounding.
[b] Harvest season beginning August 1.
[c] Bales of 480 lb net (217 kg).
Source: Prepared or estimated on the basis of official statistics of non-US governments, other source materials, reports of US agricultural attaches and other non-US service officers, results of office research and related information.

APPENDIX

II

Cotton supply 1991/92

Table AII.1 Cotton supply, use and trade by country, 1991/92: 1000 480 lb bales

	Beginning stocks	Production	Imports	Total supply	Consumption	Loss[a]	Exports	Ending stocks
Afghanistan	43	100	0	143	80	0	30	33
Albania	81	4	19	104	35	0	0	69
Algeria	6	0	200	206	200	0	0	6
Angola	7	20	0	27	15	0	5	7
Argentina	784	1 148	2	1 934	606	9	566	753
Australia	1 005	2 292	2	3 299	132	0	2 067	1 100
Austria	13	0	132	145	135	0	5	5
Baltic States	47	0	0	47	275	−261	0	33
Bangladesh	160	62	390	612	473	5	1	133
Belgium	40	0	170	210	170	0	10	30
Benin	37	345	0	382	20	0	320	42
Bolivia	10	40	25	75	65	0	0	10
Brazil	1 385	3 445	597	5 427	3 215	0	133	2 079
Bulgaria	22	78	100	200	150	0	0	50
Burkina Faso	44	320	0	364	20	0	300	44
Burma	71	101	7	179	82	3	23	71
Cambodia	1	0	5	6	5	0	0	1
Cameroon	10	230	30	270	50	0	200	20
Canada	14	0	200	214	200	0	0	14
Central Afr. Rep.	19	50	0	69	15	0	40	14
Chad	12	310	0	322	15	0	290	17
Chile	29	0	90	119	90	0	0	29
China	6 356	26 100	1 662	34 118	19 500	0	600	14 018
Colombia	23	725	79	827	482	7	160	178

Table AII.1 Continued

	Beginning stocks	Production	Imports	Total supply	Consumption	Loss[a]	Exports	Ending stocks
Costa Rica	10	4	10	24	13	0	0	11
Côte d'Ivoire	267	403	0	670	64	0	395	211
Cuba	60	5	145	210	150	0	0	60
Cyprus	1	1	1	3	2	0	0	1
Czech. & Slvk Reps	103	0	250	353	300	0	0	53
Denmark	3	0	10	13	10	0	0	3
Dominican Rep.	3	10	0	13	8	0	2	3
Ecuador	34	65	60	159	110	0	0	49
Egypt	89	1 338	341	1 768	1 465	30	90	183
El Salvador	43	15	75	133	85	0	5	43
Ethiopia	8	55	0	63	55	0	0	8
Finland	23	0	50	73	50	0	0	23
France	118	0	556	674	484	0	84	106
FSU-12	3 367	11 250	50	14 667	7 225	261	3 500	3 681
Germany	187	0	945	1 132	830	0	100	202
Ghana	10	13	40	63	50	0	0	13
Greece	235	993	124	1 352	827	0	459	66
Guatemala	31	188	14	233	120	0	63	50
Haiti	8	10	0	18	10	0	0	8
Honduras	9	7	0	16	12	0	0	4
Hong Kong	283	0	980	1 263	870	0	10	283
Hungary	51	0	150	201	150	0	0	51
India	1 765	9 430	273	11 468	8 674	0	60	2 734
Indonesia	127	14	1 800	1 941	1 750	48	0	143
Iran	74	525	0	599	500	0	50	49
Iraq	22	65	25	112	100	0	0	12
Ireland	37	0	100	137	100	0	0	37

Israel	149	90	65	304	90	0	150	64
Italy	529	1	1 460	1 990	1 470	0	0	520
Japan	653	0	2 705	3 358	2 783	0	0	575
Kenya	16	24	35	75	60	0	0	15
Korea, North	40	5	160	205	165	0	0	40
Korea, South	686	1	1 798	2 485	1 919	0	0	566
Lebanon	3	0	7	10	7	0	0	3
Madagascar	15	50	0	65	50	0	0	15
Malawi	13	25	10	48	25	0	10	13
Malaysia	36	0	175	211	185	0	0	26
Mali	59	526	0	585	50	0	500	35
Mexico	242	831	300	1 373	772	46	248	307
Morocco	69	34	140	243	155	0	2	86
Mozambique	60	60	5	125	45	0	20	60
Netherlands	9	0	40	49	40	0	2	7
Nicaragua	7	110	0	117	5	0	105	7
Niger	2	5	5	12	8	0	2	2
Nigeria	67	276	95	438	310	0	0	128
Norway	8	0	15	23	15	0	0	8
Pakistan	1 596	10 000	20	11 616	6 482	100	1 923	3 111
Panama	1	0	5	6	5	0	0	1
Paraguay	710	700	0	1 410	70	0	870	470
Peru	447	233	20	700	219	0	88	393
Philippines	65	53	294	412	253	0	0	159
Poland	138	0	250	388	275	0	0	113
Portugal	290	0	600	890	690	0	0	200
Romania	86	0	200	286	200	0	0	86
Senegal	26	100	0	126	25	0	70	31
Singapore	63	0	115	178	100	0	15	63
Somalia	1	7	5	13	12	0	0	1
South Africa	140	94	140	374	310	0	20	44
Spain	201	367	394	962	638	2	140	182

Table AII.1 Continued

	Beginning stocks	Production	Imports	Total supply	Consumption	Loss[a]	Exports	Ending stocks
Sri Lanka	6	5	40	51	45	0	0	6
Sudan	143	420	0	563	60	0	400	103
Sweden	10	0	25	35	25	0	0	10
Switzerland	200	0	280	480	280	0	10	190
Syria	223	882	0	1 105	300	0	575	230
Taiwan	594	0	1 484	2 078	1 598	0	0	480
Tanzania	116	372	0	488	55	15	257	161
Thailand	355	197	1 641	2 193	1 699	0	29	465
Togo	36	184	0	220	15	0	175	30
Tunisia	11	3	115	129	120	0	0	9
Turkey	462	2 578	420	3 460	2 641	0	255	564
Uganda	57	50	0	107	20	0	42	45
United Kingdom	120	0	100	220	100	0	0	120
United States	2 344	17 614	13	19 971	9 608	13	6 646	3 704
Uruguay	5	0	10	15	10	0	0	5
Venezuela	48	70	100	218	150	0	30	38
Vietnam	99	7	250	356	250	0	0	106
Yemen	9	65	0	74	25	0	45	4
Yugoslavia	44	1	230	275	230	0	0	45
Zaire	3	16	20	39	35	0	0	4
Zambia	114	35	0	149	46	0	50	53
Zimbabwe	93	95	155	343	200	4	25	114
Totals	28 706	95 947	23 650	148 303	84 989	282	22 372	40 660

[a] For FSU and the Baltics, adjusts for trade between them; for other foreign countries, reflects cotton lost or destroyed in the marketing channel; for the USA, reflects difference between Bureau of Census data for ending stocks and implicit stocks based on supply plus total use. A negative 'loss' is a positive number.
Source: USDA/FAS

APPENDIX

III

Production of cotton

Table AIII.1 Production of cotton (thousand tonnes)

	1984/85	1985/86	1986/87	1987/88	1988/89	1989/90	1990/91	1991/92	1992/93	1993/94
Canada	–	–	–	–	–	–	–	–	–	–
Cuba	1	1	1	1	1	1	1	1	1	1
Dom. Rep.	1	1	2	2	2	2	2	2	2	2
Mexico	242	220	139	223	308	167	175	181	50	53
USA	2 826	2 924	2 119	3 214	3 356	2 655	3 376	3 835	3 484	3 740
Sub total	3 072	3 149	2 262	3 442	3 670	2 828	3 556	4 021	3 539	3 799
Costa Rica	2	2	1	1	1	1	1	1	1	1
El Salvador	30	18	11	10	9	6	5	4	3	3
Guatemala	58	48	27	48	42	41	38	41	21	22
Honduras	6	4	4	3	3	1	1	1	1	1
Nicaragua	68	50	49	35	26	25	27	24	2	4
Sub total	165	122	92	96	81	75	72	72	29	32
Argentina	171	120	100	282	195	274	300	225	221	190
Bolivia	3	4	3	3	1	2	8	6	11	12
Brazil	965	793	633	864	709	666	717	697	751	707
Chile	–	–	–	–	–	–	–	–	–	–
Colombia	125	113	107	134	121	105	127	157	70	114
Ecuador	3	12	5	7	11	13	10	12	14	13
Paraguay	160	107	84	205	220	225	265	135	179	143
Peru	92	99	90	95	102	95	78	59	48	45
Uruguay	–	–	–	–	–	–	–	–	–	–
Venezuela	27	34	19	24	34	31	23	27	24	25
Sub total	1 546	1 283	1 041	1 614	1 392	1 410	1 527	1 318	1 318	1 248

Algeria	–	–	–	–	–	–	–	–	–	–
Egypt	399	435	403	352	311	288	296	291	281	302
Morocco	4	8	8	11	9	9	8	7	3	5
Sudan	203	142	163	136	139	128	80	95	91	104
Tunisia	–	–	–	–	0	0	0	1	1	1
Sub total	607	585	574	498	458	425	385	394	376	411
Benin	33	34	48	27	44	42	59	76	65	63
Burkina Faso	34	46	66	59	59	62	77	69	65	64
Cameroon	38	45	48	45	69	43	47	47	48	44
Cent. Afr. Rep.	17	13	10	8	11	11	14	8	12	10
Chad	36	39	34	48	53	58	60	69	70	71
Côte d'Ivoire	88	82	93	114	128	107	116	87	110	99
Madagascar	13	17	16	11	12	13	13	13	13	13
Mali	55	67	79	75	97	99	116	114	115	110
Niger	1	2	3	3	2	2	1	1	1	2
Senegal	19	11	11	15	14	12	18	20	20	20
Togo	23	26	33	28	33	34	41	41	35	33
Sub total	360	383	441	433	523	485	562	547	556	529
Angola	4	4	4	4	4	4	4	4	4	4
Burundi	2	3	3	3	3	3	2	3	3	3
Ethiopia	20	22	20	20	21	18	19	19	19	19
Ghana	0	0	3	3	2	3	5	8	7	7
Kenya	13	9	7	6	8	9	10	11	12	14
Mozambique	1	4	9	7	9	10	12	13	13	13
Nigeria	16	10	28	30	48	41	36	60	63	61
South Africa	46	47	61	78	78	60	49	21	42	44
Tanzania	52	33	72	86	63	36	47	81	70	66
Uganda	15	6	7	5	5	6	8	11	13	13

Table AIII.1 Continued

	1984/85	1985/86	1986/87	1987/88	1988/89	1989/90	1990/91	1991/92	1992/93	1993/94
Zaire	9	7	6	5	5	4	4	4	4	4
Zambia	11	12	7	23	17	10	25	16	21	19
Zimbabwe	103	89	87	116	91	67	73	20	81	82
Sub total	309	259	326	404	371	288	310	283	367	363
Bulgaria	6	5	5	4	4	4	4	4	4	4
Czech & Slovak Reps	–	–	–	–	–	–	–	–	–	–
Germany, DR	–	–	–	–	–	–	–	–	–	–
Hungary	–	–	–	–	–	–	–	–	–	–
Poland	–	–	–	–	–	–	–	–	–	–
Romania	–	–	–	–	–	–	–	–	–	–
Yugoslavia	0	0	0	0	0	0	0	–	–	–
Sub total	11	10	9	9	9	9	9	9	9	9
Austria	–	–	–	–	–	–	–	–	–	–
Finland	–	–	–	–	–	–	–	–	–	–
Norway	–	–	–	–	–	–	–	–	–	–
Sweden	–	–	–	–	–	–	–	–	–	–
Switzerland	–	–	–	–	–	–	–	–	–	–
Sub total	–	–	–	–	–	–	–	–	–	–
Belgium	–	–	–	–	–	–	–	–	–	–
Denmark	–	–	–	–	–	–	–	–	–	–
France	–	–	–	–	–	–	–	–	–	–
Germany[a]	–	–	–	–	–	–	–	–	–	–
Greece	147	163	205	174	235	268	210	211	260	219
Ireland	–	–	–	–	–	–	–	–	–	–
Italy	1	1	1	0	0	0	0	0	0	0

Netherlands	–	–	–	–	–	–	–	–	–	–
Portugal	–	–	–	–	–	–	–	–	–	–
Spain	55	69	86	83	111	61	79	74	62	62
United Kingdom	–	–	–	–	–	–	–	–	–	–
Sub total	202	232	292	256	345	329	289	285	323	282
Armenia	–	–	–	–	–	–	–	–	–	–
Azerbaijan	–	–	–	–	–	–	–	180	155	165
Byelaruss	–	–	–	–	–	–	–	–	–	–
Estonia	–	–	–	–	–	–	–	–	–	–
Georgia	–	–	–	–	–	–	–	–	–	–
Kazakhstan	–	–	–	–	–	–	–	93	95	97
Kirghizstan	–	–	–	–	–	–	–	19	20	20
Latvia	–	–	–	–	–	–	–	–	–	–
Lithuania	–	–	–	–	–	–	–	–	–	–
Moldova	–	–	–	–	–	–	–	–	–	–
Russia	–	–	–	–	–	–	–	–	–	–
Tadzhikistan	–	–	–	–	–	–	–	256	235	249
Turkmenia	–	–	–	–	–	–	–	435	400	413
Ukraine	–	–	–	–	–	–	–	–	–	–
Uzbekistan	–	–	–	–	–	–	–	1 508	1 423	1 421
Sub total[b]	2 597	2 782	2 660	2 470	2 766	2 654	2 640	2 491	2 328	2 365
China (mainland)	6 253	4 147	3 540	4 246	4 149	3 788	4 508	5 675	5 472	5 199
China (Taiwan)	0	0	0	0	0	0	0	0	0	0
Hong Kong	–	–	–	–	–	–	–	–	–	–
Sub total	6 253	4 147	3 540	4 246	4 149	3 788	4 508	5 675	5 472	5 199
Australia	249	258	213	281	292	3131	433	457	339	355
Indonesia	7	6	3	3	3	3	3	3	4	5

Table AIII.1 Continued

	1984/85	1985/86	1986/87	1987/88	1988/89	1989/90	1990/91	1991/92	1992/93	1993/94
Japan	–	–	–	–	–	–	–	–	–	–
Korea, North	3	3	3	3	3	3	3	3	3	3
Korea, South	2	1	0	0	0	0	0	0	0	0
Malaysia	–	–	–	–	–	–	–	–	–	–
Philippines	3	5	5	5	5	6	6	12	7	7
Singapore	–	–	–	–	–	–	–	–	–	–
Thailand	30	34	19	25	35	29	32	43	33	35
Vietnam	1	1	1	1	1	1	1	3	1	1
Sub total	294	308	245	319	340	355	479	521	388	407
Afghanistan	25	19	18	14	16	20	22	22	22	22
Bangladesh	3	6	5	6	7	11	17	14	17	15
India	1 820	1 964	1 579	1 555	1 802	2 308	1 989	2 040	2 190	2 125
Myanmar	13	13	12	11	20	20	21	21	19	20
Pakistan	1 008	1 216	1 319	1 468	1 425	1 455	1 637	2 193	2 200	2 210
Sri Lanka	1	2	5	5	5	6	7	7	8	9
Sub total	2 870	3 222	2 938	3 059	3 276	3 820	3 693	4 298	4 456	4 401
Iran	112	106	111	106	116	114	120	115	112	121
Iraq	14	14	15	5	4	5	5	1	2	2
Israel	88	97	67	58	63	47	52	23	30	32
Syria	153	162	126	96	114	128	145	185	183	181
Turkey	580	518	517	537	650	617	655	561	606	625
Sub total	954	904	844	811	956	920	985	892	940	969
World total[c]	19 239	17 386	15 264	17 657	18 336	17 387	19 015	20 805	20 101	20 013

[a] Federal Republic of Germany prior to 1990/91.
[b] USSR prior to 1991/92.
[c] Subtotals and total include countries not shown.

APPENDIX

IV

Harvested area

Table AIV.1 Harvested area (thousand hectares)

	1984/85	1985/86	1986/87	1987/88	1988/89	1989/90	1990/91	1991/92	1992/93	1993/94
Canada	-	-	-	-	-	-	-	-	-	-
Cuba	3	4	4	4	4	4	4	4	4	4
Dom. Rep.	1	1	1	2	3	3	3	3	3	3
Mexico	312	204	150	230	255	188	231	251	60	60
USA	4 201	4 140	3 427	4 061	4 833	3 860	4 748	5 244	4 583	5 140
Sub total	4 530	4 363	3 596	4 312	5 109	4 069	4 999	5 516	4 664	5 221
Costa Rica	2	2	2	1	2	2	2	2	2	2
El Salvador	37	27	14	13	13	10	6	5	4	4
Guatemala	63	67	31	40	42	40	36	39	23	22
Honduras	8	7	4	4	4	2	3	1	1	1
Nicaragua	112	86	65	61	40	31	44	35	2	6
Sub total	222	190	115	120	100	85	91	84	34	36
Argentina	447	339	273	492	490	540	634	535	465	419
Bolivia	9	11	7	10	1	4	18	23	24	25
Brazil	3 707	3 325	2 161	2 577	2 230	1 964	1 939	2 093	2 033	1 975
Chile	-	-	-	-	-	-	-	-	-	-
Colombia	213	183	169	234	218	193	233	271	123	202
Ecuador	9	23	17	14	21	28	24	30	30	30
Paraguay	420	340	275	420	438	509	550	480	408	347
Peru	115	152	142	133	150	145	121	112	79	75
Uruguay	-	-	-	-	-	-	-	-	-	-
Venezuela	46	61	43	53	74	68	45	50	50	50
Sub total	4 966	4 434	3 088	3 933	3 622	3 450	3 564	3 594	3 212	3 123

Algeria	–	–	–	–	–	–	–	–	–	–
Egypt	413	454	443	412	426	422	417	358	356	374
Morocco	9	12	14	15	18	15	16	12	4	8
Sudan	360	326	346	318	313	279	178	188	208	220
Tunisia	–	–	–	–	0	0	1	2	2	2
Sub total	782	792	803	745	757	716	612	560	570	603
Benin	80	83	103	72	97	91	123	151	140	133
Burkina Faso	82	94	126	170	169	150	176	180	170	162
Cameroon	73	89	94	95	112	89	94	87	85	81
Cent. Afr. Rep.	80	83	66	40	53	40	47	44	40	38
Chad	142	148	124	149	199	185	207	283	275	261
Côte d'Ivoire	146	153	159	180	213	201	199	190	190	181
Madagascar	24	33	43	22	26	29	30	31	30	29
Mali	119	146	152	149	190	178	205	216	215	204
Niger	5	6	9	12	7	9	9	12	10	10
Senegal	46	39	25	29	39	24	36	45	45	43
Togo	44	69	62	68	81	76	80	80	75	71
Sub total	841	944	965	987	1 189	1 076	1 210	1 325	1 280	1 216
Angola	14	14	10	10	10	10	10	10	10	10
Burundi	6	7	7	7	8	7	7	7	8	8
Ethiopia	53	53	53	53	37	35	36	37	37	38
Ghana	1	2	8	8	9	11	14	20	21	21
Kenya	97	68	50	40	46	51	70	74	81	89
Mozambique	15	48	108	131	108	65	78	86	86	86
Nigeria	405	220	285	320	360	380	380	430	510	510
South Africa	107	126	160	205	208	165	127	79	120	126
Tanzania	390	370	400	450	450	260	320	450	430	430
Uganda	101	45	50	35	37	42	51	78	93	93

Table AIV.1 Continued

	1984/85	1985/86	1986/87	1987/88	1988/89	1989/90	1990/91	1991/92	1992/93	1993/94
Zaire	78	62	49	41	39	33	31	30	32	33
Zambia	55	50	38	78	91	64	97	101	101	101
Zimbabwe	231	192	243	272	248	229	273	236	280	314
Sub total	1 643	1 335	1 524	1 722	1 727	1 428	1 567	1 707	1 878	1 931
Bulgaria	11	9	10	10	10	10	10	10	10	10
Czech & Slovak Reps	–	–	–	–	–	–	–	–	–	–
Germany, DR	–	–	–	–	–	–	–	–	–	–
Hungary	–	–	–	–	–	–	–	–	–	–
Poland	–	–	–	–	–	–	–	–	–	–
Romania	–	–	–	–	–	–	–	–	–	–
Yugoslavia	1	1	1	1	1	1	1	–	–	–
Sub total	29	27	27	27	27	27	27	27	27	27
Austria	–	–	–	–	–	–	–	–	–	–
Finland	–	–	–	–	–	–	–	–	–	–
Norway	–	–	–	–	–	–	–	–	–	–
Sweden	–	–	–	–	–	–	–	–	–	–
Switzerland	–	–	–	–	–	–	–	–	–	–
Sub total	–	–	–	–	–	–	–	–	–	–
Belgium	–	–	–	–	–	–	–	–	–	–
Denmark	–	–	–	–	–	–	–	–	–	–
France	–	–	–	–	–	–	–	–	–	–
Germany[a]	–	–	–	–	–	–	–	–	–	–
Greece	192	209	210	202	256	280	268	233	320	270
Ireland	–	–	–	–	–	–	–	–	–	–
Italy	2	2	3	0	0	0	0	0	0	0

Netherlands	–	–	–	–	–	–	–	–	–	–
Portugal	–	–	–	–	–	–	–	–	–	–
Spain	60	62	81	81	135	68	83	78	75	70
United Kingdom	–	–	–	–	–	–	–	–	–	–
Sub total	254	273	294	283	392	348	351	312	395	340
Armenia	–	–	–	–	–	–	–	–	–	–
Azerbaijan	–	–	–	–	–	–	–	244	235	235
Byelaruss	–	–	–	–	–	–	–	–	–	–
Estonia	–	–	–	–	–	–	–	–	–	–
Georgia	–	–	–	–	–	–	–	–	–	–
Kazakhstan	–	–	–	–	–	–	–	119	119	119
Kirghizstan	–	–	–	–	–	–	–	26	26	26
Latvia	–	–	–	–	–	–	–	–	–	–
Lithuania	–	–	–	–	–	–	–	–	–	–
Moldova	–	–	–	–	–	–	–	–	–	–
Russia	–	–	–	–	–	–	–	–	–	–
Tadzhikistan	–	–	–	–	–	–	–	295	284	290
Turkmenia	–	–	–	–	–	–	–	601	580	580
Ukraine	–	–	–	–	–	–	–	–	–	–
Uzbekistan	–	–	–	–	–	–	–	1 720	1 655	1 605
Sub total[b]	3 347	3 316	3 475	3 527	3 432	3 338	3 171	3 005	2 899	2 855
China (mainland)	6 923	5 140	4 306	4 844	5 535	5 203	5 588	6 539	6 670	6 337
China (Taiwan)	–	–	–	–	–	–	–	–	–	–
Hong Kong	–	–	–	–	–	–	–	–	–	–
Sub total	6 923	5 140	4 306	4 844	5 535	5 203	5 588	6 539	6 670	6 337
Australia	196	178	154	233	192	232	270	268	209	217
Indonesia	30	45	17	17	7	18	18	19	19	19
Japan	–	–	–	–	–	–	–	–	–	–

Table AIV.1 Continued

	1984/85	1985/86	1986/87	1987/88	1988/89	1989/90	1990/91	1991/92	1992/93	1993/94
Korea, North	8	8	8	8	8	8	8	8	8	8
Korea, South	8	8	1	1	1	1	1	1	0	0
Malaysia	–	–	–	–	–	–	–	–	–	–
Philippines	11	13	14	16	19	15	17	35	20	21
Singapore	–	–	–	–	–	–	–	–	–	–
Thailand	72	83	50	64	70	61	72	91	75	75
Vietnam	13	14	13	13	13	13	14	14	14	14
Sub total	338	349	257	351	310	347	399	436	346	355
Afghanistan	61	75	69	40	45	55	60	60	60	60
Bangladesh	12	27	23	26	17	17	20	18	21	21
India	7 382	7 533	6 948	7 061	7 343	7 331	7 402	7 684	7 915	7 598
Myanmar	160	160	155	155	160	153	156	156	156	156
Pakistan	2 242	2 364	2 505	2 568	2 620	2 599	2 662	2 881	2 900	2 929
Sri Lanka	2	7	14	16	16	18	20	22	24	26
Sub total	9 860	10 168	9 715	9 868	10 203	10 176	10 325	10 826	11 080	10 795
Iran	212	188	188	196	191	228	229	204	204	220
Iraq	50	50	50	17	12	7	7	2	4	4
Israel	63	65	46	40	48	31	31	13	19	20
Syria	178	170	144	129	171	158	156	170	192	190
Turkey	760	660	586	586	740	725	641	599	633	620
Sub total	1 285	1 155	1 036	989	1 184	1 170	1 087	1 010	1 073	1 076
World total[c]	35 019	32 486	29 201	31 708	33 587	31 434	32 992	34 939	34 129	33 915

[a] Federal Republic of Germany prior to 1990/91.
[b] USSR prior to 1991/92.
[c] Subtotals and total include countries not shown.

APPENDIX

V

Yields

Table AV.1 Yields (kg/ha)

	1984/85	1985/86	1986/87	1987/88	1988/89	1989/90	1990/91	1991/92	1992/93	1993/94
Canada	–	–	–	–	–	–	–	–	–	–
Cuba	374	272	269	269	269	269	269	269	269	269
Dom. Rep.	670	837	1 231	792	751	667	667	667	667	667
Mexico	775	1 079	926	970	1 210	889	761	720	837	884
USA	673	706	618	791	694	688	711	731	760	728
Sub total	678	722	629	798	718	695	711	729	759	728
Costa Rica	920	920	733	900	398	738	774	709	704	665
El Salvador	804	674	775	724	726	635	756	837	736	738
Guatemala	928	710	898	1 178	995	1 038	1 041	1 045	914	1 007
Honduras	812	613	892	710	708	687	462	626	639	624
Nicaragua	607	580	753	575	661	789	614	683	815	712
Sub total	741	644	799	802	807	882	789	858	846	874
Argentina	383	354	366	573	397	507	473	421	474	455
Bolivia	323	348	381	330	583	600	460	257	446	469
Brazil	260	239	293	335	318	339	370	333	370	358
Chile	–	–	–	–	–	–	–	–	–	–
Colombia	589	620	631	571	554	544	542	579	570	562
Ecuador	290	515	303	492	502	449	419	392	451	443
Paraguay	381	316	305	488	502	442	482	281	439	412
Peru	799	650	630	711	677	654	642	526	612	599
Uruguay	–	–	–	–	–	–	–	–	–	–
Venezuela	578	565	446	459	463	463	513	540	488	493
Sub total	311	289	337	410	384	409	429	367	410	400

Algeria	–	–	–	–	–	–	–	–	–	–
Egypt	966	959	909	854	729	683	709	814	789	807
Morocco	484	643	599	712	510	605	509	624	581	566
Sudan	565	434	470	426	442	458	452	504	439	474
Tunisia	–	–	–	–	499	438	608	523	517	522
Sub total	776	738	714	669	605	594	629	705	659	681
Benin	418	413	465	380	454	469	479	502	464	474
Burkina Faso	418	489	522	344	347	416	438	384	382	394
Cameroon	522	510	513	475	614	482	496	539	565	539
Cent. Afr. Rep.	214	160	145	189	215	272	299	179	300	253
Chad	250	262	273	322	264	314	288	244	255	273
Côte d'Ivoire	606	538	586	631	601	534	583	456	579	550
Madagascar	568	501	370	473	462	458	424	424	424	439
Mali	464	460	517	504	511	553	564	526	535	538
Niger	319	314	334	291	344	211	98	83	100	167
Senegal	409	280	421	531	376	503	507	453	444	457
Togo	534	384	540	412	405	440	507	505	467	465
Sub total	428	406	457	439	440	451	465	413	434	435
Angola	307	307	430	430	369	384	369	397	390	382
Burundi	358	392	468	454	401	429	302	388	395	383
Ethiopia	372	414	372	372	563	513	516	505	496	486
Ghana	179	221	337	320	272	285	348	381	321	321
Kenya	134	133	133	153	173	148	145	148	153	153
Mozambique	75	77	80	51	86	157	150	146	146	146
Nigeria	39	45	98	94	133	108	95	140	124	120
South Africa	433	372	379	378	373	364	383	268	353	348
Tanzania	133	88	180	191	141	139	146	180	163	154
Uganda	146	140	140	140	140	140	159	140	140	140

Table AV.1 Continued

	1984/85	1985/86	1986/87	1987/88	1988/89	1989/90	1990/91	1991/92	1992/93	1993/94
Zaire	115	105	124	119	118	132	136	137	128	130
Zambia	199	239	190	295	182	156	260	156	210	193
Zimbabwe	447	462	356	427	369	292	269	85	288	261
Sub total	188	194	214	235	215	202	198	166	195	188
Bulgaria	538	538	452	430	430	430	430	430	430	430
Czech & Slovak Reps	–	–	–	–	–	–	–	–	–	–
Germany, DR	–	–	–	–	–	–	–	–	–	–
Hungary	–	–	–	–	–	–	–	–	–	–
Poland	–	–	–	–	–	–	–	–	–	–
Romania	–	–	–	–	–	–	–	–	–	–
Yugoslavia	179	359	314	269	234	253	240	–	–	–
Sub total	371	367	338	330	330	328	330	330	329	329
Austria	–	–	–	–	–	–	–	–	–	–
Finland	–	–	–	–	–	–	–	–	–	–
Norway	–	–	–	–	–	–	–	–	–	–
Sweden	–	–	–	–	–	–	–	–	–	–
Switzerland	–	–	–	–	–	–	–	–	–	–
Sub total	–	–	–	–	–	–	–	–	–	–
Belgium	–	–	–	–	–	–	–	–	–	–
Denmark	–	–	–	–	–	–	–	–	–	–
France	–	–	–	–	–	–	–	–	–	–
Germany[a]	–	–	–	–	–	–	–	–	–	–
Greece	766	780	974	859	917	956	785	906	813	813
Ireland	–	–	–	–	–	–	–	–	–	–
Italy	323	323	218	333	375	343	500	500	500	500

Netherlands	–	–	–	–	–	–	–	–	–	–
Portugal	–	–	–	–	–	–	–	–	–	–
Spain	910	1 103	1 066	1 019	817	899	945	941	832	887
United Kingdom	–	–	–	–	–	–	–	–	–	–
Sub total	796	850	992	904	882	945	823	914	816	828
Armenia	–	–	–	–	–	–	–	–	–	–
Azerbaijan	–	–	–	–	–	–	–	738	660	702
Byelaruss	–	–	–	–	–	–	–	–	–	–
Estonia	–	–	–	–	–	–	–	–	–	–
Georgia	–	–	–	–	–	–	–	–	–	–
Kazakhstan	–	–	–	–	–	–	–	782	798	815
Kirghizstan	–	–	–	–	–	–	–	731	769	769
Latvia	–	–	–	–	–	–	–	–	–	–
Lithuania	–	–	–	–	–	–	–	–	–	–
Moldova	–	–	–	–	–	–	–	–	–	–
Russia	–	–	–	–	–	–	–	–	–	–
Tadzhikistan	–	–	–	–	–	–	–	868	827	859
Turkmenia	–	–	–	–	–	–	–	724	690	712
Ukraine	–	–	–	–	–	–	–	–	–	–
Uzbekistan	–	–	–	–	–	–	–	877	860	885
Sub total[b]	776	839	765	700	806	795	833	829	803	828
China (mainland)	903	807	822	876	750	728	807	868	820	820
China (Taiwan)	420	418	381	472	383	1 143	1 000	1 000	1 000	905
Hong Kong	–	–	–	–	–	–	–	–	–	–
Sub total	903	807	822	876	750	728	807	868	820	820
Australia	1 272	1 449	1 377	1 208	1 525	1 320	1 603	1 707	1 623	1 636
Indonesia	221	141	179	179	443	179	179	162	228	238
Japan	–	–	–	–	–	–	–	–	–	–

Table AV.1 Continued

	1984/85	1985/86	1986/87	1987/88	1988/89	1989/90	1990/91	1991/92	1992/93	1993/94
Korea, North	377	377	377	377	377	377	377	377	377	377
Korea, South	188	81	408	355	340	325	327	320	325	437
Malaysia	-	-	-	-	-	-	-	-	-	-
Philippines	299	353	384	327	252	367	353	329	350	330
Singapore	-	-	-	-	-	-	-	-	-	-
Thailand	413	408	379	390	499	472	452	470	440	467
Vietnam	98	98	107	107	102	82	77	199	102	104
Sub total	870	883	952	909	1 097	1 022	1 200	1 195	1 122	1 144
Afghanistan	412	257	253	350	350	364	364	364	359	360
Bangladesh	219	230	229	211	412	625	830	750	800	683
India	247	261	227	220	245	315	269	265	277	280
Myanmar	82	82	77	70	124	131	135	135	119	128
Pakistan	450	514	527	572	544	560	615	761	759	755
Sri Lanka	311	335	321	336	331	331	327	331	331	330
Sub total	291	317	302	310	321	375	358	397	402	408
Iran	527	561	592	543	609	501	525	563	548	549
Iraq	280	280	300	294	333	714	297	384	405	427
Israel	1 397	1 489	1 453	1 460	1 305	1 548	1 657	1 699	1 580	1 597
Syria	856	950	873	750	667	811	927	1 085	953	953
Turkey	763	785	883	916	878	851	1 021	938	958	1 009
Sub total	743	783	815	820	807	786	906	884	876	901
World total[c]	549	535	523	557	546	553	576	595	589	590

[a] Federal Republic of Germany prior to 1990/91.
[b] USSR prior to 1991/92.
[c] Subtotals and total include countries not shown.

APPENDIX VI

Consumption

Table AVI.1 Consumption (thousand tonnes)

	1984/85	1985/86	1986/87	1987/88	1988/89	1989/90	1990/91	1991/92	1992/93	1993/94
Canada	48	54	43	44	41	45	41	42	43	45
Cuba	44	44	49	45	46	40	39	33	26	21
Dom. Rep.	1	1	1	2	1	2	1	1	1	1
Mexico	121	138	120	149	169	164	167	178	180	189
USA	1 206	1 393	1 622	1 658	1 694	1 907	1 885	2 068	2 112	2 177
Sub total	1 422	1 632	1 838	1 899	1 953	2 159	2 135	2 324	2 364	2 435
Costa Rica	2	2	2	2	3	3	3	3	3	3
El Salvador	13	14	14	15	15	18	18	19	20	20
Guatemala	10	11	13	15	16	17	21	27	30	31
Honduras	2	2	2	2	2	2	2	2	2	2
Nicaragua	4	3	4	3	4	2	2	1	2	2
Sub total	30	32	35	38	40	42	46	52	56	58
Argentina	101	117	131	125	133	133	137	147	158	163
Bolivia	11	11	11	11	11	11	10	10	11	12
Brazil	599	692	759	811	822	764	723	719	766	807
Chile	16	22	24	25	26	24	24	25	26	26
Colombia	72	64	78	92	90	93	103	110	118	124
Ecuador	13	11	13	13	17	17	18	19	20	21
Paraguay	8	10	10	10	10	10	12	15	15	15
Peru	48	59	85	69	68	65	53	65	55	59
Uruguay	5	5	7	9	9	9	9	7	7	7
Venezuela	33	35	37	44	44	47	45	47	50	52
Sub total	907	1 026	1 153	1 209	1 229	1 172	1 134	1 164	1 225	1 286

Algeria	26	22	24	26	30	26	33	34	35	35
Egypt	297	316	302	306	288	307	315	327	331	346
Morocco	17	21	29	28	30	34	35	34	35	36
Sudan	22	23	16	18	15	12	14	13	12	12
Tunisia	14	17	17	19	19	27	26	27	27	28
Sub total	376	399	388	397	383	406	423	433	440	457
Benin	1	1	1	1	1	1	1	1	1	1
Burkina Faso	1	1	1	1	1	1	1	1	1	1
Cameroon	8	7	6	6	5	5	5	5	5	5
Cent. Afr. Rep.	0	1	1	1	1	1	1	1	1	1
Chad	2	2	2	5	3	3	3	3	3	3
Côte d'Ivoire	18	20	21	14	14	15	15	15	15	15
Madagascar	15	16	16	11	11	11	11	10	10	10
Mali	3	3	3	2	3	3	3	3	3	3
Niger	1	1	1	1	1	1	1	1	1	1
Senegal	3	3	3	2	2	2	2	2	2	2
Togo	2	2	3	3	3	3	3	2	2	2
Sub total	57	59	60	49	47	48	46	45	45	45
Angola	2	2	3	3	3	3	3	3	3	3
Burundi	1	2	2	2	3	3	2	3	3	3
Ethiopia	22	23	22	20	21	22	21	17	19	19
Ghana	4	7	9	11	9	9	9	9	10	10
Kenya	11	11	12	12	12	13	14	14	14	15
Mozambique	1	2	4	2	5	5	6	6	7	7
Nigeria	60	55	45	54	58	53	74	68	82	84
South Africa	68	77	78	76	75	80	74	63	70	74
Tanzania	16	13	11	12	10	17	14	16	14	14
Uganda	2	2	2	2	3	4	3	3	3	3

Table AVI.1 Continued

	1984/85	1985/86	1986/87	1987/88	1988/89	1989/90	1990/91	1991/92	1992/93	1993/94
Zaire	9	10	10	10	10	10	9	8	9	9
Zambia	7	7	7	7	9	10	10	10	11	11
Zimbabwe	25	25	30	31	38	39	35	49	54	60
Sub total	234	241	246	255	266	278	284	280	308	321
Bulgaria	61	57	61	66	65	65	39	26	23	21
Czech & Slovak Reps	122	121	122	121	123	123	97	63	70	88
Germany, DR	98	87	85	89	89	72	-	-	-	-
Hungary	62	64	67	68	65	58	41	30	32	36
Poland	135	144	155	163	166	137	88	72	79	86
Romania	96	93	87	82	81	81	78	65	59	65
Yugoslavia	127	128	114	118	112	98	77	-	-	-
Sub total	710	702	698	714	708	644	427	332	343	378
Austria	20	20	23	25	23	24	25	32	32	33
Finland	9	8	4	7	4	4	2	2	2	2
Norway	3	3	3	3	3	3	3	3	3	3
Sweden	7	6	5	5	5	5	5	5	5	5
Switzerland	60	67	74	64	64	61	62	61	64	67
Sub total	98	104	111	105	101	99	99	104	108	112
Belgium	40	42	54	46	43	46	40	39	40	40
Denmark	2	2	2	2	2	2	2	2	2	2
France	159	153	157	152	141	124	113	100	97	94
Germany[a]	218	213	243	230	202	206	223	178	185	193
Greece	160	165	178	170	165	170	180	180	184	187
Ireland	21	20	22	22	24	25	26	26	27	27
Italy	264	258	309	329	310	315	333	320	330	340

Netherlands	10	12	11	9	7	8	5	4	3	3
Portugal	154	166	178	187	191	190	154	150	155	159
Spain	106	138	159	151	135	154	145	144	143	145
United Kingdom	44	47	50	48	41	33	23	14	14	15
Sub total	1 177	1 218	1 364	1 346	1 263	1 275	1 245	1 158	1 180	1 207
Armenia	-	-	26	25	26	6	6	5	5	5
Azerbaijan	-	-	33	31	32	30	30	30	30	30
Byelaruss	-	-	35	33	34	34	33	32	28	28
Estonia	-	-	49	47	48	46	44	30	28	28
Georgia	-	-	25	24	24	21	19	15	14	14
Kazakhstan	-	-	37	35	36	38	39	39	40	42
Kirghizstan	-	-	23	22	22	26	26	26	25	27
Latvia	-	-	14	14	14	14	13	12	10	10
Lithuania	-	-	29	27	28	24	22	20	18	18
Moldova	-	-	39	37	38	38	36	32	28	28
Russia	-	-	1 296	1 235	1 262	1 275	1 239	1 034	934	897
Tadzhikistan	-	-	31	29	30	32	32	32	31	33
Turkmenia	-	-	6	6	6	6	7	15	20	22
Ukraine	-	-	212	205	210	213	208	171	154	148
Uzbekistan	-	-	193	189	193	200	205	207	212	220
Sub total[b]	1 879	2 003	2 047	1 960	2 003	1 995	1 960	1 700	1 578	1 550
China (mainland)	3 484	4 117	4 567	4 369	4 376	4 150	4 225	4 351	4 702	5 002
China (Taiwan)	292	431	482	405	382	360	351	369	375	381
Hong Kong	156	183	234	232	244	226	191	200	208	216
Sub total	3 932	4 731	5 284	5 005	5 001	4 735	4 767	4 920	5 285	5 599
Australia	20	21	22	22	22	22	25	28	32	32
Indonesia	136	163	180	198	250	290	336	361	386	413
Japan	693	674	747	747	759	698	650	603	591	579

Table AVI.1 Continued

	1984/85	1985/86	1986/87	1987/88	1988/89	1989/90	1990/91	1991/92	1992/93	1993/94
Korea, North	38	37	36	36	36	36	36	36	36	36
Korea, South	356	394	408	434	455	452	441	435	440	444
Malaysia	28	26	31	34	38	41	42	40	41	43
Philippines	22	29	44	54	64	67	53	50	52	54
Singapore	20	20	21	22	22	23	24	25	26	27
Thailand	159	205	247	252	280	295	328	390	400	420
Vietnam	47	52	57	59	57	57	41	44	46	46
Sub total	1 519	1 622	1 791	1 858	1 982	1 980	1 975	2 011	2 048	2 094
Afghanistan	11	8	11	11	11	11	17	17	17	17
Bangladesh	50	50	51	52	87	93	98	104	113	120
India	1 550	1 564	1 702	1 708	1 762	1 876	1 954	1 879	1 935	1 974
Myanmar	12	15	12	11	28	29	31	26	28	29
Pakistan	545	533	700	767	874	1 100	1 210	1 334	1 454	1 526
Sri Lanka	9	10	11	10	10	9	9	9	9	9
Sub total	2 178	2 181	2 488	2 560	2 773	3 118	3 320	3 370	3 557	3 676
Iran	80	91	115	120	108	119	120	120	120	120
Iraq	24	22	27	29	44	46	16	11	11	11
Israel	21	20	21	14	16	21	21	21	27	27
Syria	47	61	62	66	57	57	55	65	70	73
Turkey	414	422	521	551	552	600	557	575	604	634
Sub total	592	619	751	784	780	847	774	797	836	869
World total[c]	15 112	16 568	18 252	18 178	18 529	18 798	18 634	18 688	19 372	20 088

[a] Federal Republic of Germany prior to 1990/91.
[b] USSR prior to 1991/92.
[c] Subtotals and total include countries not shown.

Table AVI.2 Calendar year consumption (thousand tonnes)

	1984	1985	1986	1987	1988	1989	1990	1991	1992	1993
Canada	51	52	48	43	42	43	43	41	43	44
Cuba	44	44	47	46	45	43	40	35	29	23
Dom. Rep.	1	1	1	1	1	1	1	1	1	1
Mexico	119	131	127	137	161	166	166	174	179	186
USA	1 232	1 270	1 478	1 702	1 591	1 835	1 867	1 972	2 094	2 150
Sub total	1 448	1 499	1 703	1 932	1 842	2 090	2 118	2 225	2 348	2 406
Costa Rica	2	2	2	2	2	3	3	3	3	3
El Salvador	14	14	14	15	15	17	18	19	19	20
Guatemala	10	10	12	14	16	17	19	24	29	31
Honduras	2	2	2	2	2	2	2	2	2	2
Nicaragua	4	3	4	4	4	3	2	2	1	2
Sub total	32	31	34	37	39	41	44	49	54	57
Argentina	114	100	126	129	127	132	139	143	153	161
Bolivia	9	11	11	11	11	11	10	10	11	12
Brazil	552	631	737	775	838	810	729	700	746	790
Chile	15	20	23	25	26	25	24	25	26	26
Colombia	63	62	73	85	91	79	96	107	114	121
Ecuador	12	12	12	13	16	17	18	19	19	20
Paraguay	7	9	10	10	10	10	11	14	15	15
Peru	48	54	74	75	69	66	58	60	59	57
Uruguay	5	5	6	8	9	9	9	7	7	7
Venezuela	32	34	36	41	44	45	46	46	49	51
Sub total	857	938	1 107	1 171	1 239	1 204	1 139	1 130	1 199	1 261

Table AVI.2 Continued

	1984	1985	1986	1987	1988	1989	1990	1991	1992	1993
Algeria	26	24	23	25	29	28	30	33	34	35
Egypt	284	308	308	304	296	299	312	322	329	340
Morocco	17	20	26	28	29	32	35	34	34	36
Sudan	20	23	19	18	17	14	13	13	12	12
Tunisia	14	16	17	18	19	23	26	26	27	28
Sub total	362	390	393	394	389	396	416	429	437	450
Benin	1	1	1	1	1	1	1	1	1	1
Burkina Faso	1	1	1	1	1	1	1	1	1	1
Cameroon	7	7	6	6	7	5	5	5	5	5
Cent. Afr. Rep.	0	0	1	1	1	1	1	1	1	1
Chad	2	2	2	4	4	3	3	3	3	3
Côte d'Ivoire	18	19	21	17	14	14	15	15	15	15
Madagascar	15	16	16	13	11	11	11	10	10	10
Mali	3	3	3	2	3	3	3	3	3	3
Niger	1	1	1	1	1	1	1	1	1	1
Senegal	3	3	3	3	2	2	2	2	2	2
Togo	2	2	2	3	3	3	3	2	2	2
Sub total	56	58	58	53	49	47	47	46	45	45
Angola	2	2	3	3	3	3	3	3	3	3
Burundi	1	1	2	2	3	3	2	2	3	3
Ethiopia	21	22	22	21	20	21	21	19	18	19
Ghana	4	5	8	10	10	9	9	9	9	10
Kenya	10	11	12	12	12	13	14	14	14	15
Mozambique	2	2	3	3	4	5	5	6	6	7
Nigeria	59	57	49	50	56	55	65	71	76	83
South Africa	67	73	78	77	75	80	76	82	67	72

Tanzania	14	14	12	11	11	14	15	15	15	14
Uganda	3	2	2	2	3	3	3	3	3	3
Zaire	9	9	10	10	10	10	9	8	8	9
Zambia	7	7	8	6	8	9	10	10	11	11
Zimbabwe	24	25	28	31	35	39	37	43	52	58
Sub total	230	238	244	247	258	272	278	296	296	315
Bulgaria	63	58	59	64	66	65	50	32	24	22
Czech & Slovak Reps	123	121	121	126	117	127	121	80	70	81
Germany, DR	99	92	86	87	89	79	30	-	-	-
Hungary	62	63	65	67	66	61	48	35	31	34
Poland	135	136	150	158	166	160	105	69	80	83
Romania	90	94	89	84	81	81	79	71	62	62
Yugoslavia	124	128	120	116	115	104	85	-	-	-
Sub total	705	701	699	710	707	685	527	365	339	364
Austria	20	20	19	22	24	24	25	29	32	33
Finland	9	8	6	6	5	4	3	2	2	2
Norway	3	3	3	3	3	3	3	3	3	3
Sweden	6	6	5	5	5	5	5	5	5	5
Switzerland	58	64	71	68	64	62	62	62	63	66
Sub total	98	102	106	106	103	100	99	102	106	110
Belgium	39	41	46	51	42	46	43	39	39	40
Denmark	2	2	2	2	2	2	2	2	2	2
France	160	155	156	152	145	137	119	106	98	95
Germany[a]	214	216	224	235	214	205	216	197	182	189
Greece	154	160	173	176	167	168	176	180	182	186
Ireland	21	20	21	22	23	25	26	26	26	27
Italy	257	260	288	321	318	313	326	325	326	335
Netherlands	10	11	11	10	8	8	6	5	4	3

Table AVI.2 Continued

	1984	1985	1986	1987	1988	1989	1990	1991	1992	1993
Portugal	147	159	173	178	193	190	190	152	153	157
Spain	107	125	150	154	142	146	148	144	143	144
United Kingdom	44	46	47	51	43	39	28	18	14	15
Sub total	1 156	1 196	1 291	1 350	1 296	1 277	1 280	1 194	1 171	1 196
Armenia	–	–	–	–	–	–	–	–	5	5
Azerbaijan	–	–	–	–	–	–	–	–	30	30
Byelaruss	–	–	–	–	–	–	–	–	30	28
Estonia	–	–	–	–	–	–	–	–	29	28
Georgia	–	–	–	–	–	–	–	–	14	14
Kazakhstan	–	–	–	–	–	–	–	–	40	41
Kirghizstan	–	–	–	–	–	–	–	–	25	26
Latvia	–	–	–	–	–	–	–	–	11	10
Lithuania	–	–	–	–	–	–	–	–	19	18
Moldova	–	–	–	–	–	–	–	–	30	28
Russia	–	–	–	–	–	–	–	–	976	912
Tadzhikistan	–	–	–	–	–	–	–	–	31	32
Turkmenia	–	–	–	–	–	–	–	–	18	21
Ukraine	–	–	–	–	–	–	–	–	161	151
Uzbekistan	–	–	–	–	–	–	–	–	210	217
Sub total[b]	1 813	1 951	2 028	1 996	1 985	1 999	1 974	1 808	1 629	1 562
China (mainland)	3 458	3 853	4 379	4 451	4 373	4 244	4 194	4 298	4 556	4 877
China (Taiwan)	280	300	525	452	371	390	355	361	372	378
Hong Kong	160	162	203	241	233	248	248	196	205	213
Sub total	3 898	4 316	5 107	5 145	4 977	4 882	4 796	4 856	5 133	5 468
Australia	20	21	22	22	22	22	24	27	30	32

Indonesia	135	152	173	190	228	273	317	350	375	401
Japan	710	684	702	774	728	745	656	623	596	584
Korea, North	39	37	36	36	36	36	36	36	36	36
Korea, South	356	354	372	394	431	426	413	438	438	442
Malaysia	29	27	29	32	36	39	41	41	41	42
Philippines	22	26	38	50	60	66	59	51	51	53
Singapore	19	20	20	21	22	22	23	24	25	26
Thailand	158	178	222	257	268	289	314	364	396	412
Vietnam	45	50	55	58	58	57	48	43	45	46
Sub total	1 534	1 549	1 669	1 836	1 889	1 975	1 931	1 996	2 033	2 075
Afghanistan	11	9	10	11	11	11	15	17	17	17
Bangladesh	50	50	51	51	73	91	96	102	109	117
India	1 501	1 558	1 645	1 705	1 739	1 828	1 921	1 910	1 911	1 957
Myanmar	13	14	13	11	21	29	30	28	27	28
Pakistan	528	538	630	739	829	1 006	1 164	1 282	1 404	1 496
Sri Lanka	9	10	10	11	10	9	9	9	9	9
Sub total	2 112	2 180	2 360	2 530	2 684	2 974	3 236	3 349	3 479	3 627
Iran	84	87	105	118	113	114	120	120	120	120
Iraq	24	23	25	28	38	45	29	13	11	11
Israel	20	20	20	17	15	19	21	21	25	27
Syria	45	55	62	64	60	57	56	61	68	72
Turkey	403	419	480	539	552	580	575	567	592	621
Sub total	581	609	698	772	783	821	806	787	820	855
World total[c]	14 882	15 758	17 498	18 278	18 240	18 763	18 690	18 632	19 087	19 790

[a] Federal Republic of Germany prior to 1990/91.
[b] USSR prior to 1991/92.
[c] Subtotals and total include countries not shown.

APPENDIX

VII

Imports

Table AVII.1 Imports (thousand tonnes)

	1984/85	1985/86	1986/87	1987/88	1988/89	1989/90	1990/91	1991/92	1992/93	1993/94
Canada	47	62	33	43	41	44	41	43	47	45
Cuba	48	44	41	42	46	41	26	30	27	22
Dom. Rep.	–	–	–	–	–	–	–	–	–	–
Mexico	–	–	20	13	9	30	45	42	139	147
USA	3	7	1	0	1	0	1	3	3	3
Sub total	98	113	95	98	96	116	113	118	216	217
Costa Rica	1	1	1	1	2	2	2	2	2	2
El Salvador	–	1	5	7	7	7	14	16	19	19
Guatemala	2	2	3	–	1	–	–	3	10	20
Honduras	1	0	0	1	1	1	2	2	2	1
Nicaragua	–	–	–	–	–	–	–	–	–	–
Sub total	4	5	9	9	11	10	17	23	32	42
Argentina	9	20	27	17	1	10	6	10	11	11
Bolivia	5	9	9	7	11	10	7	7	5	4
Brazil	7	54	53	43	101	113	101	126	144	144
Chile	17	23	25	25	26	24	24	25	26	26
Colombia	0	1	0	0	1	4	8	17	26	26
Ecuador	10	4	4	8	6	7	5	5	6	7
Paraguay	–	–	–	–	–	–	–	–	–	–
Peru	–	–	–	2	–	–	–	5	7	9
Uruguay	7	5	9	10	8	9	8	6	7	7
Venezuela	15	10	11	29	22	16	25	27	32	34
Sub total	70	125	139	141	174	192	184	228	264	267

Table AVII.1 Continued

	1984/85	1985/86	1986/87	1987/88	1988/89	1989/90	1990/91	1991/92	1992/93	1993/94
Algeria	27	16	24	31	33	20	33	34	35	35
Egypt	31	31	15	31	31	57	52	61	61	71
Morocco	17	17	23	21	29	31	38	31	33	22
Sudan	–	–	–	–	–	–	–	–	–	–
Tunisia	13	19	18	19	18	28	27	27	27	28
Sub total	87	82	80	102	110	136	149	152	156	155
Benin	–	–	–	–	–	–	–	–	–	–
Burkina Faso	–	–	–	–	–	–	–	–	–	–
Cameroon	–	–	–	–	–	–	–	–	–	–
Cent. Afr. Rep.	–	–	–	–	–	–	–	–	–	–
Chad	–	–	–	–	–	–	–	–	–	–
Côte d'Ivoire	–	–	–	–	–	–	–	–	–	–
Madagascar	2	4	–	–	–	–	–	–	–	–
Mali	–	–	–	–	–	–	–	–	–	–
Niger	–	–	–	–	–	–	–	–	–	–
Senegal	–	–	–	–	–	–	–	–	–	–
Togo	–	–	–	–	–	–	–	–	–	–
Sub total	4	6	2	1	1	1	–	–	–	–
Angola	–	–	–	–	–	–	–	–	–	–
Burundi	–	–	–	–	–	–	–	–	–	–
Ethiopia	5	3	–	–	3	2	–	–	–	–
Ghana	2	7	10	8	7	7	4	2	2	2
Kenya	4	3	3	4	8	5	8	8	5	5
Mozambique	–	–	–	–	–	–	–	–	–	–
Nigeria	45	49	17	24	15	11	46	20	20	17

South Africa	21	30	17	5	10	9	14	27	35	37
Tanzania	–	–	–	–	–	–	–	–	–	–
Uganda	–	–	–	–	–	–	–	–	–	–
Zaire	–	5	9	1	4	5	5	4	5	5
Zambia	–	–	–	–	–	–	–	–	–	–
Zimbabwe	–	–	–	–	–	–	–	20	11	11
Sub total	78	97	57	45	48	40	78	83	80	80
Bulgaria	78	83	73	74	78	74	33	22	19	17
Czech & Slovak Reps	126	120	121	121	121	121	81	44	61	89
Germany, DR	130	116	116	116	117	109	–	–	–	–
Hungary	83	89	91	90	83	76	53	30	32	39
Poland	161	159	142	155	181	136	93	75	73	83
Romania	96	93	86	81	77	76	74	60	59	68
Yugoslavia	141	130	106	119	103	98	76	–	–	–
Sub total	821	793	738	759	764	694	413	302	319	376
Austria	23	21	22	26	22	25	28	31	33	34
Finland	8	7	4	7	4	4	2	2	2	2
Norway	3	2	3	3	3	3	3	3	3	3
Sweden	5	5	5	4	5	5	6	4	5	5
Switzerland	59	68	80	67	66	64	64	62	63	71
Sub total	100	105	115	109	100	102	104	103	107	116
Belgium	40	40	55	46	48	41	39	39	41	42
Denmark	2	2	2	2	2	2	2	2	2	2
France	162	161	177	160	149	134	118	120	100	97
Germany[a]	233	215	300	212	231	222	233	185	213	212
Greece	69	61	46	33	43	39	20	27	25	22
Ireland	18	20	22	21	23	26	27	26	28	28
Italy	260	263	354	318	323	326	336	320	331	363

Table AVII.1 Continued

	1984/85	1985/86	1986/87	1987/88	1988/89	1989/90	1990/91	1991/92	1992/93	1993/94
Netherlands	10	12	10	9	7	8	5	4	3	3
Portugal	161	171	182	197	180	198	149	160	160	181
Spain	75	80	116	113	86	116	83	73	99	105
United Kingdom	52	51	57	53	42	35	27	17	15	16
Sub total	1 082	1 077	1 322	1 166	1 137	1 146	1 041	975	1 020	1 074
Armenia	–	–	–	–	–	–	–	5	5	5
Azerbaijan	–	–	–	–	–	–	–	2	2	1
Byelaruss	–	–	–	–	–	–	–	30	27	29
Estonia	–	–	–	–	–	–	–	26	27	29
Georgia	–	–	–	–	–	–	–	14	14	14
Kazakhstan	–	–	–	–	–	–	–	22	20	20
Kirghizstan	–	–	–	–	–	–	–	16	16	16
Latvia	–	–	–	–	–	–	–	12	9	10
Lithuania	–	–	–	–	–	–	–	20	17	18
Moldova	–	–	–	–	–	–	–	31	27	28
Russia	–	–	–	–	–	–	–	1 015	940	900
Tadzhikistan	–	–	–	–	–	–	–	–	–	–
Turkmenia	–	–	–	–	–	–	–	–	–	–
Ukraine	–	–	–	–	–	–	–	165	150	147
Uzbekistan	–	–	–	–	–	–	–	2	1	1
Sub total[b]	183	108	77	87	80	59	37	1 360	1 255	1 218
China (mainland)	19	0	3	19	315	408	480	332	22	10
China (Taiwan)	341	415	512	394	384	273	358	369	405	410
Hong Kong	185	239	328	263	300	223	223	215	249	238
Sub total	545	654	844	675	999	904	1 062	916	675	659

Australia	3	1	1	0	1	1	0	1	1	1
Indonesia	127	153	203	192	242	281	324	363	414	441
Japan	683	654	816	743	771	675	634	592	566	566
Korea, North	35	34	33	30	33	33	33	33	33	33
Korea, South	358	372	419	426	461	450	447	392	440	444
Malaysia	27	24	33	30	41	38	43	38	42	44
Philippines	19	29	51	46	60	55	51	51	45	52
Singapore	23	23	27	27	27	26	27	28	31	32
Thailand	144	178	281	192	273	263	354	350	360	390
Vietnam	46	52	60	64	57	56	44	40	43	44
Sub total	1 464	1 519	1 924	1 751	1 966	1 878	1 956	1 888	1 974	2 047
Afghanistan	–	–	–	–	–	–	–	–	–	–
Bangladesh	54	50	59	51	75	110	80	85	95	105
India	–	–	–	22	41	–	–	64	–	–
Myanmar	–	–	–	–	10	10	15	7	11	11
Pakistan	2	1	0	1	1	3	0	1	1	1
Sri Lanka	8	8	7	5	4	3	3	3	3	3
Sub total	64	60	67	80	133	127	100	161	111	121
Iran	–	–	–	–	–	–	–	–	10	–
Iraq	10	8	15	22	52	33	5	2	8	11
Israel	10	10	10	8	7	5	13	14	17	15
Syria	–	–	–	–	–	–	–	–	–	–
Turkey	–	16	60	66	44	77	46	55	44	44
Sub total	22	35	88	98	106	117	66	74	81	72
World total[c]	4 620	4 779	5 554	5 120	5 725	5 524	5 320	6 383	6 290	6 441

[a] Federal Republic of Germany prior to 1990/91.
[b] USSR prior to 1991/92.
[c] Subtotals and total include countries not shown.

APPENDIX

VIII

Exports

Table AVIII.1 Exports (thousand tonnes)

	1984/85	1985/86	1986/87	1987/88	1988/89	1989/90	1990/91	1991/92	1992/93	1993/94
Canada	-	-	-	-	-	-	-	-	-	-
Cuba	-	-	-	-	-	-	-	-	-	-
Dom. Rep.	-	-	0	0	1	1	1	1	1	1
Mexico	125	97	48	79	122	46	50	54	15	11
USA	1 353	427	1 455	1 433	1 339	1 675	1 697	1 481	1 459	1 459
Sub total	1 478	524	1 504	1 513	1 462	1 723	1 748	1 536	1 476	1 471
Costa Rica	-	-	-	-	-	-	-	-	-	-
El Salvador	17	5	2	0	1	1	1	1	1	1
Guatemala	46	44	15	29	25	27	18	16	4	2
Honduras	4	3	2	3	1	1	1	1	1	1
Nicaragua	66	48	46	32	24	24	29	23	4	3
Sub total	133	100	65	65	51	53	49	41	11	7
Argentina	69	32	13	54	120	124	141	143	87	33
Bolivia	-	-	-	-	-	1	5	3	3	3
Brazil	77	78	66	130	101	144	167	60	61	20
Chile	-	-	-	-	-	-	-	-	-	-
Colombia	53	50	29	42	32	16	34	55	7	7
Ecuador	-	-	-	-	-	-	-	-	-	-
Paraguay	120	109	74	160	196	230	205	245	158	87
Peru	23	25	18	3	32	28	28	12	11	13
Uruguay	0	0	0	0	1	0	0	0	0	0
Venezuela	-	-	-	-	3	11	7	5	4	4
Sub total	342	294	201	389	484	553	586	523	332	167

Table AVIII.1 Continued

	1984/85	1985/86	1986/87	1987/88	1988/89	1989/90	1990/91	1991/92	1992/93	1993/94
Algeria	-	-	-	-	-	-	-	-	-	-
Egypt	152	148	121	88	60	43	18	20	26	30
Morocco	-	-	1	5	8	1	3	1	3	2
Sudan	145	115	222	149	162	141	103	109	91	95
Tunisia	-	-	-	0	0	0	0	0	0	0
Sub total	297	263	344	242	230	186	124	129	121	128
Benin	18	22	38	42	40	42	46	75	69	62
Burkina Faso	32	41	60	57	60	61	72	63	66	62
Cameroon	24	39	35	49	59	38	45	48	48	39
Cent. Afr. Rep.	17	13	8	7	8	10	8	10	12	9
Chad	33	36	33	43	48	48	57	66	67	69
Côte d'Ivoire	66	64	64	70	101	111	86	90	104	84
Madagascar	-	-	-	-	1	2	3	3	4	3
Mali	49	60	75	73	89	96	113	116	112	102
Niger	0	0	2	2	2	1	0	0	0	0
Senegal	15	7	8	13	13	11	16	16	18	18
Togo	16	22	31	30	31	31	38	32	33	29
Sub total	271	303	356	386	452	452	484	521	534	477
Angola	1	1	1	1	2	2	0	0	0	0
Burundi	1	2	1	0	0	0	0	0	0	0
Ethiopia	1	-	-	0	0	-	-	-	-	-
Ghana	-	-	-	-	-	-	-	-	-	-
Kenya	2	1	1	-	1	1	2	2	1	1
Mozambique	0	2	5	5	5	5	4	8	8	8
Nigeria	-	-	-	-	-	2	-	-	-	-
South Africa	-	-	-	4	5	10	3	4	-	-

Tanzania	19	27	43	39	62	37	39	66	45	44
Uganda	9	6	5	3	2	2	4	7	7	7
Zaire	–	–	0	–	–	–	–	–	–	–
Zambia	11	6	1	1	2	4	9	14	11	11
Zimbabwe	60	79	61	53	82	63	49	5	33	33
Sub total	107	131	123	114	171	133	120	110	109	108
Bulgaria	23	30	20	12	16	14	4	–	–	–
Czech & Slovak Reps	–	–	–	–	–	–	–	–	–	–
Germany, DR	33	28	30	30	30	38	–	–	–	–
Hungary	–	–	–	–	–	–	–	–	–	–
Poland	13	11	–	–	6	2	1	2	2	2
Romania	–	–	–	–	–	–	–	–	–	–
Yugoslovia	–	–	–	–	–	0	–	–	–	–
Sub total	91	93	70	64	74	76	22	2	2	2
Austria	2	1	1	1	1	2	0	0	1	1
Finland	–	–	–	–	–	–	–	–	–	–
Norway	–	–	–	–	–	–	–	–	–	–
Sweden	–	–	–	–	–	–	–	–	–	–
Switzerland	0	0	0	1	2	2	2	2	2	2
Sub total	2	1	1	2	4	3	2	2	3	3
Belgium	–	1	2	3	4	3	2	2	2	2
Denmark	–	–	–	–	–	–	–	–	–	–
France	8	4	8	10	11	9	6	15	12	3
Germany[a]	14	13	20	16	29	24	44	24	20	20
Greece	38	16	73	48	85	84	85	100	107	33
Ireland	0	0	0	0	0	0	0	0	0	0
Italy	1	1	2	2	2	1	2	2	2	2
Netherlands	–	0	–	0	–	–	–	–	–	–
Portugal	–	5	6	3	3	3	3	3	3	3

Table AVIII.1 Continued

	1984/85	1985/86	1986/87	1987/88	1988/89	1989/90	1990/91	1991/92	1992/93	1993/94
Spain	19	30	33	22	58	22	29	33	19	22
United Kingdom	0	–	1	2	3	3	5	3	1	1
Sub total	81	70	144	107	196	150	176	182	166	86
Armenia	–	–	–	–	–	–	–	–	–	–
Azerbaijan	–	–	–	–	–	–	–	120	115	125
Byelaruss	–	–	–	–	–	–	–	–	–	–
Estonia	–	–	–	–	–	–	–	–	–	–
Georgia	–	–	–	–	–	–	–	–	–	–
Kazakhstan	–	–	–	–	–	–	–	70	80	80
Kirghizstan	–	–	–	–	–	–	–	13	12	9
Latvia	–	–	–	–	–	–	–	–	–	–
Lithuania	–	–	–	–	–	–	–	–	–	–
Moldova	–	–	–	–	–	–	–	–	–	–
Russia	–	–	–	–	–	–	–	–	–	–
Tadzhikistan	–	–	–	–	–	–	–	230	199	220
Turkmenia	–	–	–	–	–	–	–	383	383	390
Ukraine	–	–	–	–	–	–	–	–	–	–
Uzbekistan	–	–	–	–	–	–	–	1 168	1 219	1 300
Sub total[b]	650	683	744	760	758	746	327	1 984	2 008	2 124
China (mainland)	206	609	690	506	356	188	202	149	315	484
China (Taiwan)	–	–	–	2	–	–	–	–	–	–
Hong Kong	43	68	77	73	45	27	32	20	22	22
Sub total	249	678	767	580	402	215	234	169	337	506
Australia	150	255	272	164	286	300	309	501	327	310
Indonesia	–	–	–	–	–	–	–	–	–	–
Japan	–	–	–	–	–	–	–	–	–	–

Korea, North	–	–	–	–	–	–	–	–	–	–
Korea, South	–	–	–	–	–	–	–	–	–	–
Malaysia	–	–	–	–	–	–	–	–	–	–
Philippines	–	–	–	–	–	–	–	–	–	–
Singapore	3	3	3	3	3	3	3	3	3	3
Thailand	12	7	9	13	8	4	5	6	6	4
Vietnam	2	3	6	6	5	3	–	–	–	–
Sub total	167	269	290	187	302	310	318	510	337	317
Afghanistan	11	–	8	5	5	6	7	7	7	7
Bangladesh	0	0	0	0	0	0	–	–	–	–
India	27	63	231	23	18	184	255	11	93	215
Myanmar	–	–	–	–	2	1	5	2	2	2
Pakistan	275	685	631	513	823	296	248	430	599	628
Sri Lanka	–	–	–	–	–	–	–	–	–	–
Sub total	312	749	870	541	849	487	515	450	701	852
Iran	0	5	19	2	0	–	–	–	–	–
Iraq	–	–	–	–	–	–	–	–	–	–
Israel	77	89	56	53	56	26	42	16	24	19
Syria	103	100	60	33	58	70	91	100	100	120
Turkey	155	68	112	40	145	45	164	45	25	50
Sub total	341	268	252	134	265	146	304	167	155	195
World total[c]	4 522	4 426	5 732	5 084	5 698	5 234	5 009	6 328	6 290	6 441

[a] Federal Republic of Germany prior to 1990/91.
[b] USSR prior to 1991/92.
[c] Subtotals and total include countries not shown.

APPENDIX

IX

End stocks

Table AIX.1. Ending stocks (thousand tonnes)

	1984/85	1985/86	1986/87	1987/88	1988/89	1989/90	1990/91	1991/92	1992/93	1993/94
Canada	11	19	9	8	8	7	7	9	12	12
Cuba	24	26	19	17	18	20	7	6	8	10
Dom. Rep.	1	1	1	1	1	0	0	0	0	0
Mexico	27	12	22	20	45	33	32	22	15	15
USA	893	2 035	1 094	1 256	1 544	653	510	827	762	870
Sub total	958	2 095	1 146	1 304	1 618	715	558	865	798	908
Costa Rica	1	1	1	1	1	1	1	1	1	1
El Salvador	8	9	6	8	9	4	3	3	5	5
Guatemala	9	5	7	10	12	9	8	9	6	15
Honduras	3	3	3	1	1	1	0	1	1	1
Nicaragua	15	14	12	12	10	9	5	5	1	1
Sub total	36	31	30	32	34	24	18	20	14	23
Argentina	91	86	71	191	129	156	184	129	116	121
Bolivia	0	2	3	2	2	2	2	1	3	3
Brazil	782	779	640	606	493	364	291	335	404	428
Chile	5	6	7	7	6	6	6	6	6	6
Colombia	37	37	37	37	37	37	35	45	15	29
Ecuador	3	8	4	6	5	8	5	3	3	2
Paraguay	95	83	83	118	132	117	165	40	46	87
Peru	56	71	52	100	80	48	50	37	26	8
Uruguay	3	2	5	5	4	4	3	2	2	2
Venezuela	23	32	25	9	19	8	4	6	8	11
Sub total	1 094	1 106	927	1 081	906	750	746	605	629	696
Algeria	10	4	4	9	11	5	5	5	5	5
Egypt	81	105	74	77	48	36	51	57	41	41

Table AIX.1 Continued

	1984/85	1985/86	1986/87	1987/88	1988/89	1989/90	1990/91	1991/92	1992/93	1993/94
Morocco	1	2	3	3	4	9	17	22	20	8
Sudan	247	255	184	156	117	91	55	28	16	14
Tunisia	1	3	3	3	2	3	4	4	5	5
Sub total	341	369	269	248	183	146	133	116	88	73
Benin	23	34	43	27	30	29	41	41	36	36
Burkina Faso	13	18	23	24	21	21	26	30	28	28
Cameroon	26	26	33	23	33	33	30	24	20	20
Cent. Afr. Rep.	4	5	5	5	7	7	12	9	8	8
Chad	11	12	13	17	19	26	26	26	26	26
Côte d'Ivoire	30	29	37	66	80	62	77	59	50	50
Madagascar	5	10	9	9	9	9	7	7	6	6
Mali	37	42	42	41	46	46	46	41	41	46
Niger	1	1	1	1	1	1	1	0	0	1
Senegal	7	7	6	7	6	5	5	7	7	7
Togo	9	12	11	6	6	6	6	12	12	14
Sub total	167	195	223	227	259	245	277	257	235	242
Angola	3	5	5	5	3	1	2	2	3	3
Burundi	1	0	0	1	1	0	0	0	0	0
Ethiopia	4	5	3	2	5	4	2	3	3	3
Ghana	0	0	4	4	4	5	5	5	5	4
Kenya	6	7	4	2	4	4	6	8	10	14
Mozambique	8	8	8	8	8	8	10	9	7	6
Nigeria	5	9	9	9	14	11	19	31	32	27
South Africa	52	63	74	67	75	54	38	19	27	34
Tanzania	57	50	20	49	40	21	15	14	25	34
Uganda	3	1	1	3	3	3	4	5	8	12
Zaire	1	3	7	3	2	1	1	1	1	2

Zambia	1	1	0	15	21	17	24	15	14	12
Zimbabwe	79	77	79	109	84	52	42	29	33	33
Sub total	227	235	219	284	269	189	171	145	174	187
Bulgaria	15	16	13	13	14	13	7	7	7	7
Czech & Slovak Reps	66	64	64	64	63	61	44	25	16	17
Germany, DR	30	31	31	28	26	24	-	-	-	-
Hungary	16	17	22	22	18	14	9	9	9	12
Poland	36	36	22	25	35	31	31	32	23	18
Romania	28	28	26	26	22	17	13	8	8	11
Yugoslavia	32	34	26	20	17	17	17	-	-	-
Sub total	225	229	206	199	196	179	124	100	83	87
Austria	9	9	7	6	5	4	7	6	6	6
Finland	2	2	2	2	1	1	1	1	1	1
Norway	2	1	1	1	1	1	1	1	1	1
Sweden	2	2	3	2	2	2	2	2	2	2
Switzerland	26	26	32	34	33	34	34	33	29	31
Sub total	40	41	45	45	44	43	45	42	38	40
Belgium	11	11	16	17	21	14	11	9	9	9
Denmark	0	0	1	1	0	0	0	0	0	0
France	28	28	39	37	33	44	42	47	38	38
Germany[a]	74	63	100	67	66	58	49	32	40	40
Greece	72	102	92	79	96	148	113	71	65	87
Ireland	4	4	4	4	3	3	4	4	5	6
Italy	51	55	99	86	97	92	93	91	91	113
Netherlands	2	2	1	1	1	1	1	1	1	1
Portugal	31	31	29	46	32	37	29	35	38	56
Spain	32	13	23	46	71	72	60	31	31	31
United Kingdom	5	5	6	6	5	3	2	2	1	1
Sub total	311	313	411	389	426	471	404	323	320	383

Table AIX.1 Continued

	1984/85	1985/86	1986/87	1987/88	1988/89	1989/90	1990/91	1991/92	1992/93	1993/94
Armenia	–	–	–	–	–	–	–	1	1	1
Azerbaijan	–	–	–	–	–	–	–	77	89	100
Byelaruss	–	–	–	–	–	–	–	5	4	4
Estonia	–	–	–	–	–	–	–	4	3	4
Georgia	–	–	–	–	–	–	–	2	2	2
Kazakhstan	–	–	–	–	–	–	–	32	27	22
Kirghizstan	–	–	–	–	–	–	–	4	3	3
Latvia	–	–	–	–	–	–	–	2	1	1
Lithuania	–	–	–	–	–	–	–	3	2	2
Moldova	–	–	–	–	–	–	–	4	3	3
Russia	–	–	–	–	–	–	–	127	133	136
Tadzhikistan	–	–	–	–	–	–	–	54	59	55
Turkmenia	–	–	–	–	–	–	–	88	85	85
Ukraine	–	–	–	–	–	–	–	26	22	21
Uzbekistan	–	–	–	–	–	–	–	578	571	473
Sub total[b]	407	610	557	393	478	449	841	1 007	1 004	913
China (mainland)	4 330	3 751	2 038	1 400	1 137	995	1 554	3 061	3 538	3 260
China (Taiwan)	131	115	145	133	135	48	56	56	85	115
Hong Kong	28	35	58	41	64	39	39	34	53	53
Sub total	4 489	3 902	2 241	1 574	1 336	1 081	1 648	3 150	3 676	3 428
Australia	192	175	109	205	191	183	300	205	186	201
Indonesia	8	4	49	46	41	35	28	34	66	99
Japan	163	114	221	218	235	203	192	182	157	143
Korea, North	8	8	8	5	5	5	5	4	4	4
Korea, South	115	93	113	122	106	123	128	85	85	85

Malaysia	6	5	7	4	7	5	6	4	5	6
Philippines	2	8	21	18	18	12	16	28	28	32
Singapore	2	2	5	7	9	9	9	9	11	12
Thailand	40	39	62	12	31	24	76	73	60	62
Sub total	551	462	608	650	653	606	771	633	61	653
Afghanistan	17	29	27	14	14	17	14	12	9	6
Bangladesh	28	9	19	24	18	46	45	40	38	38
India	660	996	640	455	519	784	558	772	934	871
Myanmar	6	4	4	4	4	4	4	4	4	4
Pakistan	320	319	309	498	228	290	469	900	1 049	1 106
Sri Lanka	2	2	2	2	2	3	4	6	8	11
Sub total	1 034	1 360	1 002	998	786	1 146	1 097	1 737	2 045	2 039
Iran	35	45	22	7	15	10	10	5	6	7
Iraq	12	12	15	12	25	17	11	3	1	3
Israel	8	5	6	6	4	10	11	11	20	22
Syria	27	28	32	29	29	31	30	50	62	51
Turkey	132	176	121	125	122	171	150	146	166	151
Sub total	217	269	198	182	199	243	215	218	260	237
World total[c]	10 096	11 218	8 082	7 607	7 386	6 285	7 048	9 219	9 976	9 908

[a] Federal Republic of Germany prior to 1990/91.
[b] USSR prior to 1991/92.
[c] Subtotals and total include countries not shown.

Bibliography

A detailed bibliography on all cotton issues can be found in the library of the Tropical Products Institute at Chatham, UK.

Beasley J O (1940) 'The origin of the American tetraploid Gossypium species', *Am Nat* 74: 285–6.

Beasley J O (1942) 'Meiotic chromosome behaviour in species, species hybrids, haploids and induced polyploids of Gossypium', *Genetics* 27: 25–54.

Bell T (1990) *The Story of Cotton*, New York.

Bell T (1993) *From Field to Fabric*, New York.

Fryxell P (1979) *The Natural History of the Cotton Tribe*.

Helfrich H H ed (1970) *The Environmental Crisis*, Yale University Press, New Haven, Conn.

Hutchison J (1947) *New Phytol* 'Notes on the classification of genera related to Gossypium', 46: 123–41.

Hutchison J (1959) *The Application of Genetics to Cotton Improvement*, CUP, London.

Hutchison J, Silow R and Stephens S (1947) *The Evolution of Gossypium*, OUP, London.

ICAC, *Monthly Cotton Analysis*, Washington, DC.

ITMF (1991, 1992) *Annual Review of Textile Statistics*, Geneva.

Liverpool Raw Cotton Annual (1958).

Morris D (1988) *Cotton*, Economist Intelligence Unit, London.

Munro J M (1987) *Cotton*, 2nd edition, Longman, Harlow.

New York Cotton Exchange (1992) *An Introduction to World Cotton Futures and Options*, New York.

Nofi A A (1974) *The American Civil War*, Simulation Publications, New York.
Prentice A (1972) *Cotton*, Longman, London.
Rambousek W *et al* (1986) *Volkart: The History of a Trading Company*, Volkart, Geneva.
Reich C (1986) 'The Greening of America', *The Economist.*
Starbird I *et al* (1987) *The US Cotton Industry*, USDA Economic Research Service, Washington.
Stewart J (1991) *Cotton Biotechnology*, Research Mimeo.
Tocom (1991) *Credibility and Globalisation*, Tokyo.
USDA (1986) *The US Cotton Industry*, US Department of Agriculture, Washington.

Index